Bayesian Disease Mapping

Third Edition

CHAPMAN & HALL/CRC
Interdisciplinary Statistics Series

Series editors: N. Keiding, B.J.T. Morgan, C.K. Wikle, P. van der Heijden

Bayesian Disease Mapping

Hierarchical Modeling in Spatial Epidemiology

Third Edition

By
Andrew B. Lawson

CRC Press
Taylor & Francis Group
Boca Raton London New York

CRC Press is an imprint of the
Taylor & Francis Group, an **informa** business

A CHAPMAN & HALL BOOK

CRC Press
Taylor & Francis Group
6000 Broken Sound Parkway NW, Suite 300
Boca Raton, FL 33487-2742

First issued in paperback 2020

ISBN-13: 978-1-138-57542-4 (hbk)
ISBN-13: 978-0-367-78122-4 (pbk)

Contents

List of Tables

Preface to Third Edition

Bayesian methods for disease mapping have reached a maturity with a wide variety of software and associated methods. With the development of a wider range of posterior sampling algorithms and various posterior approximation methods, some sophistication in analysis can be within reach of a wide range of potential users.

In this work I have tried to provide an overview of both the main areas of Bayesian hierarchical modeling in its application to geographical analysis of disease, and to cover the newer developments. To this end, as well as including chapters on more conventional topics such as relative risk estimation, clustering, spatial survival, and longitudinal analysis, I have also added extended chapters on computational methods, infectious disease modeling, zoonoses and surveillance. These latter topics are of particular currency due to the increasing awareness of the interconnectedness of human and animal health.

One Health (http://www.onehealthinitiative.com/) is a global initiative to provide support for cross-disciplinary research and public health practice between human and veterinary specialists in the face of many emerging zoonotic threats (e.g. Zika, Dengue, West Nile, Ebola, Leishmaniasis, Echinococcus).

There are many people who have helped in the production of this work. In particular, I have to thank a range of postdoctoral fellows and graduate students who have provided help at various times: Chawarat Rotejanaprasert, Rachel Carroll, Mehreteab Aregay, Raymond Boaz and Georgiana Onicescu. I must also thank those at CRC Press for great help in finalizing the work, in particular, Rob Calver and Lara Spieker for general production support, and Shashi Kumar for LaTeX help.

Finally, I again acknowledge the enduring support of my family, and, in particular, Pat for her understanding during the sometimes difficult activity of book writing.

Andrew B. Lawson
Charleston, United States
2018

MATLAB® Is a registered trademark of The Math Works, Inc. For product information, please contact:

The MathWorks, Inc.
3 Apple Hill Drive
Natick, MA, 01760-2098 USA
Tel: 508-647-7000
Fax: 508-647-7001
E-mail: info@mathworks.com
Web: www.mathworks.com

Preface to Second Edition

Since the first edition of this book was published there has been considerable development of new Bayesian tools/methods for disease mapping. These new tools have arisen in a range of application areas, but featured prominently are space-time data and their analysis, and the linear and non-linear modeling of predictor effects on health outcomes. The primary area, that of space-time analysis, has seen developments in infectious disease modeling, space-time prospective surveillance, latent structure, and Kalman filter modeling. One area in particular, survival analysis, has seen many examples of novel approaches to incorporation of spatial contextual effects in different types of survival models.

In this work I have tried to provide an overview of both the main areas of Bayesian hierarchical modeling in its application to geographical analysis of disease, and to cover the newer developments. To this end, as well as including chapters on more conventional topics such as relative risk estimation, clustering, spatial survival, and longitudinal analysis, I have also added an extended chapter on case event modeling, and spatio-temporal analysis, as well as new chapters on predictor-outcome modeling, and finally spatio-temporal health surveillance.

There are many people who have helped in the production of this work. In particular, I have to thank a range of postdoctoral fellows and graduate students who have provided help at various times: Ana Corberan, Jungsoon Choi, Monir Hossain, Ahmed Al-Hadrami, and Georgiana Onicescu. I must also thank those at CRC Press for great help in finalizing the work, in particular, Rob Calver and Rachel Holt for general production support and Shashi Kumar for LaTeX help.

Finally I would like to again acknowledge the continual and patient support of my family, and, in particular, Pat for her understanding during the sometimes fraught activity of book writing.

Andrew B. Lawson
Charleston, United States
2013

Preface to First Edition

Bayesian approaches to biostatistical problems have become commonplace in epidemiological, medical, and public health applications. Indeed the use of Bayesian methodology has seen great advances since the introduction of, first, BUGS, and then WinBUGS. WinBUGS is a free software package that allows the development and fitting of relatively complex hierarchical Bayesian models. The introduction of fast algorithms for sampling posterior distributions in the 1990s has meant that relatively complex Bayesian models can be fitted in a straightforward manner. This has led to a great increase in the use of Bayesian approaches not only to medical research problems but also in the field of public heath. One area of important practical concern is the analysis of the geographical distribution of health data found commonly in both public health databases and in clinical settings. Often population level data are available via government data sources such as online community health systems (e.g., for the state of South Carolina this is http://scangis.dhec.sc.gov/scan/, while in the state of Georgia it is http://oasis.state.ga.us/) or via centrally organized data registries where individual patient records are held. Cancer registry data (such as SEER in the United States) usually include individual diagnosis type and date as well as demographic information at a finer level of resolution.

Most government sources hold publicly accessible *aggregated* health data due to confidentiality requirements. The resulting count data, usually available at county or postal/census region level, can yield important insights into the general spatial variation of disease in terms of incidence or prevalence. Data can also be analyzed with respect to health inequalities or disparities related to health service provision. While this form of data and its analysis are relatively well documented, there are other areas of novel application of spatial methodology that are less well recognized currently. For example, common sources of *individual level* data are disease registries where notification of a disease case leads to registering the individual and his or her demographic details. In addition, some diagnostic information is usually held. This is typically found on cancer registries, but other diseases have similar registration processes. In clinical trials, or community-based behavioral intervention trials, individual patient information is often held and disease progression is noted over the duration of the trial. In two ways, it may be important or relevant to consider spatial information in such applications. First, the recruitment or dropout process for trials may have a spatial component. Second, there may be unobserved confounding variables that have a spatial expression over the course of the trial. These issues may lead to the consideration of longitudinal or survival analyses where geo-referencing is admitted as a confounding factor. In general, the focus area of this work is, in effect, *spatial biostatistics*, as the inclusion of clinical and registry-level analysis, as well as population level analyses, lies within the range of applications for the methods covered.

In this work I have tried to provide an overview of the main areas of Bayesian hierarchical modeling in its application to geographical analysis of disease. I have tried to orient the coverage to deal with both population level analyses and also individual level analyses resulting from cancer registry data and also the possibility of the use of data on health service utilization (disease progression via health practitioner visits, etc.), and designed studies (clinical or otherwise). To this end, as well as including chapters on more conventional topics such as relative risk estimation and clustering, I have included coverage of spatial survival and longitudinal analysis, with a section on repeated event analysis.

There are many people who have helped in the production of this work. In particular, I would like to recognize sources of encouragement from Andrew Cliff, Sudipto Banerjee, Emmanuel Lesaffre, Peter Rogerson, and Allan Clark. In addition, I have to thank a range of postdoctoral fellows and graduate students who have provided help at various times: Hae-Ryoung Song, Ji-in Kim, Huafeng Zhou, Kun Huang, Junlong Wu, Yuan Liu, and Bo Ma. I must also thank those at CRC Press for great help in finalizing the work, in particular, Rob Calver for general production support and Shashi Kumar for LaTeX help.

Finally I would like to acknowledge the continual and patient support of my family, and, in particular, Pat for her understanding during the sometimes fraught activity of book writing.

Andrew Lawson
Charleston, United States
2008

Part I

Background

1

Introduction

Some Basic Ideas and History Concerning Bayesian Methods

Bayesian methods have become commonplace in modern statistical applications. The acceptance of these methods is a relatively recent phenomenon however. This acceptance has been facilitated in large measure by the development of fast computational algorithms that were simply not commonly available or accessible as recently as the late 1980s. The widespread adoption of Markov chain Monte Carlo (MCMC) methods for posterior distribution sampling has led to a large increase in Bayesian applications. Most recently Bayesian methods have become commonplace in epidemiology, and the pharmaceutical industry, and they are becoming more widely accepted in public health practice. As early as 1993, review articles appeared extolling the virtues of MCMC in medical applications (Gilks et al., 1993). This increase in use has been facilitated by the implementation of software which provides a platform for the posterior distribution sampling which is necessary when relatively complex Bayesian models are employed. The development of the BUGS (Bayesian inference using Gibbs sampling) package and its Windows incarnation WinBUGS (Spiegelhalter et al., 2007) have had a huge effect on the dissemination and acceptance of these methods. To quote Cowles (2004): "A brief search for recently published papers referencing WinBUGS turned up applications in food safety, forestry, mental health policy, AIDS clinical trials, population genetics, pharmacokinetics, pediatric neurology, and other diverse fields, indicating that Bayesian methods with WinBUGS indeed are finding widespread use."

Basic ideas in Bayesian modeling stem from the extension of the likelihood paradigm to allow parameters within the likelihood model to have distributions called prior distributions. Thus parameters are allowed to be stochastic. By making this allowance, in turn, parameters in the prior distributions of the likelihood parameters can also be stochastic. Hence a natural parameter hierarchy is established. These hierarchical models form the basis of inference under the Bayesian paradigm. By combining the likelihood (data) model with suitable prior distributions for the parameters a so-called posterior distribution is formed which describes the behavior of the parameters after reviewing the data. There is a natural sequence underlying this approach that allows it to well describe the progression of scientific advance. The prior distribution is the current idea about the variation in the parameter set; data is collected (and modeled within the likelihood) and this updates our understanding of the parameter set variation via a posterior distribution. This posterior distribution can become the prior distribution for the parameter set before the next data experiment.

Various reviews of the different aspects of the Bayesian paradigm and modeling are now available. Among these a seminal work on Bayesian theory has been provided by Bernardo and Smith (1994). Recent general reviews of Bayesian methods appear in Leonard and Hsu (1999), Carlin and Louis (2008) and Gelman et al. (2004). Overviews of Bayesian modeling are also provided in Congdon (2003), Congdon (2005), Gelman and Hill (2007), Congdon (2007), Congdon (2010), Lesaffre and Lawson (2012) and Congdon (2014). Ntzoufras (2009) has provided an excellent and comprehensive resource for Bayesian methods with WinBUGS applications. In addition, Lunn et al. (2012) provide a briefer overview. For overview of MCMC methods, Gamerman and Lopes (2006) and Marin and Robert (2007) are useful starting points. For fuller coverage Gilks et al. (1996), Robert and Casella (2005), and Brooks et al. (2011) are useful resources.

Some Basic Ideas and History Concerning Disease Mapping

Disease mapping goes under a variety of names, some of which are: spatial epidemiology, environmental epidemiology, disease mapping, small area health studies. However at the center of these different names are two characteristics. First a spatial or geographical distribution is the focus and so the relative location of events is important. This brings the world of geographical information systems into play, while also including spatial statistics as a key component. The second ingredient is disease and the spatial distribution of disease is the focus. Hence the fundamental issue is how to analyze disease incidence or prevalence when we have geographical information. Sometimes this is called *geo-referenced* disease data, specifying the labeling of outcomes with spatial tags.

It is apparent that none of the names listed above include the term 'statistics'. This is unfortunate as it is often the case that statistical methodology (especially methodology from spatial statistics) is involved in the analysis of maps of disease. A more appropriate description of the area of focus of this work is *spatial biostatistics* as this emphasizes the broad nature of the focus. In later chapters, I focus on both population level analysis and analysis of clinical studies where longitudinal and survival data arise and so more conventional biostatistical applications are also stressed here.

The area of disease mapping has had a long but checkered history. Some of the first epidemiological studies were geographic in nature. For example, the study of the spatial distribution of cholera victims around the Broad Street pump by John Snow (Snow, 1854) was one of the earliest epidemiological studies and it was innately geographical. The use of geo-referenced data in observational studies was overtaken and subordinated to more rigorous clinical studies in medicine and often the geo-referencing is assumed to be irrelevant. In more recent decades, the development of fast computational platforms and geographical information system (GIS) capabilities has allowed a much greater sophistication in the handling of geo-referenced data. This coupled with advances in computational algorithms has allowed many spatial problems to be addressed effectively with accessible software. The recent rise on open source (free) software has provided wide access to students and professionals. The existence of free software, such as Quantum GIS (QGIS), GRASS, R and WinBUGS/OpenBUGS, and CARBayes and INLA, enhances this access.

Within the area of spatial biostatistics the major advances have been relatively recent. In the area of risk estimation and modeling the development of Bayesian models with random effects fitted via MCMC was first proposed by Besag et al. (1991). Since that time there has been a large increase the use of such methods. The use of scanning methods for disease cluster detection was also developed by Kulldorff and Nagarwalla (1995). There is now widespread use of scanning methods in cluster detection and surveillance. The widespread use of Bayesian methods in most areas of disease mapping is now well established and there is a need to review and summarize these disparate strands in one place. The focus here is on the use of Bayesian models and computational methods in application to studies in spatial biostatistics. Useful resources for the application of MCMC in this area are found in Lawson (2013) and Onicescu and Lawson (2016), Lawson and Choi (2016) and Blangiardo and Cameletti (2016), while an overview of nested Laplace approximation in this area is given by Blangiardo and Cameletti (2015).

Recent reviews of the general area of application of statistics in disease mapping can be found in Lawson et al. (1999), Elliott et al. (2000), Waller and Gotway (2004), Lawson (2006b) and Lawson et al. (2016). For a more epidemiologic slant, the edited work by Elliott et al. (1992) is very useful, while recent works by Peng and Dominici (2008), Thomas (2009) and Nieuwenhuijsen (2015) emphasize environmental epidemiology applications. For a GIS slant on both human and veterinary health see for example Maheswaran and Craglia (2004), Cromley and McLafferty (2011), Durr and Gatrell (2004) or Pfeiffer et al. (2008).

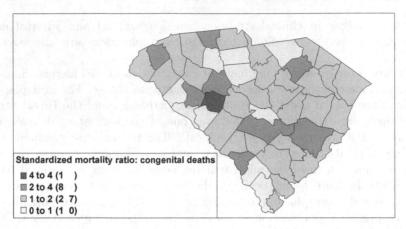

Standardized mortality ratio: congenital deaths

- ■ 4 to 4 (1   )
- ▨ 2 to 4 (8   )
- ▢ 1 to 2 (2   7)
- ☐ 0 to 1 (1   0)

FIGURE 1.1: South Carolina congenital deaths by county, 1990: standardized mortality ratio.

1.1   Data Sets

In the following chapters, a range of data sets are analyzed. Most of these are available publicly and can be downloaded from public domain web sites. In a few cases the data are confidential and cannot be accessed widely without approval. These data sets are not made available here. All other data sets are available (along with a selection of relevant programs) at the web site http://academicdepartments.musc.edu/phs/research/lawson/.

The data sets listed here are in order of appearance in the book, and allow access to the data. Some data sets are restricted by confidentiality agreements. Some data sets are well known and are not displayed as they are viewable elsewhere.

1. South Carolina (SC) county level congenital anomaly deaths, 1990. This data set consists of counts of deaths from congenital anomalies within the year 1990 in 46 counties of South Carolina, along with expected death rates computed from the statewide rate, without age-gender standardisation, and applied to the county total population. The data is available from the SCAN system at South Carolina Department of Health and Environmental Control (SCDHEC): (http://scangis.dhec.sc.gov/scan/). Both maps and tabulations of this data are available online from that source. Figure 1.1 displays the standardized mortality ratios for this example computed as the ratio of the count of disease to the expected rate within each county. The expected rate is calculated from the standardized rate from the statewide incidence rate of the condition.

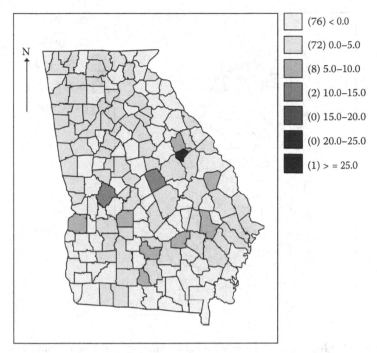

FIGURE 1.2: Georgia oral cancer standardized mortality ratio by county for 2004, using statewide rate.

2. Georgia oral cancer mortality, 2004. This data set consist of counts of oral cancer deaths within the 159 counties of the state of Georgia. It also includes expected rates computed from the statewide overall rate for 2004, and applied to the total county population. The data is available from the OASIS online system of the Georgia Division of Public Health (https://oasis.state.ga.us/). Both maps and tabulations of this data are available online from that source. Figure 1.2 displays the standardized mortality ratios for this example computed as the ratio of the count of disease to the expected rate within each county. Expected rates are computed from the statewide incidence rate.

3. Ohio respiratory cancer mortality count data set. The full data set covers 1968 through 1988 and consists of count data classified by age, gender and race for the state. This full data set has been analyzed many times (see e.g. Knorr-Held and Besag, 1998; Carlin and Louis, 2000). Subsets of the data set using total counts in counties, or functions of counts, are used here: for example 1968 and 1979 in Chapter 5 and 1979 through 1988 in Chapter 12. The data are accompanied by expected rates computed for each county from the statewide rate stratified by age and gender groups and applied to these groups in the county population and then summed.

☐	(97) < 0.0
☐	(51) 0.0–5.0
◫	(7) 5.0–10.0
◼	(3) 10.0–15.0
■	(1) > = 50.0

FIGURE 1.3: Georgia county level asthma mortality for year 2000: standardized mortality ratios based on the statewide standard population rate.

4. Ohio county data for the autologistic model. Counts of first order and second order neighbors of each county and their totalled binary outcomes after thresholding by exceedance for the state of Ohio. These data are used in autologistic modeling when binary outcomes are observed.

5. Georgia asthma mortality, 2000. This data set consist of counts of asthma deaths within the 159 counties of the state of Georgia. It also includes expected rates computed from the statewide overall rate for 2000 applied to the total county population. The data is available from the OASIS online system of the Georgia Division of Public Health (http://oasis.state.ga.us/). Both maps and tabulations of this data are available online from that source. Figure 1.3 displays the standardized mortality ratios for this example computed as the ratio of the count of deaths to the expected death rate within each county. Expected rates are computed from the statewide incidence rate.

6. Larynx cancer incidence, Lancashire, North west England (1973 through 1984). This data set was made available by Peter Diggle. Variants of the data set have appeared at different times. The data set consists of the residential addresses of cases of larynx cancer (58) reported for 1973 through 1984 in the Charnock Richards area of Lancashire, England,

UK. Within the map area is an incinerator (location: easting 35450, northing 41400) and the data was originally collected to help in the analysis of larynx cancer incidence around this location. Besides the case address locations there are 978 control disease addresses (respiratory cancer incident cases) within the same study region.

7. South Carolina congenital anomaly deaths, 1990: additional covariate information. For each county, the percent poverty listed under the US census of 1990 and also the average household income for the same census are given.

8. Georgia oral cancer 2004 multi-level data. This data set consists of counts of oral cancer and expected rates for the 159 counties of Georgia as well as the counts and expected rates for the 18 public health districts of Georgia for the same period. The public health districts are groupings of counties and are aggregations of the county level data. The expected rates are computed from statewide rates and applied to the local unit population (district or county). Figure 1.4 displays the geographies for the 18 public health districts and 159 counties of Georgia. In Chapter 9 these geographies are used with associated count data to examine multiple scale models.

9. Anonymized binary outcome (misalignment example). This data set consists of 140 binary indicators (0: control, 1: case) and their address locations and the measured soil chemical concentrations of arsenic (As) found in a network of 119 sampling sites. The soil chemical values must be interpolated to the sites of the binary outcome variables.

10. Georgia chronic multiple disease example. For the state of Georgia for the year of 2005, this data set consists of three ambulatory care sensitive chronic diseases: asthma, chronic obstructive pulmonary disease (COPD) and angina. These diseases could be affected by poor air quality and so could have common patterning or correlation. The specific data was counts of disease at county level in Georgia for 2005 for all age and gender groups. This data is publicly available from the OASIS online system of the Georgia Division of Public Health (http://oasis.state.ga.us/). Both maps and tabulations of this data are available online from that source. Figure 1.5 displays the standardized incidence ratios for the three diseases. The expected rates are computed from the statewide rate for each disease.

11. United Kingdom industrial town multiple disease example. The data set consists of residential locations cited on death certificates for respiratory disease (bronchitis) and air-way cancers (respiratory, gastric and oesophageal) for 1966 through 1976. These diseases were chosen as a set of diseases potentially related to adverse air pollution. A control disease

FIGURE 1.4: Georgia public health district geographies (top panel) and county geographies (bottom panel).

(lower body cancer as composite control) was also obtained. Another control (coronary heart disease) was available but it displays a confounding with smoking. The data consist of 630 coordinates of the residential locations of the composite control and the three diseases comprising the

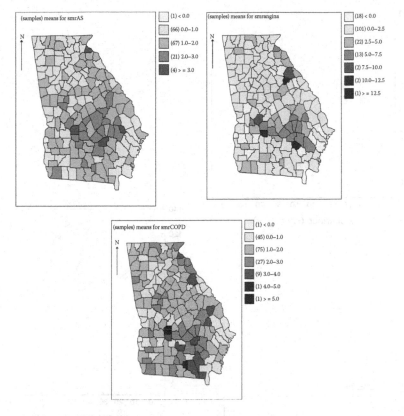

FIGURE 1.5: Georgia county level standardized incidence ratios for three diseases: asthma (top left), COPD (bottom), angina (top right).

data set. As an example of the data from this study, Figure 1.6 displays three plots of the spatial distribution of the case diseases of interest: gastric and oesophageal cancer, respiratory cancer, and bronchitis.

12. Seizure data example. This data set consists of seizure counts on 59 participants in a clinical trial of an anti-convulsive therapy for epilepsy. Each participant had available a group indicator (0,1 - control, treatment), seizure count at four time points, baseline seizure count, and age in years. The data set had been analyzed by Breslow and Clayton (1993), and is discussed in detail by Diggle et al. (2002). I have added a randomly assigned spatial county indicator for South Carolina. The full data set, for each individual, consists of variables: seizure count, county indicator, baseline count, age, and group.

13. Burkitt's lymphoma data set: This data set appears in the splancs package in R. It consists of locations of Burkitt's lymphoma cases in the

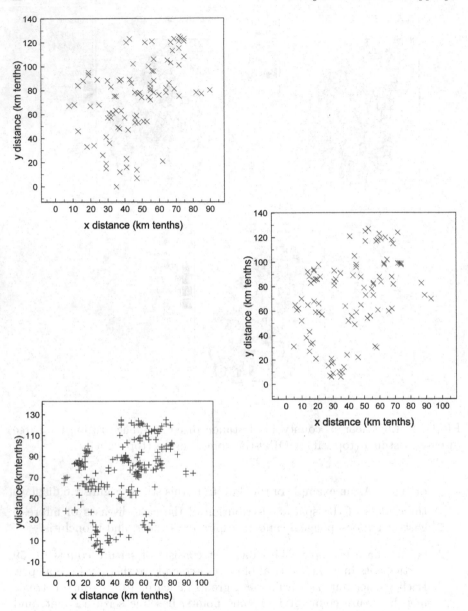

FIGURE 1.6: United Kingdom industrial town mortality study: gastric and oesophageal cancer (top); respiratory cancer (middle); bronchitis (bottom).

Western Nile district of Uganda for 1960 through 1975. In the data set the locations of cases and diagnosis dates (days from January 1st 1960) are given as well as the ages of the cases. There is no background population information in this example. There are a total of 188 cases.

14. Georgia very low birth weight ST example. This data set consists of counts of very low birth weight births for the counties of Georgia for the sequence of 11 years from 1994 to 2004. The total birth counts for the same period and counties are also available. The data is available from the OASIS online system of the Georgia Division of Public Health (http://oasis.state.ga.us/). Both maps and tabulations of this data are available online from that source. Figure 1.7 displays the rate ratios for the 11 years of very low birth weight births in relation to the county birth rates over the 11 years for all counties.

15. Ohio respiratory cancer data set. This data collection is similar to data set 4 described earlier except that it covers 21 years (1968 through 1988) with a binary outcome created by threshold exceedance. A one-time unit lag and first and second order spatial neighborhoods served as covariates.

16. Georgia asthma ST data set. This data set consists of ambulatory sensitive asthma case counts at county level for 8 years for <1 year age group (1999 through 2006). Expected rates are also available calculated from the statewide rate over the 8-year period. The data is available from the OASIS online system of the Georgia Division of Public Health (http://oasis.state.ga.us/). Both maps and tabulations of this data are available online from that source. The expected rates were computed from the overall rate (= total rate/total population <1year) times the local population count <1 year. Figures 1.8 and 1.9 display the standardized incidence ratios for the ambulatory asthma data set for 1999 through 2004 and 2005 through 2006, respectively.

17. South Carolina flu season C+ notifications data. This consists of count data for laboratory C+ notifications for the 46 counties of South Carolina over 13 time periods (biweekly recording) for the 2004-2005 flu season. This data is publicly available from the SCAN system at South Carolina Department of Health and Environmental Control (http://scangis.dhec.sc.gov/scan/). Both maps and tabulations of this data are available online from that source. Figure 1.10 displays the count of notifications for four selected counties of South Carolina for the flu season over 13 biweekly periods.

18. Foot and mouth data set. This data set consists of parish level counts of foot and mouth disease (FMD) in Cumbria, England, during the outbreak in 2001 (Figure 1.11). The time period is biweekly, starting in February 2nd and ending on June 1st (8 biweekly periods). This data set was made available by Dr Mark Stevenson, University of Melbourne, Australia. Figure 1.11 displays the standardized incidence ratios for FMD by parish.

19. The spatial distribution of incidence of visceral leishmaniasis (VL) in municipalities of Brazil for 2007 through 2015. VL is considered a dis-

FIGURE 1.7: Georgia county level very low birth weight (VLBW) risk ratios.

FIGURE 1.8: Standardized incidence ratios for Georgia county level ambulatory asthma for 1999 through 2004. Top row: 1999, 2000, 2001. Bottom row: 2002, 2003, 2004.

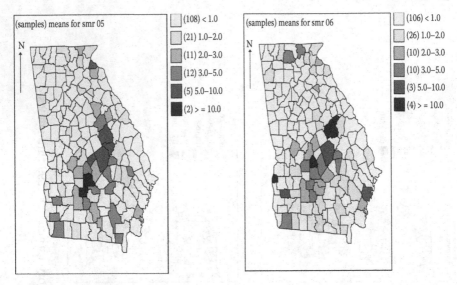

FIGURE 1.9: Standardized incidence ratios for Georgia county level ambulatory asthma for 2005 (left) and 2006 (right).

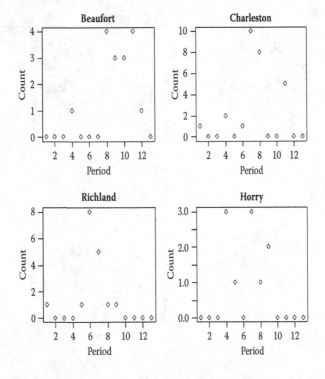

FIGURE 1.10: South Carolina influenza C+ notifications for 2004-2005 flu season: counts for 13 biweekly time periods.

FIGURE 1.11: Northwest England foot and mouth disease (FMD) during the 2001 epidemic: parish level standardized incidence ratios for eight biweekly periods.

ease of significant public health concern due to its increasing incidence and geographical expansion, especially into urban areas where the dog appears as the main reservoir. These data consist of dog case and human case counts by municipality for 2007 through 2015 in 5,564 municipalities in Brazil. The data were provided by the Ministry of Health and are considered in Chapter 14, Section 14.4.

20. The spatial distribution of tularemia in health districts of Finland 1995 through 2013. Tularemia (https://www.cdc.gov/tularemia/index.html) is an infectious disease caused by an intracellular bacterium, *Francisella tularensis* which is tick- or deer fly-borne and is zoonotic. The disease is endemic in North America and parts of Europe, with recurrent outbreaks in Sweden and Finland. *Francisella tularensis* has a wide range of hosts with transmission most commonly via arthropod vectors. Rodents could play a role in the zoonotic transmission of the disease after findings of a relationship between vole population cycles and human tularemia incidence in Sweden and Finland. These data consist of incidence of disease in voles (binary or categorical indicator) and human case counts for health districts in Finland for 1995 through 2013. These data are analysed in Chapter 14, Section 14.4.

2

Bayesian Inference and Modeling

The development of Bayesian inference has as its kernel the data likelihood. The likelihood is the joint distribution of the data evaluated at the sample values. It can also be regarded as a function describing the dependence of a parameter or parameters on sample values. Hence there can be two interpretations of this function. In Bayesian inference it is this latter interpretation that is of prime importance. In fact the *likelihood principle*, by which observations come into play through the likelihood function, and only through the likelihood function, is a fundamental part of the Bayesian paradigm (Bernardo and Smith, 1994, Section 5.1.4). This implies that the information content of the data is entirely expressed by the likelihood function. Furthermore, the likelihood principle implies that any event that did not happen has no effect on an inference, since if an unrealized event does affect an inference then there is some information not contained in the likelihood function.

2.1   Likelihood Models

The likelihood for data $\{y_i\}, i = 1, ..., m$, is defined as

$$L(\mathbf{y}|\theta) = \prod_{i=1}^{m} f(y_i|\theta) \qquad (2.1)$$

where θ is a p length vector $\theta : \{\theta_1, \theta_2, ..., \theta_p\}$ and $f(.|.)$ is a probability density (or mass) function. The assumption is made here that the "sample" values of $\mathbf{y}$ given the parameters are independent, and hence it is possible to take the product of individual contributions in (2.1). Hence the data are assumed to be conditionally independent. Note that in many spatial applications the data would not be unconditionally independent and would in fact be correlated. This conditional independence is an important assumption fundamental to many disease mapping applications. The logarithm of the likelihood is also useful in model development and is defined as:

$$l(\mathbf{y}|\theta) = \sum_{i=1}^{m} \log f(y_i|\theta). \qquad (2.2)$$

19

2.1.1   Spatial Correlation

Within spatial applications it is often found that correlation will exist between spatial units. This correlation is geographical and relates to the basic idea that locations close together in space often have similar values of outcome variables while locations far apart are often different. This spatial correlation (or autocorrelation as it's sometimes called) must be allowed for in spatial analyses. This may have an impact on the structure and form of likelihood models that are assumed for spatial data. The assumption made in the construction of conventional likelihoods is that the individual contribution to the likelihood is independent and this independence allows the likelihood to be derived as a product of probabilities. However, if this independence criterion is not met, then a different approach would be required.

2.1.1.1   Conditional Independence

In some circumstances it is possible to consider *conditional* independence of the data given parameters at a higher level of the hierarchy. For instance in count data examples y_i from the i th area might be thought to be independent of other outcomes given knowledge of the model parameters. In the simple case of dependence on a parameter vector θ, then conditioning on the parameters can allow $[y_i | \theta]$ to be assumed to be an independent contribution. This simply states that dependence only exists unconditionally (i.e. unobserved effects can induce dependence). This is often true in disease mapping examples where confounders that have spatial expression may or may not be measured in a study and their exclusion may leave residual correlation in the data. Note that this approach to correlation does not completely account for spatial effects as there can be residual correlation effects after inclusion of confounders. These effects could be due to unobserved or unknown confounders. Alternatively they could be due to intrinsic correlation in the process. Hence the assumption of conditional independence may only be valid if correlation is accounted for somewhere within the model.

The idea of inclusion of spatial correlation at a hierarchical level *above* the likelihood is a fundamental assumption often made in Bayesian small area health modeling. This means that the correlation appears in prior distributions rather than in the likelihood itself. Often parameters are given such priors and it is assumed that conditional independence applies in the likelihood. This is valid for many situations and will be the focus of most of this book.

2.1.1.2   Joint Densities with Correlations

Situations exist where spatial correlation can be incorporated within a joint distribution of the data. For example if a continuous spatial process is observed at measurement sites (such as air pollutants, soil chemical concentration, water quality) then often a spatial Gaussian process (SGP) will be assumed (Ripley, 1981). This process assumes that any realization of the process is multivariate normal with spatially-defined covariance, within its specification. Hence, if these data were observed outcome data, then the joint density would include spatial correlation (see Section 5.4.2).

Alternatively, it is possible to consider discrete outcome data where correlation is explicitly modelled. The autologistic and auto Poisson models were developed for lattice data with spatial correlation included via dependence on a spatial neighborhood (Besag and Tantrum, 2003). In this approach, the normalization of the likelihood is computationally prohibitive and resort is often made to likelihood approximation (see Section 2.1.1.3).

2.1.1.3 Pseudolikelihood Approximation

Pseudolikelihood has been proposed as an option to exact likelihood analysis when correlation exists. It has a number of variants (composite, local, pairwise: Lindsay (1988), Tibshirani and Hastie (1987), Kauermann and Opsomer (2003), Nott and Rydén (1999), Varin et al. (2005)). Pseudolikelihood has been used for autologistic models both in space and time (most recently by Besag and Tantrum (2003)). In space, the likelihood is given by

$$L_p(\mathbf{y}|\theta) = \prod_{i-1}^{m} f(y_i|y_{j \neq i}, \theta).$$

For the autologistic model, with binary outcome y_i a simple version could be:

$$f(y_i|y_{j \neq i}) = \frac{\exp[m(\beta, \{y_j\}_{j \in \delta_i})]}{1 + \exp[m(\beta, \{y_j\}_{j \in \delta_i})]}$$

where δ_i is a neighborhood set of the i th location or area, and $m(.)$ is a specified function (such as mean or median) and β is a parameter controlling the spatial smoothing or degree of correlation. For non-lattice data the neighborhood can be defined by adjacency (for count data this could be adjacent regions and for case event data this could be tesselation neighbors). In the simple case of adjacency then $m(\beta, \{y_j\}_{j \in \delta_i})$ could simply reduce to $m(\beta, \{y_j\}_{j \in \delta_i}) = \beta \sum_{j \in \delta_i} y_j$ which is the total of neighboring ones. Thus in this case, under pseudo-likelihood, the spatial correlation is accommodated by a logistic regression on neighbor count (see Section 5.7.1). It is known that pseudolikelihood is least biased when relatively low spatial correlation exists (see e.g. Diggle et al., 1994). While the autologistic model has seen some application, the auto Poisson model is limited by its awkward negative correlation structure. An autobinomial model is also available for the situation where y_i is a count of disease out of a finite local population n_i (see e.g. Cressie, 1993, p. 431 and Cressie and Wikle, 2011, chapter 4 for more examples).

2.2 Prior Distributions

All parameters within Bayesian models are stochastic and are assigned appropriate probability distributions (unless their value is fixed). Hence a single parameter value is simply one possible realization of the possible values of the parameter, the probability of which is defined by the prior distribution. The prior distribution is assigned to the parameter before seeing the data. Note also that one interpretation of prior distributions is that they provide additional "data" for a problem and so they can be used to improve estimation or identification of parameters. For a single parameter, θ, the prior distribution can be denoted $\mathbf{g}(\theta)$, while for a parameter vector, θ, the joint prior distribution is $\mathbf{g}(\theta)$.

2.2.1 Propriety

It is possible that a prior distribution can be *improper*. Impropriety is defined as the condition that integration of the prior distribution of the random variable θ over its range (Ω) is not finite:

$$\int_\Omega g(\theta) d\theta = \infty.$$

A prior distribution is improper if its normalizing constant is infinite. While impropriety is a limitation of any prior distribution, it is not necessarily the case that an improper prior will lead to impropriety in the posterior distribution. The posterior distribution can often be proper even with an improper prior specification.

2.2.2 Non-Informative Priors

Often prior distributions are assumed that do not make strong preferences over values of the variables. These are sometimes known as *vague*, or *reference* or *flat* or *non-informative* prior distributions. Usually, they have a relatively flat form yielding close-to-uniform preference for different values of the variables. This tends to mean that in any posterior analysis (see Section 2.3) the prior distribution(s) will have little impact compared to the likelihood of the data. *Jeffrey's* priors were developed in an attempt to find such reference priors for given distributions. They are based on the Fisher information matrix. For example, for the binomial data likelihood with common parameter p, then the Jeffrey's prior distribution is $p \sim Beta(0.5, 0.5)$. This is a proper prior distribution. However it is not completely non-informative as it has asymptotes close to 0 and 1. Jeffrey's prior for the Poisson data likelihood with common mean θ is given by $g(\theta) \propto \theta^{-\frac{1}{2}}$ which is *improper*. This also is not particularly non-informative. The Jeffrey's prior is locally uniform, however.

Choice of non-informative priors can often be made with some general understanding of the range and behavior of the variable. For example, variance parameters must have prior distributions on the positive real line. Non-informative distributions in this range are often in the gamma, inverse gamma, or uniform families. For example, $\tau \sim G(0.001, 0.001)$ will have a small mean (1) but a very large variance (1000) and hence will be relatively flat over a large range. Another specification chosen is $\tau \sim G(0.1, 0.1)$ with variance 10 for a more restricted range. On the other hand, a uniform distribution on a large range has been advocated for the standard deviation (Gelman, 2006): $\sqrt{\tau} \sim U(0, 1000)$. For parameters on an infinite range, such as regression parameters, then a distribution centered on zero with a large variance will usually suffice. The zero-mean Gaussian or Laplace distribution could be assumed. For example, a zero-mean Gaussian with variance τ_β, such as

$$\beta \sim N(0, \tau_\beta)$$
$$\tau_\beta = 100000$$

is typically assumed in applications. The Laplace distribution is favored in large scale Bayesian regression to encourage removal of covariates (Balakrishnan and Madigan, 2006).

Of course sometimes it is important to be informative with prior distributions. Identifiability is an issue relating to the ability to distinguish between parameters within a parametric model (see e.g. Bernardo and Smith (1994), p. 239). In particular, if a restricted range must be assumed to allow a number of variables to be *identified*, then it may be important to specify distributions that will provide such support. Ultimately if the likelihood has little or no information about the separation of parameters then separation or identification can only come from prior specification. In general, if proper prior distributions are assumed for parameters then they may often be identified in the posterior distribution. However how far they are identified may depend on the assumed variability. An example of identification which arises in disease mapping is where a linear predictor is defined to have two random effect components:

$$\log \theta_i = v_i + u_i,$$

and the components have different normal prior distributions with variances (say, τ_v, τ_u). These variances can have gamma prior distributions such as:

$$\tau_v \sim G(0.001, 0.001)$$
$$\tau_u \sim G(0.1, 0.1).$$

The difference in the variability of the second prior distribution allows there to be some degree of identification. Note that this means that a priori τ_v will be allowed greater variability in the variance of v_i than that found in u_i.

2.3   Posterior Distributions

Prior distributions and likelihood provide two sources of information about any problem. The likelihood informs about the parameter via the data, while the prior distributions inform via prior beliefs or assumptions. When there are large amounts of data, i.e. the sample size is large, the likelihood will contribute more to the relative risk estimation. When the example is data poor then the prior distributions will dominate the analysis.

The product of the likelihood and the prior distributions is called the posterior distribution. This distribution describes the behavior of the parameters after the data are observed and prior assumptions are made. The posterior distribution is defined as:

$$p(\theta|\mathbf{y}) = L(\mathbf{y}|\theta)\mathbf{g}(\theta)/C \tag{2.3}$$

$$\text{where } C = \int_p L(\mathbf{y}|\theta)\mathbf{g}(\theta)d\theta$$

where $\mathbf{g}(\theta)$ is the joint distribution of the θ vector. Alternatively this distribution can be specified as a proportionality: $p(\theta|\mathbf{y}) \propto L(\mathbf{y}|\theta)\mathbf{g}(\theta)$.

A simple example of this type of model in disease mapping is where the data likelihood is Poisson and there is a common relative risk parameter with a single gamma prior distribution:

$$p(\theta|\mathbf{y}) \propto L(\mathbf{y}|\theta)g(\theta)$$

where $g(\theta)$ is a gamma distribution with parameters α, β i.e. $G(\alpha, \beta)$, and $L(\mathbf{y}|\theta) = \prod_{i=1}^{m}\{(e_i\theta)^{y_i} \exp(-e_i\theta)\}$ bar a constant only dependent in the data. A compact notation for this model is:

$$y_i|\theta \sim Pois(e_i\theta)$$
$$\theta \sim G(\alpha, \beta).$$

This leads to a posterior distribution for fixed α, β of:

$$[\theta|\{y_i\}, \alpha, \beta] = L(\mathbf{y}|\theta, \alpha, \beta).p(\theta)/C$$

$$\text{where } C = \int L(\mathbf{y}|\theta, \alpha, \beta).p(\theta)d\theta.$$

In this case the constant C can be calculated directly and it leads to another gamma distribution:

$$[\theta|\mathbf{y}, \alpha, \beta] = \frac{\beta^{*\alpha^*}}{\Gamma(\alpha^*)}\theta^{\alpha^*-1}\exp(-\theta\beta^*) \text{ where } \alpha^* = \sum y_i + \alpha, \ \beta^* = \sum e_i + \beta.$$

2.3.1 Conjugacy

Certain combinations of prior distributions and likelihoods lead to the same distribution family in the posterior as for the prior distribution. This can lead to advantages in inference as the posterior form will follow from the prior specification. For instance, for the Poisson likelihood with mean parameter θ then with a gamma prior distribution for θ, the posterior distribution of θ is also gamma. Similar results hold for binomial likelihood and beta prior distribution and for a normal data likelihood with a normal prior distribution for the mean. The table below gives a small selection of results of this conjugacy. Conjugacy can often be found by examining the kernel of the prior-likelihood product. The un-normalized kernel should have a recognizable form related to the conjugate distribution. For example, a beta form has un-normalized kernel $\theta^{\alpha-1}(1-\theta)^{\beta-1}$. Conjugacy always guarantees a proper posterior distribution. Note that conjugacy may not be possible within a large parameter hierarchy but conditional conjugacy could be useful to exploit when examining model adequacy. It is also the case that for the sophisticated hierarchical models found in disease mapping, simple conjugacy is less likely to be available.

Likelihood	Prior	Posterior
$\mathbf{y} \sim \text{Poisson}(\theta)$	$\theta \sim G(\alpha, \beta)$	$\theta\|\mathbf{y} \sim G(\sum y_i + \alpha, m + \beta)$
$\mathbf{y} \sim \text{binomial}(\mathbf{p}, 1)$	$\mathbf{p} \sim Beta(\alpha_1, \alpha_2)$	$\mathbf{p}\|\mathbf{y} \sim Beta(\sum y_i + \alpha_1, m - \sum y_i + \alpha_2)$
$\mathbf{y} \sim \text{normal}(\mu, \tau)$, τ fixed	$\mu \sim N(\alpha_0, \tau_0)$	$\mu\|\mathbf{y} \sim N(\frac{\tau_0 \sum y_i + \alpha_0 \tau}{m\tau_0 + \tau}, \frac{\tau_0 \tau}{m\tau_0 + \tau})$
$\mathbf{y} \sim \text{gamma}(1, \beta)$	$\beta \sim G(\alpha_0, \beta_0)$	$\beta\|\mathbf{y} \sim G(1 + \alpha_0, \beta_0 + \sum y_i)$

2.3.2 Prior Choice

Choice of prior distributions is very important as the prior distributions of parameters can affect the posterior significantly. The balance between prior and posterior evidence is related to the dominance of the likelihood and is a sample size issue. For example, with large samples the likelihood usually dominates the prior distributions. This effectively means that current data are given priority in their weight of evidence. Prior distributions that dominate the likelihood are informative, but have less influence as sample size increases. Hence, with additional data, the data speak more. Of course when parameters are not identified within a likelihood then additional data are unlikely to change the importance of informative priors in identification. Propriety of posterior distributions is important as only under propriety can the absolute statements about probability of posterior parameter values be made. Some typical prescriptions for prior choice are as follows:

2.3.2.1   Regression Parameters

Within a mean model often a linear combination of predictors and regression parameters is found: $x_i'\beta$, where x_i' is a row vector of covariate values for the i th unit and β is a corresponding regression parameter vector. Here the βs are often assumed to have zero mean Gaussian prior distribution with a τ variance parameter: $\beta_* \sim N(0, \tau_*)$.

2.3.2.2   Variance or Precision Parameters

Gaussian distributions are controlled by variances. In Bayesian models it is often the inverse of the variance (the precision) that is specified. In simple Gaussian models, the precision, τ^{-1}, has a conjugate gamma prior distribution: i.e. $\tau^{-1} \sim Ga(a, b)$. This implies that the variance has an inverse gamma prior distribution. This conjugacy arguement has led to the common assumption that precisions have gamma prior distributions and variances have inverse gamma distributions. Note that both these distributions have similar forms and so it is possible to consider $\tau \sim Ga(a, b)$ and τ^{-1} as inverse gamma distributed. An alternative proposal for a non-informative prior distribution for these parameters was made by Gelman (2006), whereby a uniform prior was assumed for the standard deviation i.e. $\tau^{1/2} \sim U(0, c)$ where c is a fixed limit (such as $c = 10$ or 100). Recently, it has been suggested that weakly informative prior distributions should be adopted for different situations to honor the feasible range of the parameter and the 'null' behavior (penalised complexity, Simpson et al., 2015). For wider discussion of priors for different situations also see https://github.com/stan-dev/stan/wiki/prior-choice-recommendations. The recommendation of $\tau \sim Ga(1, 0.1)$ i.e. and exponential prior with mean 10 is proposed for the variance. Other recommendations include $\tau \sim Ga(2, 1/A)$ which bounds away from 0. A can be 1 or scaled to the range of the parameter.

2.3.2.3   Correlation Parameters

A correlation parameter usually lies on the range (-1,1) and often a uniform prior can be assumed : $\rho \sim U(-1, 1)$. Sometimes a transformation is also used. For example you could use $z = \log\left\{\frac{1+\rho}{1-\rho}\right\} \sim N(0, 2.25)$ and $\rho = \frac{\exp(z)-1}{\exp(z)+1}$. For strictly positive correlation, which arises in spatial applications, $\rho \sim U(0, 1)$ or $-\ln(\rho) \sim Ga(1, 1)$ i.e. $Exp(1)$, could be used.

2.3.2.4   Probabilities

Often probabilities are assumed to have beta prior distributions as they lie on the 0-1 range. $p \sim beta(1, 1)$ provides a uniform prior distribution. An alternative is the Jeffrey's prior $beta(0.5, 0.5)$ which is uniform over a large central range. A transformation to a prior distribution on $\text{logit}(p) \sim N(0, 3)$ could also be made.

2.3.2.5 Correlated Parameters

Often it is unrealistic to assume that all parameters are independent under prior specification. For instance, a vector of regression parameters could be assumed to have correlation between the parameters. In that case it is common to assume a multivariate Gaussian distribution with a defined covariance matrix Σ, say, or precision matrix $P = \Sigma^{-1}$. Note that under the posterior distribution the regression parameters will be correlated even when they are assumed independent under prior assumptions. However assuming a correlated joint prior distribution allows the direct estimation of parameters related to this correlation. A common assumption for a prior distribution for the precision matrix P is a Wishart distribution, which is the mutivariate generalisation of the gamma distribution and is conjugate. The covariance matrix then has an inverse Wishart distribution.

2.4 Predictive Distributions

The posterior distribution summarizes our understanding about the parameters given observed data and plays a fundamental role in Bayesian modeling. However we can also examine other related distributions that are often useful when prediction of new data (or future data) is required. Define a new observation of y as y^*. We can determine the predictive distribution of y^* in two ways. In general the predictive distribution is defined as

$$p(y^*|\mathbf{y}) = \int L(y^*|\theta)p(\theta|\mathbf{y})d\theta. \qquad (2.4)$$

Here the prediction is based on marginalizing over the parameters in the likelihood of the new data ($L(y^*|\theta)$) using the posterior distribution $p(\theta|y)$ to define the contribution of the observed data to the prediction. This is termed the posterior predictive distribution. A variant of this definition uses the prior distribution instead of the posterior distribution:

$$p(y^*|\mathbf{y}) = \int L(y^*|\theta)p(\theta)d\theta. \qquad (2.5)$$

This emphasizes the prediction based only on the prior distribution (before seeing any data). Note that this distribution (the prior predictive (2.5)) is just the marginal distribution of y^*.

2.4.1 Poisson-Gamma Example

A classic example of a predictive distribution that arises in disease mapping is the negative binomial distribution. Let y_i, $i = 1, ..., m$ be counts of disease in arbitrary small areas (e.g. census tracts, zip codes, districts). Also define, for the same areas, expected rates $\{e_i\}$, and relative risks $\{\theta_i\}$. We assume that independently $y_i \sim Poisson(e_i\theta_i)$ given θ_i. Assume that $\theta_i = \theta$ $\forall i$ and that the prior distribution of θ, $p(\theta)$, is $\theta \sim gamma(\alpha, \beta)$ where $E(\theta) = \alpha/\beta$, and $var(\theta) = \alpha/\beta^2$. The posterior distribution of θ is

$$[\theta|\mathbf{y}, \alpha, \beta] = \frac{\beta^{*\alpha^*}}{\Gamma(\alpha^*)}\theta^{\alpha^*-1}\exp(-\theta\beta^*) \text{ where } \alpha^* = \sum y_i + \alpha, \ \beta^* = \sum e_i + \beta.$$

It follows that the (prior) predictive distribution is

$$[\mathbf{y}^*|\mathbf{y}, \alpha, \beta] = \int f(\mathbf{y}^*|\theta)f(\theta|\alpha, \beta)d\theta \tag{2.6}$$

$$= \prod_{i=1}^{m}\left[\frac{\beta^\alpha}{\Gamma(\alpha)}\frac{\Gamma(y_i^* + \alpha)}{(e_i + \beta)^{(y_i^* + \alpha)}}\right].$$

The posterior predictive distribution is given by

$$[\mathbf{y}^*|\mathbf{y}, \alpha, \beta] = \int f(\mathbf{y}^*|\theta)f(\theta|\mathbf{y}, \alpha, \beta)d\theta \tag{2.7}$$

$$= \int f(\mathbf{y}^*|\theta)f(\theta|\mathbf{y}, \alpha, \beta)d\theta$$

$$= \prod_{i=1}^{m}\left[\frac{\beta^{*\alpha^*}}{\Gamma(\alpha^*)}\frac{\Gamma(y_i^* + \alpha^*)}{(e_i + \beta^*)^{(y_i^* + \alpha^*)}}\right].$$

2.5   Bayesian Hierarchical Modeling

In Bayesian modeling the free parameters have distributions. These distributions control the form of the parameters and are specified by the investigator based, usually, on prior belief concerning their behavior. These distributions are prior distributions and I will denote such distributions by $g(\theta)$. In the disease mapping context a commonly assumed prior distribution for θ in a Poisson likelihood model is the gamma distribution and the resulting model is the gamma-Poisson model.

2.6   Hierarchical Models

A simple example of a hierarchical model that is commonly found in disease mapping is where the data likelihood is Poisson and there is a common relative risk parameter with a single gamma prior distribution:

$$p(\theta|\mathbf{y}) \propto L(\mathbf{y}|\theta)g(\theta)$$

where $g(\theta)$ is a gamma distribution with parameters α, β, i.e. $G(\alpha, \beta)$, and $L(y|\theta) = \prod_{i=1}^{m}\{(e_i\theta)^{y_i}\exp(e_i\theta)\}$ bar a constant only dependent in the data. A compact notation for this model is:

$$y_i|\theta \sim Pois(e_i\theta)$$
$$\theta \sim G(\alpha, \beta).$$

In the previous section a simple example of a likelihood and prior distribution was given. In that example the prior distribution for the parameter also had parameters controlling its form. These parameters (α, β) can have assumed values, but more usually an investigator will not have a strong belief in the prior parameters' values. The investigator may want to estimate these parameters from the data. Alternatively and more formally, as parameters within models are regarded as stochastic (and thereby have probability distributions governing their behavior), then these parameters must also have distributions. These are known as hyperprior distributions, and the parameters are known as hyperparameters.

The idea that the values of parameters could arise from distributions is a fundamental feature of Bayesian methodology and leads naturally to the use of models where parameters arise within hierarchies. In the Poisson-gamma example there is a two level hierarchy: θ has a $G(\alpha, \beta)$ distribution at the first level of the hierarchy and α will have a hyperprior distribution (h_α) as will β (h_β), at the second level of the hierarchy. This can be written as:

$$y_i|\theta \sim Pois(e_i\theta)$$
$$\theta|\alpha, \beta \sim G(\alpha, \beta)$$
$$\alpha|\nu \sim h_\alpha(\nu)$$
$$\beta|\rho \sim h_\beta(\rho).$$

For these types of models it is also possible to use a graphical tool to display the linkages in the hierarchy. This is known as a directed acyclic graph or DAG for short. On such a graph lines connect the levels of the hierarchy and parameters are nodes at the ends of the lines. Clearly it is important to terminate a hierarchy at an appropriate place, otherwise one could always assume an infinite hierarchy of parameters. Usually the cut-off point is chosen to lie where further variation in parameters will not affect the lowest level model. At this point the parameters are assumed to be fixed. For example, in the gamma-Poisson model if you assume α and β were fixed then the Gamma prior would be fixed and the choice of α and β would be uninformed. The data would not inform about the distribution at all. However by allowing a higher level of variation, i.e. hyperpriors for α, β, then we can fix the values of ν and ρ without heavily influencing the lower level variation. Figure 2.1 displays the DAG for the simple two level gamma-Poisson model just described.

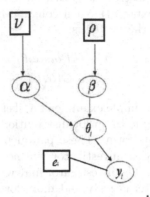

FIGURE 2.1: Directed acyclic graph for Poisson gamma hierarchical model.

2.7   Posterior Inference

When a simple likelihood model is employed, often maximum likelihood is used to provide a point estimate and associated variability for parameters. This is true for simple disease mapping models. For example, in the model $y_i|\theta \sim Pois(e_i\theta)$ the maximum likelihood estimate of θ is the the overall rate for the study region, i.e. $\sum y_i / \sum e_i$. On the other hand, the SMR $(=y_i/e_i)$ is the maximum likelihood estimate for the model $y_i|\theta_i \sim Pois(e_i\theta_i)$.

When a Bayesian hierarchical model is employed it is no longer possible to provide a simple point estimate for any of the θ_is. This is because the parameter is no longer assumed to be fixed but to arise from a distribution of possible values. Given the observed data, the parameter or parameters of interest will be described by the posterior distribution, and hence this distribution must be found and examined. It is possible to examine the expected value (mean) or the mode of the posterior distribution to give a point estimate for a parameter or parameters: e.g. for a single parameter θ, say, then $E(\theta|y) = \int \theta\, p(\theta|y)d\theta$, or $\arg\max_{\theta} p(\theta|y)$. Just as the maximum likelihood estimate is the mode of the likelihood, then the maximum a posteriori estimate is that value of the parameter or parameters at the mode of the posterior distribution. More commonly the expected value of the parameter or parameters is used. This is known as the posterior mean (or Bayes estimate). For simple unimodal symmetrical distributions, the modal and mean estimates coincide.

For some simple posterior distributions it is possible to find the exact form of the posterior distribution and to find explicit forms for the posterior mean or mode. However, it is commonly the case that for reasonably realistic models within disease mapping, it is not possible to obtain a closed form for the posterior distribution. Hence it is often not possible to derive simple estimators for parameters such as the relative risk. In this situation resort must be made to posterior approximation either via numerical approximation or by posterior sampling. In the latter case, simulation methods are used to obtain samples from the posterior distribution which then can be summarized to yield estimates of relevant quantities. In the next section the use of sampling algorithms for this purpose is discussed. In Chapter 15, numerical approximation is also considered.

An exception to this situation where a closed form posterior distribution can be obtained is the gamma-Poisson model where α, β are fixed. In that case, the relative risks have posterior distribution given by:

$$\theta_i | y_i, e_i, \alpha, \beta \sim G(y_i + \alpha, e_i + \beta)$$

and the posterior expectation of θ_i is $(y_i + \alpha)/(e_i + \beta)$. The posterior variance is also available: $(y_i + \alpha)/(e_i + \beta)^2$, as is the modal value which is

$$\arg\max_{\theta} p(\theta|\mathbf{y}) = \begin{cases} [(y_i + \alpha) - 1]/(e_i + \beta) & \text{if } (y_i + \alpha) \geq 1. \\ 0 & \text{if } (y_i + \alpha) < 1 \end{cases}$$

Of course, if α and β are not fixed and have hyperprior distributions then the posterior distribution is more complex. Clayton and Kaldor (1987) use an approximation procedure to obtain estimates of α and β from a marginal likelihood apparently on the assumption that α and β had uniform hyperprior distributions. These estimates are displayed in Figure 2.2. Note that these are not the full posterior expected estimates of the parameters from within a two level model hierarchy.

2.7.1 Bernoulli and Binomial Examples

Another example of a model hierarchy that arises commonly is the small area health data in which a finite population exists within an area and within that population binary outcomes are observed. A fuller discussion of these models is given in Section 5.1.3. In the case event example, define the case events as $s_i : i = 1, ..., m$ and the control events as $s_i : i = m + 1, ..., N$ where $N = m + n$ the total number of events in the study area. Associated with each location is a binary variable (y_i) which labels the event either as a case $(y_i = 1)$ or a control $(y_i = 0)$. A conditional Bernoulli model is assumed for the binary outcome where p_i is the probability of an individual being a case, given the location of the individual. Hence we can specify that $y_i \sim Bern(p_i)$. Here the probability will usually have either a prior distribution associated with it, or will be linked to other parameters and covariate or random effects, possibly via a linear predictor. Assume that a logistic link is appropriate for the probability and that two covariates are available for the individual: x_1 :

FIGURE 2.2: Twenty-six census enumeration districts (tracts) in Falkirk, Scotland: respiratory cancer mortality counts, 1978 through 1983. Top panel is the standardised mortality ratio map using external age × sex standardized expected rates and the bottom panel shown the gamma Poisson estimates of risk using the empirical Bayes approach of Clayton and Kaldor (1987).

age, x_2 : exposure level (of a health hazard). Hence,

$$p_i = \frac{\exp(\alpha_0 + \alpha_1 x_{1i} + \alpha_2 x_{2i})}{1 + \exp(\alpha_0 + \alpha_1 x_{1i} + \alpha_2 x_{2i})}$$

is a valid logistic model for this data with three parameters $(\alpha_0, \alpha_1, \alpha_2)$. Assume that the regression parameters will have independent zero-mean Gaussian prior distributions. The hierarchical model is specified in this case as:

$$y_i|p_i \sim Bern(p_i)$$
$$logit(p_i) = \mathbf{x}_i'\alpha$$
$$\alpha_j|\tau_j \sim N(0, \tau_j)$$
$$\tau_j \sim G(\psi_1, \psi_2).$$

In this case, x_i' is the i th row of the design matrix (including an intercept term), α is the (3×1) parameter vector, τ_j is the variance for the j th

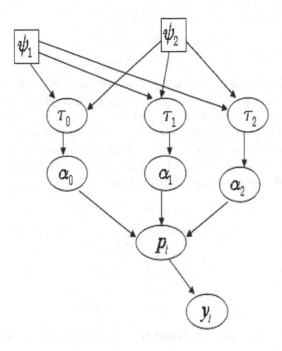

FIGURE 2.3: The Bernoulli hierarchical model where the logit link to a linear predictor is assumed.

parameter, and ψ_1, and ψ_2 are fixed scale and shape parameters. Figure 2.3 displays the hierarchy for this model.

In the binomial case we would have a collection of small areas within which we observe events. Define the number of small areas as m and the total population as n_i. Within the population of each area individuals have a binary label which denotes the case status of the individual. The number of cases are denoted as y_i and it is often assumed that the cases follow an independent binomial distribution, conditional on the probability that an individual is a case, defined as p_i: $y_i \sim Bin(p_i, n_i)$.

The likelihood is given by $L(y_i|p_i, n_i) = \prod_{i=1}^{m} \binom{n_i}{y_i} p_i^{y_i} (1 - p_i)^{(n_i - y_i)}$. Here the probability will usually have either a prior distribution associated with it, or will be linked to other parameters and covariate or random effects, possibly via a linear predictor such as $\text{logit}(p_i) = x_i'\alpha + z_i'\gamma$. In this general case, the z_i' is a vector of individual level random effects and the γ is a unit vector. Assume that a logistic link is appropriate for the probability and that a random effect at the individual level is to be included: v_i. Hence,

$$p_i = \frac{\exp(\alpha_0 + v_i)}{1 + \exp(\alpha_0 + v_i)}$$

would represent a basic model with intercept to capture the overall rate and prior distribution for the intercept and the random effect could be assumed to be $\alpha_0 \sim N(0, \tau_{\alpha_0})$, and $v_i \sim N(0, \tau_v)$. The hyperprior distribution for the variance parameters could be a distribution on the positive real line such as the gamma, inverse gamma, or uniform. The uniform distribution has been proposed for the standard deviation ($\sqrt{\tau_*}$) by Gelman (2006). Here for illustration, I define a gamma distribution:

$$y_i \sim Bin(p_i, n_i)$$
$$logit(p_i) = \alpha_0 + v_i$$
$$\alpha_0 \sim N(0, \tau_{\alpha_0})$$
$$v_i \sim N(0, \tau_v)$$
$$\tau_{\alpha_0} \sim G(\psi_1, \psi_2)$$
$$\tau_v \sim G(\phi_1, \phi_2).$$

The hierarchy for this case would be as displayed in Figure 2.4.

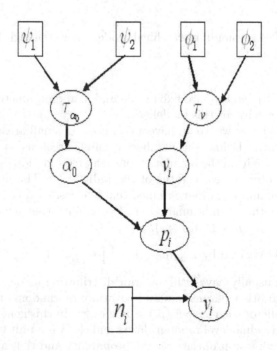

FIGURE 2.4: The hierarchical model for the binomial example with a logit link to a single intercept term and an individual level random effect.

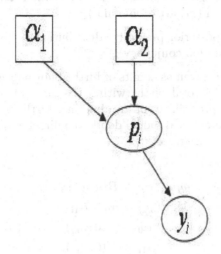

FIGURE 2.5: The hierarchical model for the beta Bernoulli hierarchy with fixed α_1, α_2 parameters.

An alternative approach to the Bernoulli or binomial distribution at the second level of the hierarchy is to assume a distribution directly for the case probability p_i. This might be appropriate when limited information about p_i is available. This is akin to the assumption of a gamma distribution as prior distribution for the Poisson relative risk parameter. Here one choice for the prior distribution could be a beta distribution:

$$p_i \sim Beta(\alpha_1, \alpha_2).$$

In general, the parameters α_1 and α_2 could be assigned hyperprior distributions on the positive real line, such as gamma or exponential. However if a uniform prior distribution for p_i is favored then $\alpha_1 = \alpha_2 = 1$ can be chosen. The hierarchy for this last situation with a Bernoulli model is displayed in Figure 2.5.

2.8 Exercises

1) Derive the posterior distribution for θ where $L(y|\theta) = \prod_{i=1}^{m}\{(e_i\theta)^{y_i}$ $\exp(-e_i\theta)\}$ and the prior distribution for θ is $Exp(\beta)$, where $Exp(\beta)$ denotes an exponential distribution with mean β.

2) For the Poisson-gamma distribution in Section 2.4.1 derive the prior predictive distribution (negative binomial).

3) Show that the posterior predictive distribution is also negative binomial (hint: use gamma-gamma conjugacy).

4) Observed data is given as counts of birth abnormalities in m small areas: $\{y_i\}$ $i = 1, ..., m$. The total births within the same areas are $\{n_i\}$ and are assumed fixed. The probability of an abnormal birth in the i th area is ψ_i. For the following hierarchical model define the directed acyclic graph (DAG) assuming that $\tau_{\alpha_0}, \tau_{\alpha_1}$ are fixed:

$$
\begin{aligned}
[y_i|n_i, \psi_i] &\sim Bin(n_i, \psi_i) \quad \forall i \\
\text{logit}(\psi_i) &= \alpha_0 + \alpha_{1i} \\
\alpha_0 &\sim N(0, \tau_{\alpha_0}) \\
\alpha_{1i} &\sim N(0, \tau_{\alpha_1}),
\end{aligned}
$$

where $N(0, \tau)$ denotes a Gaussian distribution with zero mean and variance τ.

3

Computational Issues

3.1 Posterior Sampling

Once a posterior distribution has been derived, from the product of likelihood and prior distributions, it is important to assess how the form of the posterior distribution is to be evaluated. If single summary measures are needed then it is sometimes possible to obtain these directly from the posterior distribution either by direct maximization (mode: maximum a posteriori estimation) or analytically in simple cases (mean or variance for example)(see Section 2.3). If a variety of features of the posterior distribution are to be examined then often it will be important to be able to access the distribution via posterior sampling. Posterior sampling is a fundamental tool for exploration of posterior distributions and can provide a wide range of information about their form. Define a posterior distribution for data $\mathbf{y}$ and parameter vector θ as $p(\theta|\mathbf{y})$. We wish to represent features of this distribution by taking a sample from $p(\theta|\mathbf{y})$. The sample can be used to estimate a variety of posterior quantities of interest. Define the sample size as m_p. For analytically tractable posterior distributions may be available to directly simulate the distribution. For example the gamma-Poisson model with α, β known, in Section 2.7, leads to the gamma posterior distribution: $\theta_i \sim G(y_i + \alpha, e_i + \beta)$. This can either be simulated directly (on R: rgamma) or sample estimation can be avoided by direct computation from known formulas. For example, in this instance, the moments of a gamma distribution are known: $E(\theta_i) = (y_i + \alpha)/(e_i + \beta)$ etc.

Define the sample values generated as: θ_{ij}^*, $j = 1, ..., m_p$. As long as a sample of reasonable size has been taken then it is possible to approximate the various functionals of the posterior distribution from these sample values. For example, an estimate of the posterior mean would be $\widehat{E}(\theta_i) = \widehat{\theta}_i = \sum_{j=1}^{m_p} \theta_{ij}^*/m_p$, while the posterior variance could be estimated as $\widehat{var}(\theta_i) = \frac{1}{m_p-1} \sum_{j=1}^{m_p} (\theta_{ij}^* - \widehat{\theta}_i)^2$, the sample variance. In general, any real function of the j th parameter $\gamma_j = t(\theta_j)$ can also be estimated in this way. For example, the mean of γ_j is given by $\widehat{E}(\gamma_j) = \widehat{\gamma}_j = \sum_{j=1}^{m_p} t(\theta_{ij}^*)/m_p$. Note that credibility intervals can also be found for parameters by estimating the respective sample quantiles. For example if $m_p = 1000$ then the 25th and 975th largest values would yield an equal tail 95% credible interval for γ_j. The median is also available as the 50% percentile of the sample, as are other percentiles.

The empirical distribution of the sample values can also provide an estimate of the marginal posterior density of θ_i. Denote this density as $\pi(\theta_i)$. A smoothed estimate of this marginal density can be obtained from the histogram of sample values of θ_i. Improved estimators can be obtained by using conditional distributions. A Monte Carlo estimator of $\pi(\theta_i)$ is given by

$$\widehat{\pi}(\theta_i) = \frac{1}{n} \sum_{j=1}^{n} \pi(\theta_i | \theta_{j,-i})$$

where the $\theta_{j,-i} \; j = 1, ..., n$ is a sample from the marginal distribution $\pi(\theta_{-i})$.

Often m_p is chosen to be ≥ 500, more often 1000 or 10, 000. If computation is not expensive then large samples such as these are easily obtained. The larger the sample size, the closer the posterior sample estimate of the functional will be.

Generally, the complete sample output from the distribution is used to estimate functionals. This is certainly true when independent sample values are available (such as when the distribution is analytically tractable and can be sampled from directly, such as in the gamma-Poisson case). In other cases, where iterative sampling must be used, it is sometimes necessary to subsample the output sample. In the next section, this is discussed more fully.

3.2   Markov Chain Monte Carlo (MCMC) Methods

Often in disease mapping, realistic models for maps have two or more levels and the resulting complexity of the posterior distribution of the parameters requires the use of sampling algorithms. In addition, the flexible modeling of disease could require switching between a variety of relatively complex models. In this case, it is convenient to have an efficient and flexible posterior sampling method which could be applied across a variety of models. Efficient algorithms for this purpose were developed within the fields of physics and image processing to handle large scale problems in estimation. In the late 1980s and early 1990s these methods were developed further, particularly for dealing with Bayesian posterior sampling for more general classes of problems (Gilks et al., 1993, 1996). Now posterior sampling is commonplace and a variety of packages (including WinBUGS, its descendant OpenBUGS, MlwiN, R, with MCMCpack, CARBayes and R-NIMBLE) have incorporated these methods. In Chapter 15 we will explore further these packages in disease mapping applications. For general reviews of this area the reader is referred to Cassella and George (1992), Robert and Casella (2005), Gamerman and Lopes (2006) and Brooks et al. (2011). Markov chain Monte Carlo (MCMC) methods use iterative simulation of parameter values within a Markov chain. The convergence of this chain to a stationary distribution, which is assumed to be the posterior distribution, must be assessed.

Prior distributions for the p components of θ are usually defined independently, as $g_i(\theta_i)$ for $i = 1, ..., p$. The posterior distribution of $\boldsymbol{\theta}$ and $\mathbf{y}$ is defined as:

$$P(\boldsymbol{\theta}|\mathbf{y}) \propto L(\mathbf{y}|\boldsymbol{\theta}) \prod_i g_i(\theta_i). \tag{3.1}$$

The aim is to generate a sample from the posterior distribution $P(\boldsymbol{\theta}|\mathbf{y})$. Suppose we can construct a Markov chain with state space $\boldsymbol{\theta}_c$, where $\boldsymbol{\theta} \in \boldsymbol{\theta}_c \subset \Re^k$. The chain is constructed so that the equilibrium distribution is $P(\boldsymbol{\theta}|\mathbf{y})$, and the chain should be easy to simulate from. If the chain is run over a long period, then it should be possible to reconstruct features of $P(\boldsymbol{\theta}|\mathbf{y})$ from the realized chain values. This forms the basis of the MCMC method, and algorithms are required for the construction of such chains. A selection of literature on this area is found in Ripley (1987), Gelman and Rubin (1992), Smith and Roberts (1993), Besag and Green (1993), Smith and Gelfand (1992), Tanner (1996), Chen et al. (2000), Robert and Casella (2005). MCMC methods are now commonly applied in spatial statistical modeling: see for example Cressie and Wikle (2011), Gelfand et al. (2010), and Banerjee et al. (2014).

The basic algorithms used for this construction are:

1. Metropolis and its extension Metropolis-Hastings algorithm

2. Gibbs sampler algorithm

3.3 Metropolis and Metropolis-Hastings Algorithms

In all MCMC algorithms, it is important to be able to construct the correct *transition probabilities* for a chain which has $P(\boldsymbol{\theta}|\mathbf{y})$ as its equilibrium distribution. A Markov chain consisting of $\theta^1, \theta^2, ...\theta^t$ with state space Θ and equilibrium distribution $P(\boldsymbol{\theta}|\mathbf{y})$ has transitions defined as follows.

Define $q(\boldsymbol{\theta}, \boldsymbol{\theta}')$ as a transition probability function, such that, if $\theta^t = \boldsymbol{\theta}$, the vector $\boldsymbol{\theta}^t$ drawn from $q(\boldsymbol{\theta}, \boldsymbol{\theta}')$ is regarded as a proposed possible value for θ^{t+1}.

3.3.1   Metropolis Updates

In this case choose a symmetric proposal $q(\boldsymbol{\theta}, \boldsymbol{\theta}')$ and define the transition probability as

$$p(\boldsymbol{\theta}, \boldsymbol{\theta}') = \begin{cases} \alpha(\boldsymbol{\theta}, \boldsymbol{\theta}')q(\boldsymbol{\theta}, \boldsymbol{\theta}') & \text{if } \boldsymbol{\theta}' \neq \boldsymbol{\theta} \\ 1 - \sum_{\boldsymbol{\theta}''} q(\boldsymbol{\theta}, \boldsymbol{\theta}'')\alpha(\boldsymbol{\theta}, \boldsymbol{\theta}'') & \text{if } \boldsymbol{\theta}' = \boldsymbol{\theta} \end{cases}$$

where $\alpha(\boldsymbol{\theta}, \boldsymbol{\theta}') = \min\left\{1, \frac{P(\boldsymbol{\theta}'|\mathbf{y})}{P(\boldsymbol{\theta}|\mathbf{y})}\right\}$.

In this algorithm a proposal is generated from $q(\boldsymbol{\theta}, \boldsymbol{\theta}')$ and is accepted with probability $\alpha(\boldsymbol{\theta}, \boldsymbol{\theta}')$. The acceptance probability is a simple function of the ratio of posterior distributions as a function of the ratio of posterior distributions as a function of $\boldsymbol{\theta}$ values. The proposal function $q(\boldsymbol{\theta}, \boldsymbol{\theta}')$ can be defined to have a variety of forms but must be an irreducible and aperiodic transition function. Specific choices of $q(\boldsymbol{\theta}, \boldsymbol{\theta}')$ lead to specific algorithms.

3.3.2   Metropolis-Hastings Updates

In this extension to the metropolis algorithm the proposal function is not confined to symmetry and

$$\alpha(\boldsymbol{\theta}, \boldsymbol{\theta}') = \min\left\{1, \frac{P(\boldsymbol{\theta}'|\mathbf{y})q(\boldsymbol{\theta}', \boldsymbol{\theta})}{P(\boldsymbol{\theta}|\mathbf{y})q(\boldsymbol{\theta}, \boldsymbol{\theta}')}\right\}.$$

Some special cases of chains are found when $q(\boldsymbol{\theta}, \boldsymbol{\theta}')$ has special forms. For example, if $q(\boldsymbol{\theta}, \boldsymbol{\theta}') = q(\boldsymbol{\theta}', \boldsymbol{\theta})$ then the original metropolis method arises and further, with $q(\boldsymbol{\theta}, \boldsymbol{\theta}') = q(\boldsymbol{\theta}')$, (i.e. when no dependence on the previous value is assumed) then

$$\alpha(\boldsymbol{\theta}, \boldsymbol{\theta}') = \min\left\{1, \frac{w(\boldsymbol{\theta}')}{w(\boldsymbol{\theta})}\right\}$$

where $w(\boldsymbol{\theta}) = P(\boldsymbol{\theta}|\mathbf{y})/q(\boldsymbol{\theta})$ and $w(.)$ are importance weights. One simple example of the method is $q(\boldsymbol{\theta}') \sim \text{uniform}(\boldsymbol{\theta}_a, \boldsymbol{\theta}_b)$ and $g_i(\theta_i) \sim \text{uniform}(\theta_{ia}, \theta_{ib})$ $\forall i$; this leads to an acceptance criterion based on a likelihood ratio. Hence the original metropolis algorithm with uniform proposals and prior distributions leads to a stochastic exploration of a likelihood surface. This, in effect, leads to the use of prior distributions as proposals. However, in general, when the $g_i(\theta_i)$ are not uniform this leads to inefficient sampling. The definition of $q(\boldsymbol{\theta}, \boldsymbol{\theta}')$ can be quite general in this algorithm and, in addition, the posterior distribution only appears within a ratio as a function of $\boldsymbol{\theta}$ and $\boldsymbol{\theta}'$. Hence, the distribution is only required to be known up to proportionality.

3.3.3   Gibbs Updates

The Gibbs sampler has gained considerable popularity, particularly in applications in medicine, where hierarchical Bayesian models are commonly applied (see, Gilks et al., 1993). This popularity is mirrored in the availability of software which allows its application in a variety of problems (e.g. Win/OpenBUGS, MLwiN, JAGS, MCMCpack, NIMBLE on R). This sampler is a special case of the metropolis-Hastings algorithm where the proposal is generated from the conditional distribution of θ_i given all other θ's, and the resulting proposal value is accepted with probability 1.

More formally, define

$$q(\theta_j, \theta'_j) = \begin{cases} p(\theta^*_j | \theta^{t-1}_{-j}) & if \ \theta^*_{-j} = \theta^{t-1}_{-j} \\ 0 & \text{otherwise} \end{cases}$$

where $p(\theta^*_j | \theta^{t-1}_{-j})$ is the conditional distribution of θ_j given all other $\boldsymbol{\theta}$ values $(\boldsymbol{\theta}_{-j})$ at time $t-1$. Using this definition it is straightforward to show that

$$\frac{q(\boldsymbol{\theta}, \boldsymbol{\theta}')}{q(\boldsymbol{\theta}', \boldsymbol{\theta})} = \frac{P(\boldsymbol{\theta}'|\mathbf{y})}{P(\boldsymbol{\theta}|\mathbf{y})}$$

and hence $\alpha(\boldsymbol{\theta}, \boldsymbol{\theta}') = 1$.

3.3.4 Metropolis-Hastings (M-H) versus Gibbs Algorithms

There are advantages and disadvantages to M-H and Gibbs methods. The Gibbs sampler provides a *single* new value for each θ at each iteration, but requires the evaluation of a conditional distribution. On the other hand the M-H step does not require evaluation of a conditional distribution but does not guarantee the acceptance of a new value. In addition, block updates of parameters are available in M-H, but not usually in Gibbs steps (unless joint conditional distributions are available). If conditional distributions are difficult to obtain or computationally expensive, then M-H can be used and is usually available.

In summary, the Gibbs sampler may provide faster convergence of the chain if the computation of the conditional distributions at each iteration is not time consuming. The M-H step will usually be faster at each iteration, but will not necessarily guarantee exploration. In straightforward hierarchical models where conditional distributions are easily obtained and simulated from, then the Gibbs sampler is likely to be favoured. In more complex problems, such as many arising in spatial statistics, resort to the M-H algorithm may be required.

A simple M-H example Assume that for m regions, the count $n_i \ i = 1, ..., m$ is observed. In addition, the expected count in the i th region, e_i is also observed. Assume also that the counts are independently distributed and have a Poisson distribution with $E(n_i) = \theta.e_i$, where θ is a constant parameter describing the relative risk over the whole study window. The likelihood in this case, bar a constant, is given by

$$L(\theta) = \exp(-\theta \sum_{i=1}^{m} e_i). \prod_{i=1}^{m} (\theta e_i)^{n_i}. \tag{3.2}$$

Assuming a flat prior distribution for θ, then the M-H sampler for this problem reduces to a stochastic exploration of the likelihood surface. Hence the following sampler criterion is found for the θ parameter in this case:

$$\frac{L(\theta')}{L(\theta)} = \exp\{s_e(\theta - \theta')\} \cdot \left(\frac{\theta'}{\theta}\right)^{s_n}$$

$$where \ s_e = \sum_{i=1}^{m} e_i \text{ and } s_n = \sum_{i=1}^{m} n_i.$$

3.3.5   Special Methods

Alternative methods exist for posterior sampling when the basic Gibbs or M-H updates are not feasible or appropriate. For example, if the range of the parameters is restricted then slice sampling can be used (Robert and Casella, 2005, Chapter 7; Neal, 2003). When exact conditional distributions are not available but the posterior is log-concave then adaptive rejection sampling algorithms can be used. The most general of these algorithms (ARS algorithm; Robert and Casella, 2005, pp. 57-59) has wide applicability for continuous distributions, although it may not be efficient for specific cases. Block updating can also be used to effect in some situations. When generalized linear model components are included then block updating of the covariate parameters can be effected via multivariate updating.

A variant known as metropolis adjusted Langevin algorithm (MALA) has been proposed and includes a gradient search (Langevin dynamic) before a M-H update. This can lead to more efficient searching of the posterior space. This is a special case of Hamiltonian Monte Carlo (HMC; Neal, 2011). The R package GeoRglm uses this approach. In addition, the programming package STAN also implements this HMC approach. In general, the acceptance rate of these samplers is higher than for standard M-H samplers.

3.3.6   Convergence

MCMC methods require the use of diagnostics to assess whether the iterative simulations have reached the equilibrium distribution of the Markov chain. Sampled chains must be run for an initial burn-in period until they can be assumed to provide approximately correct samples from the posterior distribution of interest. This burn-in period can vary considerably between different problems. In addition, it is important to ensure that the chain manages to explore the parameter space properly so that the sampler does not 'stick' in local maxima of the surface of the distribution. Hence, it is crucial to ensure that a burn-in period is adequate for the problem considered. Judging convergence has been the subject of much debate and can still be regarded as art rather than science: a qualitative judgement has to be made at some stage as to whether the burn-in period is long enough.

There are a wide variety of methods now available to assess convergence of chains within MCMC. Robert and Casella (2005) and Liu (2001) provide reviews. The available methods are largely based on checking the distributional properties of samples from the chains. In general they define an output stream for a parameter vector $\boldsymbol{\theta}$ as $\{\boldsymbol{\theta}^1, \boldsymbol{\theta}^2, ...\boldsymbol{\theta}^m, \boldsymbol{\theta}^{m+1}....\boldsymbol{\theta}^{m+m_p}\}$. Here the m th value is the end of the burn-in period and a (converged) sample of size m_p is taken. Hence the converged sample is $\{\boldsymbol{\theta}^{m+1}....\boldsymbol{\theta}^{m+m_p}\}$. Define a function of the output stream as $\gamma = t(\boldsymbol{\theta})$ so that $\gamma^1 = t(\boldsymbol{\theta}^1)$.

3.3.6.1 Single-Chain Methods

First, global methods for assessing convergence have been proposed which involve monitoring functions of the posterior output at each iteration. Globally this output could be the log posterior value ($\log p(\widehat{\boldsymbol{\theta}}|\mathbf{y})$ where $\widehat{\boldsymbol{\theta}}$ are the estimated parameters at a given iteration), or the deviance of the model ($-2[l(y|\widehat{\boldsymbol{\theta}}) - l(y|\widehat{\boldsymbol{\theta}}_{ref})]$) where $\widehat{\boldsymbol{\theta}}_{ref}$ is a saturated or other reference model estimate). (In Win/OpenBUGS the deviance is assumed to be $-2l(y|\widehat{\boldsymbol{\theta}})$). These methods look for stabilization of the probability value. This value forms a time series, and special cusum methods have been proposed (Yu and Mykland, 1998). This approach emphasizes the overall convergence of the chain rather than individual parameter convergence. Two basic statistical tools that can be used to check sequences of output have been proposed by Geweke (1992) and Yu and Mykland (1998). For the Geweke statistic, the sequence of output is broken up into two segments following a burn-in of m length. The first and last segments of length n_b and n_a respectively are defined. Averages of the first and last segments of output are obtained:

$$\overline{\gamma}_b = \frac{1}{n_b} \sum_{j=m+1}^{m+n_b} \gamma^j$$

$$\overline{\gamma}_a = \frac{1}{n_a} \sum_{j=m+m_p-n_a+1}^{m+m_p} \gamma^j.$$

As m_p gets large then the statistic

$$G = \frac{\overline{\gamma}_a - \overline{\gamma}_b}{\sqrt{\widehat{var}(\gamma_a) + \widehat{var}(\gamma_b)}} \rightarrow N(0, 1) \text{ in distribution,}$$

where $\widehat{var}(\gamma_a), \widehat{var}(\gamma_b)$ are empirical variance estimates. Usually it is assumed that $n_b = 0.1n$ and $n_a = 0.5n$. Note that we can set $\gamma^j = -2l(y|\boldsymbol{\theta}^j)$ or $\gamma^j = \log p(\boldsymbol{\theta}^j|\mathbf{y})$ and so the deviance or log posterior can be monitored as an overall measure. This test is available on R in the CODA package (geweke.diag). CODA is the R package for convergence diagnostics and contains a variety of test diagnostics for MCMC convergence. It is also available in the R package CARBayes as the main single chain diagnostic criterion. The second test for

FIGURE 3.1: Cusum plot of S_t against t for 1000 length converged sample of gamma$(1, 1)$ posterior output.

single sequences was proposed by Yu and Mykland (1998) and later modified by Brooks (1998). For a post-convergence sequence of length m_p an average is computed

$$\widehat{\mu} = \frac{1}{m_p} \sum_{j=m+1}^{m+m_p} \gamma^j.$$

This average is used within a cusum calculation by defining a cusum of the sequence:

$$\widehat{S}_t = \sum_{j=m+1}^{t} [\gamma^j - \widehat{\mu}] \quad \text{for } t = m+1, .., m+m_p.$$

In the original proposal, a plot of $\widehat{S}_t$ against t was proposed. The interpretation of the plot relies on the identification of the hairiness or spikiness of the cusum: a smooth cusum suggesting under-exploration of the posterior distribution, while a spiky plot represents rapid mixing. Figure 3.1 displays this plot for a 1000 length output sample from a gamma(1,1) posterior distribution. Brooks (1998) further quantified this approach by deriving a statistic that measures the spikeyness of $\widehat{S}_t$.

Define

$$d_i = \begin{cases} 1 & \text{if} \quad S_{i-1} > S_i \text{ and } S_i < S_{i+1} \\ & \text{or } S_{i-1} < S_i \text{ and } S_i > S_{i+1}, \\ 0 & \text{else} \end{cases}$$

FIGURE 3.2: Cusum plot (left) and D_t statistic graph (right) for a 1000 size realization from a Gaussian distribution with mean 5.6 and variance 1.

for all $i = m + 1, \ldots, m + m_p - 1$. Further define

$$D_t = \frac{1}{t - m - 1} \sum_{i=m+1}^{t-1} d_i \quad m + 2 \leq t \leq m + m_p.$$

This statistic can be used in a number of ways. For an *i.i.d* sequence symmetric about the mean then the expected value of d_i would be $1/2$. Further, D_t can be treated as a binomial variate with $E(D_t) = 1/2$ and $var(D_t) = \frac{1}{4(t-m-1)}$ and D_t will be approximately Gaussian with $100(1 - \alpha/2)\%$ bounds

$$\frac{1}{2} \pm Z_{\alpha/2} \sqrt{\frac{1}{4(t - m - 1)}}.$$

These bounds can be used as a formal tool to detect convergence, but it should be noted that a large sample is required for the Gaussian assumption to hold. Figure 3.2 displays an example of this form of plot for a Gaussian distribution sample with mean 5.6 and variance 1. This sample displays reasonable convergence behavior. However, for asymmetric distributions or non-convergent output, the D_t diagnostic could violate the limits. Note that for 'sticky' samplers, where values may stay for long periods (such as is possible with metropolis-Hastings samplers), then the d_i can be modified to allow for such static behavior (see e.g. Brooks (1998) for details and Section 3.3.7.1 below).

Second, graphical methods have been proposed which allow the comparison of the whole distribution of successive samples. Quantile-quantile plots of successive lengths of single variable output from the sampler can be used for

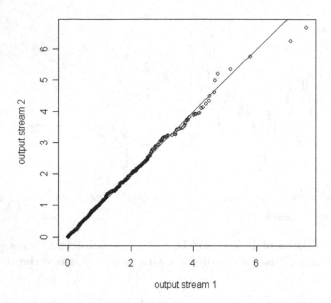

FIGURE 3.3: Quantile-quantile plot of two sequences of 1000 length of converged sample output from a gamma posterior distribution with parameters $\alpha = 1, \beta = 1$. The equality line is marked.

this purpose. Figure 3.3 displays an example of such a plot. On R, with vectors out1 and out2 this can be created via commands:

```
>plot(sort(out1),sort(out2),xlab="output stream 1",ylab="output stream 2")
>lines(x,y)
> cor(sort(out1),sort(out2))
```

Further assessment of the degree of equality can be made via use of a correlation test. The Pearson correlation coefficient between the sorted sequences can be examined and compared to special tables of critical values. This adds some formality to the relatively arbitrary nature of visual inspection.

3.3.6.2   Multi-Chain Methods

Single chain methods can, of course, be applied to each of a multiple of chains. In addition, there are methods that can only be used for multiple chains. The Brooks-Gelman-Rubin (BGR) statistic was proposed as a method for assessing the convergence of multiple chains via the comparison of summary measures across chains (Gelman and Rubin, 1992, Brooks and Gelman, 1998, Robert and Casella, 2005, Chapter 8, Gelman and Shirley, 2011).

This statistic is based on between-and within-chain variances. For the univariate case we have p chains and a sample of size n and a sample value of γ_i^j $j = 1, ..., n$; $i = 1, ..., p$. Denote the average over the sample for the i th chain as $\overline{\gamma}_i = \frac{1}{n} \sum_{j=1}^{n} \gamma_i^j$ and the overall average as $\overline{\gamma}. = \frac{1}{p} \sum_{i=1}^{p} \overline{\gamma}_i$ and the variance of the i th chain is $\tau_i^2 = \frac{1}{n-1} \sum_{j=1}^{n} (\gamma_i^j - \overline{\gamma}_i)^2$. Then the between- and within-sequence variances are

$$B = \frac{n}{p-1} \sum_{i=1}^{p} (\overline{\gamma}_i - \overline{\gamma}.)^2$$

$$W = \frac{1}{p} \sum_{i=1}^{p} \tau_i^2.$$

The marginal posterior variance of the γ is estimated as $\frac{n-1}{n} W + \frac{1}{n} B$ and this is unbiased asymptotically($n \to \infty$) . Monitoring the statistic

$$R = \sqrt{\frac{n-1}{n} + \frac{1}{n} \frac{B}{W}}$$

for convergence to 1 is recommended. If the R for all parameters and functions of parameters is between 1.0 and 1.1 (Gelman et al., 2004) this is acceptable for most studies. Note that this depends on the sample size taken and closeness will be more easily achieved for a large post-convergence m_p. Brooks and Gelman (1998) extended this diagnostic to a multiparameter situation. On R the statistic is available in the CODA package as gelman.diag. On Win/OpenBUGS the Brooks-Gelman-Rubin (BGR) statistic is available in the Sample Monitor Tool. On Win/OpenBUGS, the width of the central 80% interval of the pooled runs and the average width of the 80% intervals within the individual runs are color-coded (green, blue), and their ratio R is red; for plotting purposes the pooled and within-interval widths are normalized to have an overall maximum of one. On Win/OpenBUGS the statistics are calculated in bins of length 50. R would generally be expected to be greater than 1 if the starting values are suitably over-dispersed. Brooks and Gelman (1998) emphasize that one should be concerned both with convergence of R to 1, and with convergence of both the pooled and within interval widths to stability. One caveat should be mentioned concerning the use of between- and within-chain diagnostics. If the posterior distribution being approximated were to be highly multimodal, which could be the case in many mixture and spatial problems then the variability across chains could be large even when close to the posterior distribution and it could be that very large bins would need to be used for computation.

There is some debate about whether it is useful to run one long chain as opposed to multiple chains with different start points. The advantage of multiple chains is that they provide evidence for the robustness of convergence across different subspaces. However, as long as a single chain samples the parameter space adequately, it provides benefits. The reader is referred to Robert and Casella (2005), chapter 8 and Gelman and Shirley (2011) for a thorough discussion of diagnostics and their use.

3.3.7 Subsampling and Thinning

MCMC samplers often produce correlated samples of parameters. That is, a parameter value γ_i^j is likely to be similar to γ_i^{j-1}. This is likely to be true if successful proposals are based on proposal distributions with small variances, or where acceptances are localized to small areas of the posterior surface. In the former case, it may be that only small subsections of the posterior surface are being explored and so the sampler will not reach equilibrium for some time. Hence there may be an issue of lack of convergence when this occurs. The latter case could arise when a very spiky likelihood dominates. These correlated samples do not create problems for subsequent use of output streams, unless the sample size is very small (m_p small), or convergence has not been reached. Summary statistics could be affected by such auto-correlation. While measures of central tendency may not be much affected, the variance and other spread measures could be downward biased due to the (positive) autocorrelation in the stream. One possible remedy for this correlation is to take subsamples of the output. The simplest approach to this is to *thin* the stream by taking systematic samples at every k th iteration. By lengthening the gap between sampled units, then the more likely the correlation will be reduced or eliminated.

3.3.7.1 Monitoring Metropolis-Like Samplers

Samplers that don't necessarily accept a new value at each iteration cannot be monitored as easily as those that do produce new values (such as the Gibbs sampler). With, for example, a metropolis-Hastings algorithm the acceptance rate of new proposals is an important measure of the performance of the algorithm. The acceptance rate is defined as the number of iterations where new values are accepted out of a batch of iterations. Let's assume we have a batch size of $n_l = 100$ iterations and during that period we observe m_l accepted proposals. We assume that the number of parameters is small ($p << n_l$) so that there are potentially many transitions that could be made within n_l. The acceptance rate is just $A_r = \frac{m_l}{n_l}$. This rate could be a useful indicator of the behavior of the sampler. For example if the sampler is not mixing well then it may stick in various places and fail to find acceptable proposals. This would lead to a low acceptance rate. However, on the other hand, a high acceptance rate may signify good proposals but could also mean that the sampler is 'stuck' in the vicinity of a peak in the posterior surface and not searching the space in general. In both cases, the proposals may either be too small or too large to adequately search the space. Usually as a guide to a reasonable acceptance rate for a M-H algorithm with small dimension (1 to 2 parameters) then $A_r \approx$

0.5 would be reasonable (Robert and Casella, 2005). For higher dimensions $(p > 2)$ then $A_r \approx 0.25$ is reasonable. Hence for reversible jump algorithms (which are based on M-H steps with high dimension) then $A_r \approx 0.25$ might be expected. For metropolis-Langevin or Langevin-Hastings algorithms (such as used in the R package geoRglm or STAN) that incorporate gradient terms then higher rates are optimal ($A_r \approx 0.6$). It should be borne in mind that achievement of an optimal A_r alone does not necessarily imply convergence to a stationary distribution, although poor A_r could be due to lack of mixing and hence lack of convergence. It is also possible for chains to have high acceptance and very low convergence. On WinBUGS when a metropolis update is used then the acceptance rate can be set using the Monitor Met button in the Model Menu. This generates a plot of the acceptance rate over iteration for batches of $n_l =100$ iterations. For user defined likelihood models using the zeroes or ones trick, A_r is always available.

The D_t statistic of Brooks (1998) can be modified for application to M-H algorithms where extended periods of stickiness arise:

$$d_i = \begin{cases} 1 & \text{if} \quad S_{i-1} > S_i \text{ and } S_i < S_{i+1} \\ & or \ S_{i-1} < S_i \text{ and } S_i > S_{i+1} \\ & or \ S_{i-1} < S_i, S_{i+k} < S_i \text{ and} \\ & \quad S_i = S_{i+1} = \ldots = S_{i+k}, \\ & or \ S_{i-1} > S_i \ , \ S_{i+k} > S_i \text{ and} \\ & \quad S_i = S_{i+1} = \ldots = S_{i+k} \\ \frac{1}{2} & \text{if} \qquad S_{i-1} - S_i = S_{i+1} \\ 0 & \text{else} \end{cases}$$

for all $i = m + 1, \ldots, m + n - 1$.

In addition, for complex reversible jump samplers there may be a need for stratified convergence checking. For example, the dimension of the parameter set may lead to stratifying the number of parameters and this can lead to χ^2 tests and Kolmogorov-Smirnov statistics comparing a number of chains by their cumulative distribution functions (Brooks et al., 2003). Monitoring of dimension-changing algorithms is still an open issue.

3.4 Perfect Sampling

The idea of MCMC is that simulation from a posterior distribution can be achieved over time and iterations are followed until convergence to the equilibrium distribution is found. Propp and Wilson (1996) proposed a different approach whereby instead of iteration towards this equilibrium, a search is made to find a path from the past which will lead to coalescence at the current time. In essence a stopping time for the chain is found which corresponds to the equilibrium distribution. This is known as *coupling from the past* (CFTP). Examples of the application of such exact sampling have been made to point processes and Ising models (van Lieshout and Baddeley, 2002, Móller and Waagpetersen, 2004), case event data cluster modeling (McKeague and Loiseaux, 2002) where special MCMC (reversible jump birthdeath sampling) must be used, and to autologistic models for spatial and space-time data (Besag and Tantrum, 2003).

However CFTP is not guaranteed to work for MCMC transitional kernels that are not uniformly ergodic (Robert and Casella, 2005). However perfect *slice* sampling may help towards a general algorithm that has general appeal (Mira et al., 2001).

Currently, the main problem with perfect sampling is that it is not possible to provide a general algorithm from which modeling of particular situations is immediately available. In fact, for most applications, the algorithm has to be specially designed and it is often therefore relatively difficult to adapt to changes of model form: for example, inclusion or exclusion of covariates may not be possible without significant alteration to the algorithm.

3.5 Posterior and Likelihood Approximations

From the point of view of computation it is now straightforward to examine a range of posterior distributional forms. This is certainly true for most applications of disease mapping where relative risk is estimated. However there are situations where it may be easier or more convenient to use a form of approximation to the posterior distribution or to the likelihood itself. Some approximations have been derived originally when posterior sampling was not possible and where the only way to obtain fully Bayesian estimates was to approximate (Bernardo and Smith, 1994). However other approximations arise due to the intractability of spatial integrals (for example in point process models).

3.5.1 Pseudolikelihood and Other Forms

In Section 2.1.1.3 the idea of pseudolikelihood was briefly introduced. In certain spatial problems, found in imaging and elsewhere, normalizing constants arise which are highly multidimensional. A simple example is the case of a Markov point process. Define the realization of m events within a window T as $\{s_1,......s_m\}$. Under a Markov process assumption the normalized probability density of a realization is

$$f_\theta(\mathbf{s}) = \frac{1}{c(\theta)} h_\theta(\mathbf{s})$$

$$\text{where } c(\theta) = \sum_{k=0}^{\infty} \frac{1}{k!} \int_{T^k} h_\theta(\mathbf{s}) \lambda^k(d\mathbf{s}).$$

Conditioning on the number of events (m), the normalization of $f_m(\mathbf{s}) \propto h_m(\mathbf{s})$ is over the m-dimensional window:

$$c(\theta) = \int_T \cdots \int_T h_m(\{\mathbf{s}_1, \ldots \mathbf{s}_m\}) d\mathbf{s}_1, \ldots d\mathbf{s}_m.$$

For a conditional Strauss process, $f_m(\mathbf{s}) \propto \gamma^{n_R(\mathbf{s})}$ and $n_R(\mathbf{s})$ is the number of R-close pairs of points to $\mathbf{s}$.

It is also true that a range of lattice models developed for image processing applications also produces awkward normalization constants (auto-Poisson and autologistic models and Gaussian-Markov random field models: Besag and Tantrum (2003), Rue and Held (2005)).

This has led to the use of approximate likelihood models in many cases. For example, for Markov point processes it is possible to specify a conditional intensity (Papangelou) which is independent of the normalization. This conditional intensity $\lambda^*(\xi, \mathbf{s}|\theta) = h(\xi \cup \mathbf{s})/h(\xi)$ can be used within a pseudo-likelihood function. In the case of the above Strauss process this is just $\lambda^*(\xi, \mathbf{s}|\theta) = \lambda^*(\mathbf{s}|\theta) = \gamma^{n_R(\mathbf{s})}$ and the pseudo-likelihood is:

$$L_p(\{\mathbf{s}_1, \ldots \mathbf{s}_m\}|\theta) = \prod_{i=1}^{m} \lambda^*(\mathbf{s}_i|\theta) \exp(-\int_T \lambda^*(\mathbf{u}|\theta) d\mathbf{u}).$$

As this likelihood has the form of an inhomogeneous Poisson process likelihood, then this is relatively straightforward to evaluate. The only issue is the integral of the intensity over the window T. This can be handled via special numerical integration schemes (Berman and Turner, 1992, Lawson, 1992a, Lawson, 1992b, and Section 5.1.1). Bayesian extensions are generally straightforward. Note that once a likelihood contribution can be specified it can be incorporated within a posterior sampling algorithm such as metropolis-Hastings. This can be implemented on WinBUGS via a zeroes trick if the Berman-Turner weighting is used. For example the model with the i th likelihood component: $l_i = \log \lambda^*(\mathbf{s}_i|\theta) - w_i \lambda^*(\mathbf{s}_i|\theta)$ can be fitted using this method, where the weight w_i is based on the Dirichlet tile area of the i th point or a function of the Delauney triangulation around the point (see Berman and Turner, 1992, Baddeley and Turner, 2000 and Appendix C.5.3 of Lawson, 2006b).

In application to lattice models, Besag and Tantrum (2003) give the example of a Markov random field of m dimension where the pseudolikelihood

$L_p = \prod\limits_{i=1}^{m} p(y_i^0|y_{-i}^0;\theta)$ is the product of the full conditional distributions. In the (auto)logistic binary case

$$p(y_i^0|y_{-i}^0;\theta) = \frac{\exp(f(\alpha_0,\{y_{-i}^0\}_{\in\partial i},\theta)}{1+\exp(f(\alpha_0,\{y_{-i}^0\}_{\in\partial i},\theta)}$$

where ∂_i denotes the adjacency set of the i th site.

Other variants of these likelihoods have been proposed. Local likelihood (Tibshirani and Hastie (1987)) is a variant where a contribution to likelihood is defined within a local domain of the parameter space. In spatial problems this could be a spatial area. This has been used in a Bayesian disease mapping setting by Hossain and Lawson (2005). Pairwise likelihood (Nott and Rydén (1999), Heagerty and Lele (1998)) has been proposed for image restoration and for general spatial mixed models (Varin et al. (2005)). Approximations of point process likelihoods based on aggregations have been explored by Hossain and Lawson (2008). All these variants of full likelihoods will lead to models that are approximately valid for real applications. It should be borne in mind however that they ignore aspects of the spatial correlation and if these are not absorbed in some part of the model hierarchy the appropriateness of the model may be affected.

3.5.2 Asymptotic Approximations

It is possible to approximate a posterior distribution with a simpler distribution which is found asymptotically. The usefulness of approximations lies in their often common form and also the ease with which parameters may be estimated under the approximations. Often the asymptotic approximating distribution will be a normal distribution. Here two possible approaches are examined: the asymptotic quadratic form approximation and integral approximation via Laplace's method.

3.5.2.1 Asymptotic Quadratic Form

Large sample convergence in form of the likelihood or posterior distribution is considered here. In many cases the limiting form of a likelihood or posterior distribution in large samples can be used as an approximation. The Taylor series expansion of the function $f(.)$ around vector $\mathbf{a}$ is *

$$f(\mathbf{a}) + U(\mathbf{a})^T(\mathbf{x}-\mathbf{a}) + \frac{1}{2}(\mathbf{x}-\mathbf{a})^T H(\mathbf{a})(\mathbf{x}-\mathbf{a}). + R$$

where $U(\mathbf{a})$ is the score vector evaluated at $\mathbf{a}$, R is a remainder, and $H(\mathbf{a})$ is the Hessian matrix of second derivatives of $f(.)$ evaluated at $\mathbf{a}$. For an arbitrary

---

*more here

log likelihood with p length vector of parameters θ, then an expansion around a point is required. Usually the mode of the distribution is chosen. Define the modal vector as θ^{mo} and $l(\mathbf{y}|\theta) \equiv l(\theta)$ for brevity. The expansion is defined as:

$$l(\theta) = l(\mathbf{y}|\theta^{mo}) + U(\theta^{mo})^T (\theta - \theta^{mo}) - \frac{1}{2}(\theta - \theta^{mo})^T H(\theta^{mo})(\theta - \theta^{mo}).$$

Here $U(\theta^{mo}) = \mathbf{0}$ as we have expanded around the maxima and so this reduces to

$$l(\theta) = l(\mathbf{y}|\theta^{mo}) - \frac{1}{2}(\theta - \theta^{mo})^T H(\theta^{mo})(\theta - \theta^{mo}). \qquad (3.3)$$

Note that $H(\theta^{mo})$ describes the local curvature of the likelihood at the maxima and is defined by

$$H(\theta^{mo}) = \left(-\frac{\partial^2 l(\theta)}{\partial \theta_i \partial \theta_j} \right)\bigg|_{\theta = \theta^{mo}}.$$

This approximation, given θ^{mo}, consists of a constant and a quadratic form around the maxima. In a likelihood analysis the θ^{mo} might be replaced by maximum likelihood estimates $\widehat{\theta}^{mo}$.

For a posterior distribution, it is possible to also approximate the prior distribution with a Taylor expansion, in which case a full posterior approximation would be obtained. In the case of the joint prior distribution, defined by $p(\theta|\Gamma)$ where Γ is a parameter vector or matrix, assuming that Γ is fixed, the approximation around the modal vector θ^p, again assuming the score vector is zero at the maxima, is given by

$$\log p(\theta|\Gamma) = \log p(\theta^p|\Gamma) - \frac{1}{2}(\theta - \theta^p)^T H_p(\theta^p)(\theta - \theta^p) + R_0$$

where R_0 is the remainder term and

$$H_p(\theta^p) = \left(-\frac{\partial^2 \log p(\theta|\Gamma)}{\partial \theta_i \partial \theta_j} \right)\bigg|_{\theta = \theta^p}.$$

Again given θ^p, this is simply a quadratic form around the maxima. There are then two posterior approximations that might be considered:

i) Likelihood approximation only:

$$p(\theta|\mathbf{y}) \propto p(\theta|\Gamma) \exp\left\{ -\frac{1}{2}(\theta - \theta^{mo})^T H(\theta^{mo})(\theta - \theta^{mo}) \right\}$$

or

ii) Full posterior approximation:

$$p(\theta|\mathbf{y}) \propto \exp\left\{ -\frac{1}{2}(\theta - \theta^p)^T H_p(\theta^p)(\theta - \theta^p) - \frac{1}{2}(\theta - \theta^{mo})^T H(\theta^{mo})(\theta - \theta^{mo}) \right\}$$

$$\propto \exp\left\{ -\frac{1}{2}(\theta - \mathbf{m}^n)^T H_n(\theta - \mathbf{m}^n) \right\}$$

where $H_n = H_p(\boldsymbol{\theta}^p) + H(\boldsymbol{\theta}^{mo})$ and $\mathbf{m}^n = H_n^{-1}(H_p(\boldsymbol{\theta}^p)\boldsymbol{\theta}^p + H(\boldsymbol{\theta}^{mo})\boldsymbol{\theta}^{mo})$. Note that $H(\boldsymbol{\theta}^{mo})$ is the observed information matrix. As the sample size increases this quadratic form approximation improves its accuracy and two important results follow:

i) The posterior distribution tends towards a normal distribution i.e.

$$as \ m \to \infty \ then \ p(\boldsymbol{\theta}|\mathbf{y}) \to N_p(\boldsymbol{\theta}|\mathbf{m}^n, H_n).$$

ii) The information matrix tends towards the Fisher (expected) information matrix in the sense that $H(\boldsymbol{\theta}^{mo}) \to mI(\boldsymbol{\theta}^{mo})$ where the ijth element is

$$I(\boldsymbol{\theta})_{ij} = \int p(y|\boldsymbol{\theta}) \left(-\frac{\partial^2 l(\boldsymbol{\theta})}{\partial \theta_i \partial \theta_j} \right) dy.$$

This means that it is possible to consider further asymptotic distributional forms. For instance if the variability in the prior distribution is negligible compared to the likelihood then

$$p(\boldsymbol{\theta}|\mathbf{y}) \to N_p(\boldsymbol{\theta}|\boldsymbol{\theta}^{mo}, H(\boldsymbol{\theta}^{mo}))$$

or

$$p(\boldsymbol{\theta}|\mathbf{y}) \to N_p(\boldsymbol{\theta}|\boldsymbol{\theta}^{mo}, mI(\boldsymbol{\theta}^{mo})).$$

Often the maximum likelihood (ML) estimates would be substituted for $\boldsymbol{\theta}^{mo}$. If $\boldsymbol{\theta}^{mo}$ are given or estimated via ML the posterior distribution will be multivariate normal in large samples.

Hence a normal approximation to the posterior distribution is justified at least asymptotically (as $m \to \infty$). This approximation should be reasonably good for continuous likelihood models and may be reasonable for discrete models when the rate parameter (Poisson) is large or the binomial probability is not close to 0 or 1. Of course this is likely not to hold when there is sparseness in the count data, as can arise when rare diseases are studied. Further discussion of different asymptotic results can be found in Bernardo and Smith (1994).

An example of such a likelihood approximation would be where a binomial likelihood has been assumed and $y_i|p_i \sim Bin(p_i, n_i)$ with $p_i \sim beta(2, 2)$. In this case, assume $p(\boldsymbol{\theta}|\mathbf{y}) \sim N_p(\boldsymbol{\theta}|\mathbf{m}^n, H_n)$ and $\mathbf{m}^n = \frac{\widehat{p}_i(1-\widehat{p}_i)}{n_i} \left[\frac{n_i}{\widehat{p}_i(1-\widehat{p}_i)}\widehat{p}_i \right] = \widehat{p}_i$, $H_n = 0 + \frac{n_i}{\widehat{p}_i(1-\widehat{p}_i)}$ where $\widehat{p}_i = \frac{y_i}{n_i}$ and so the distribution is $N_m(p_i|\widehat{p}_i, diag\{\frac{n_i}{\widehat{p}_i(1-\widehat{p}_i)}\})$. Hence the approximate distribution is centered around the saturated maximum likelihood estimator. In this case the prior distribution has little effect on the mean or the variance of the resulting Gaussian distribution. If, on the other hand an asymmetric prior distribution favouring low rates of disease were assumed such as $p_i \sim Beta(1.5, 5)$, then the approximation is given by $\mathbf{m}^n = (H_p(\boldsymbol{\theta}^p) + \frac{n_i}{\widehat{p}_i(1-\widehat{p}_i)})^{-1} [H_p(\boldsymbol{\theta}^p)\boldsymbol{\theta}^p + \frac{n_i\widehat{p}_i}{\widehat{p}_i(1-\widehat{p}_i)}]$ and $H_n = H_p(\boldsymbol{\theta}^p) + \frac{n_i}{\widehat{p}_i(1-\widehat{p}_i)}$ where $H_p(\boldsymbol{\theta}^p) = 81.383$ and $\boldsymbol{\theta}^p = 0.11$. Here the mean and variance are influenced considerably.

Note that it is also possible to approximate posterior distributions with mixtures of normal distributions and this could lead to closer approximation to complex (multi-modal) distributions. Hierarchies with more than two levels have not been discussed here. However in principle, if a normal approximation can be made to each prior in turn (perhaps via mixtures of normals) then a quadratic form would result with a more complex form.

3.5.2.2 Laplace Integral Approximation

In some situations ratios of integrals must be evaluated and it is possible to employ an integral approximation method suggested by Laplace (Tierney and Kadane, 1986). For example the posterior expectation of a real valued function $g(\boldsymbol{\theta})$ is given by

$$E(g(\boldsymbol{\theta})|\mathbf{y}) = \int g(\boldsymbol{\theta})p(\boldsymbol{\theta}|\mathbf{y})d\theta.$$

This can be considered as a ratio of integrals, given the normalization of the posterior distribution. The approximation is given by

$$\widehat{E}(g(\boldsymbol{\theta})|\mathbf{y}) \approx \left(\frac{\sigma^*}{\sigma}\right) \exp\{-m[h^*(\boldsymbol{\theta}^*) - h(\boldsymbol{\theta})]\}$$

where $-mh(\boldsymbol{\theta}) = \log p(\boldsymbol{\theta}) + l(\mathbf{y}|\boldsymbol{\theta})$ and $-mh^*(\boldsymbol{\theta}) = \log g(\boldsymbol{\theta}) + \log p(\boldsymbol{\theta}) + l(\mathbf{y}|\boldsymbol{\theta})$ and

$$-h(\widehat{\boldsymbol{\theta}}) = \max_{\boldsymbol{\theta}}\{-h(\boldsymbol{\theta})\}, \ -h^*(\boldsymbol{\theta}^*) = \max_{\boldsymbol{\theta}}\{-h^*(\boldsymbol{\theta})\},$$

$$\widehat{\sigma} = |m\nabla^2 h(\widehat{\boldsymbol{\theta}})|^{-1/2} \text{ and } \widehat{\sigma} = |m\nabla^2 h^*(\boldsymbol{\theta}^*)|^{-1/2}$$

where

$$[\nabla^2 h(\widehat{\boldsymbol{\theta}})]_{ij} = \frac{\partial^2 h(\boldsymbol{\theta})}{\partial \theta_i \partial \theta_j}\bigg|_{\theta=\widehat{\theta}}.$$

3.5.2.3 INLA and R-INLA

A development in the use of approximations to Bayesian models has been proposed in a sequence of papers by Rue and coworkers (Rue et al., 2009; Lindgren et al., 2011; Simpson et al., 2012; Blangiardo et al., 2013; Lindgren and Rue, 2015; Rue et al., 2016). The basic idea is that a wide range of models which have a latent Gaussian structure can be approximated via integrated nested Laplace approximation (INLA). These approximations can be seen as successive approximations of functions within integrals. The integrals are then approximated by fixed integration schemes and not using Monte Carlo integration. If we consider a Poisson model for observed counts: y_i, $i = 1, ..., m$ then with the set of hyperparameters given by ϕ and a log link to an additive set of effects (random effects),

$$y_i|\lambda_i \sim Pois(e_i\theta_i)$$
$$\theta_i = \exp\{\alpha + v_i + u_i\}$$
$$\text{where } \boldsymbol{\lambda}_i = \{\alpha, v_i, u_i\}^T.$$

Note that the parameters in $\boldsymbol{\lambda}$ all have Gaussian distributions and have prior distribution $P(\boldsymbol{\lambda}|\phi)$. In this case it is possible to approximate the posterior marginal distribution $P(\boldsymbol{\lambda}_i|\boldsymbol{y})$ by

$$P(\boldsymbol{\lambda}_i|\boldsymbol{y}) = \int_{\phi} P(\boldsymbol{\lambda}_i|\boldsymbol{y}, \phi)P(\phi|\boldsymbol{y})d\phi$$

for each component λ_i of the latent fields. The terms $P(\lambda_i|\boldsymbol{y}, \phi)$ and $P(\phi|\boldsymbol{y})$ can each be approximated by Laplace approximation. The simplest of these is the Gaussian approximation where matching of the mode and curvature to a normal distribution is used. Finally the integral approximation leads to

$$\widetilde{P}(\boldsymbol{\lambda}_i|\boldsymbol{y}) = \sum_k \triangle_k \widetilde{P}(\boldsymbol{\lambda}_i|\boldsymbol{y}, \phi_k)\widetilde{P}(\phi_k|\boldsymbol{y}).$$

This approximation approach is now available in R (package R-inla: www.r-inla.org), and can be used for a wide variety of applications. Application of these approximations has been made by Schrodle et al. (2011) to veterinary spatial surveillance data. In that work they demonstrate the closeness of the final estimates to that achieved using posterior sampling within MCMC. However they do not show any simulated comparisons where a ground truth is compared. Further demonstration of the capabilities in spatio-temporal modeling is given by Schrodle and Held (2011). A recent simulated comparison of the performance of INLA to MCMC for disease mapping can be found in Carroll et al. (2015).

In a recent extension, the parallel with finite element solutions to differential equations is exploited by Lindgren et al. (2011) whereby the spatial field is a solution to a stochastic partial differential equation (SPDE) with form $\lambda(s_i) = \sum_k \phi_k(s)w_k$ where the $\phi_k(s)$ are basis functions and w_k are weights. This is formally close in form to the kernel process convolution models of Higdon (2002). A comparison is made by Simpson et al. (2012). Some examples of the use of INLA for Bayesian disease mapping examples are given in Chapter 15.

3.6   Alternative Computational Aproaches

3.6.1   Maximum A Posteriori Estimation (MAP)

While MCMC has been extensively used for relatively small data problems, when data size becomes prohibitive (in 1000s rather than 100s) then computational speed becomes critical. The approximations of INLA present a fast answer to this situation. However, other proposals have been suggested that do not require complete posterior sampling. These have often been developed for imaging applications where large pixel or voxel arrays must be scanned and updated. Maximum a posteriori estimation does not seek to sample the posterior but instead provides a modal estimate of the posterior parameter vector. Hence this is a maximisation albeit stochastic algorithm. Various special algorithms have been developed: simulated annealing (Kirkpatrick et al., 1983) seeks to use a temperature schedule to optimize the exploration for the global maximum. The Swenson-Wang algorithm (Wang and Swendsen, 1990) is a variant of metropolis-Hastings algorithm with a temperature schedule that controls the optimisation. A development called *simulated tempering* allows the temperature to be dynamically changed.

3.6.2   Iterated Conditional Modes (ICMs)

Iterated conditional modes is a deterministic algorithm for obtaining a configuration of a local maximum of the joint probability of a posterior distribution. It does this by iteratively maximizing the probability of each variable conditioned on the rest. It is therefore the maximisation equivalent of the Gibbs sampler but not stochastic. Besag (1986) first described the use of this approach for analysis of image data, but it could be implemented in the context of disease mapping analysis also.

3.6.3   MC³ and Parallel Tempering

Parallel tempering (Geyer, 2011), also known as replica exchange MCMC sampling, and also MC³, relies on running multiple chains in parallel with different temperature schedules and swapping or exchanging states between the chains. MH updates based on joint temperature schedules are used. The method has been implemented in the R package mcmc. Coupling MC chains is useful as it allows the sampling of multiple modes that are very relevant to mixture or latent component type problems. Some other examples and R code can be found at https://darrenjw.wordpress.com/tag/mcmcmc/.

3.6.4   Variational Bayes

Another method which provides an approximation to posterior inference is variational Bayes. In this approach the posterior is defined as $P(\boldsymbol{\theta}|Y) = P(\boldsymbol{\theta}, Y)/P(Y)$ and $log(P(Y)) = \int \log\{\frac{P(\boldsymbol{\theta}, Y)}{P(\boldsymbol{\theta}|Y)}\}q(\boldsymbol{\theta})d\boldsymbol{\theta}$ where $q(\boldsymbol{\theta})$ has the same support as the posterior distribution. The resulting mean field approximation is given by maximising $F = \int \log\{\frac{P(\boldsymbol{\theta}, Y)}{q(\boldsymbol{\theta})}\}q(\boldsymbol{\theta})d\boldsymbol{\theta}$ with respect to $q(\boldsymbol{\theta})$. Nathoo et al. (2013) made a simulated comparison of MCMC and variational Bayes within an EEG imaging example. They found that the efficiency of pursuing a variational approach is problem dependent and could be beneficial. However, they also found that variational Bayes underestimated posterior parameter variability compared to MCMC.

3.6.5   Sequential Monte Carlo

Sequential Monte Carlo (SeqMC) was developed as an alternative to MCMC. It is most suited to systems that evolve over time as it involves sequential updating (Doucet et al., 2001, Morel et al., 2012). However it can be applied to static data via data splitting and subsetting (see e.g. Chopin, 2002). This is sometimes known as particle filtration as it involves the iterative generation of parameter values (particles) which are iteratively filtered. Vidal-Rodeiro and Lawson (2006b) provide an example of applying seqMC to a spatio-temporal disease mapping example.

3.7   Approximate Bayesian Computation (ABC)

Approximate Bayesian computation (ABC) is the term given to Monte Carlo methods applied to situations where the likelihood cannot be evaluated easily. Usually the algorithm proceeds by simulating parameters from prior distributions and then data are simulated from the likelihood conditional on the simulated parameters. The data is compared with observed data and the process is repeated until the generated data is 'close to' the observed data, within a certain tolerance. The basic algorithm looks like

  1) Generate $\boldsymbol{\theta}'$ from the prior distribution $\pi(.)$.

  2) Generate $\mathbf{z}$ from the likelihood $L(.|\boldsymbol{\theta}')$ until $\rho\{\eta(\mathbf{z}), \eta(\mathbf{y})\} \leq \epsilon$ set $\theta_i = \boldsymbol{\theta}'_i$. and

  3) Repeat 1) and 2) to convergence for a sample of $i = 1, ..., P$. The $\rho\{\eta(\mathbf{z}), \eta(\mathbf{y})\}$ is a loss function measuring how far the statistic $\eta(\mathbf{z})$ is from $\eta(\mathbf{y})$. The resulting sample of $\{\boldsymbol{\theta}_1, ..., \boldsymbol{\theta}_p\}$ should be an i.i.d. sample from the posterior distribution. Clearly it is important that both the prior distribu-

tion and the likelihood can be simulated from easily. In addition, if the prior distribution is flat then the sampling will be very inefficient. A remedy is to include a transition kernel (as in MH sampling) and also a randomized acceptance step combined with the loss function. This leads to a likelihood-free MCMC sampler (see e.g. Marin et al., 2012). Adaptations have also been made whereby sequential Mont Carlo is added to the ABC to give adaptive Seq MC-ABC (Morel et al., 2012). Various speed-ups have been suggested and Bayesian optimization for likelihood-free models (BOLFI) has been developed (Gutmann and Corander, 2016). An example of a likelihood-free MCMC sampler for a gamma-Poisson model for 46 areas could be

---

generate initial θ^{t-1} from $Ga(1,1)$
for i in $1:1000$
repeat
1) generate a, b each from $Exp(1)$
2) generate θ^t from $G(.|\theta^{t-1})$ a Markov kernel
3) generate $\{z_j\}_{j=1,\ldots,46}$ from Poiss$(e_j\theta_j)$
$U1 \sim U(0,1)$
$rat = \frac{Ga(\theta^t,a,b)}{Ga(\theta^{t-1},a,b)} \frac{N(\theta^{t-1};\theta^t,\kappa)}{N(\theta^t;\theta^{t-1},\kappa)}$
if($U1 < rat$ and $|\mathbf{z} - \mathbf{y}|^2 \leq 0.1$)
set $\theta_i = \theta$
break
end for

Note that the Markov kernel is a symetric Gaussian and $\frac{N(\theta^{t-1};\theta^t,\kappa)}{N(\theta^t;\theta^{t-1},\kappa)}$ cancels in the ratio. An example of applying this ABC sampler to the South Carolina congenital abnormality 1990 county level data is given in Chapter 15. If the likelihood is known and can be evaluated then it is much more efficient to directly sample the posterior distribution via conventional MCMC.

3.8 Exercises

1) Assume a generalized linear model with Poisson likelihood: $[y_i|\theta_i] \sim Poiss(e_i\theta_i)$ with log link to linear predictor $\log\theta_i = \eta_i = \mathbf{x}'_i\beta$ where $\mathbf{x}'_i$ is a row covariate vector of p length and β is a p length parameter vector. The parameter vector is assumed to have a Gaussian prior distribution with $\mathbf{0}$ mean vector and the parameters are assumed independent; hence $\beta \sim \mathbf{N}_p(\mathbf{0}, \mathbf{\Gamma})$ where $\mathbf{\Gamma} = \mathbf{diag}\{\tau_1....,\tau_p\}$. Show that a normal approximation

to the posterior distribution of this model, given the maximum likelihood estimates $\widehat{\boldsymbol{\beta}}$, is given by $N_p(\boldsymbol{\theta}|\mathbf{m}^n, H_n)$ where $H_n = H_p(\boldsymbol{\theta}^p) + H(\boldsymbol{\theta}^m)$ and $\mathbf{m}^n = H_n^{-1}[H_p(\boldsymbol{\theta}^p)\boldsymbol{\theta}^p + H(\boldsymbol{\theta}^m)\boldsymbol{\theta}^m]$ and $\boldsymbol{\theta}^p = \mathbf{0}$, $H_p(\boldsymbol{\theta}^p) = diag\{\tau_{\beta_1}^{-1/2}....\tau_{\beta_p}^{-1/2}\}$ where σ_{β_*} is the standard deviation in the Gaussian prior distribution for β_*.

$$H(\boldsymbol{\theta}^m)_{jk} = \sum_{i=1}^{m} [A_{jk} + B_{jk}]$$
$$where$$
$$A_{jk} = \frac{y_i x_{ji} x_{ki}}{\widehat{\eta}_i (\ln \widehat{\eta}_i)^2}, \ B_{jk} = e_i x_{ji} x_{ki} \exp\{\widehat{\eta}_i\}$$
$$and \ \widehat{\eta}_i = \mathbf{x}_i' \widehat{\boldsymbol{\beta}}.$$

4

Residuals and Goodness-of-Fit

Attainment of convergence of MCMC algorithms does not necessarily yield good models. If a model is mis-specified then it will be of limited use. There are many issues relating to model goodness-of-fit that should be of concern when evaluating models for geo-referenced disease data. In this section, I treat general issues related to the use of goodness-of-fit (GOF) measures, residual diagnostics and the use of posterior output to yield risk exceedance probabilities.

4.1   Model GOF Measures

Goodness-of fit-criteria vary depending on the properties of the criteria and the nature of the model. In conventional generalized linear modeling with fixed effects, the deviance is an important measure (McCullagh and Nelder (1989)). Usually this measure of model adequacy compares a fitted model to a saturated model. It is based on the difference between the log likelihood of the data under either model:

$$D = -2[l(\boldsymbol{y}|\widehat{\boldsymbol{\theta}}_{fit}) - l(\boldsymbol{y}|\widehat{\boldsymbol{\theta}}_{sat})].$$

The saturated model has a single parameter per observation. Often a relative measure of fit is used so that deviances are compared and the change in deviance between model 1 and model 2 is used:

$$\Delta D = -2[l(\boldsymbol{y}|\widehat{\boldsymbol{\theta}}_1) - l(\boldsymbol{y}|\widehat{\boldsymbol{\theta}}_2)].$$

Hence the saturated likelihood cancels in this relative comparison. The deviance is used in goodness-of-fit measures in Bayesian modeling, but usually without reference to a saturated model.

One disadvantage of using the deviance directly is that it does not allow for the degree of parameterization in the model: a model can be made to more closely approximate data by increasing the number of parameters. Hence attempts have been made to penalize model complexity. One example of this is the Akaike information criterion (AIC). This is defined as:

$$AIC = -2l(\boldsymbol{y}|\widehat{\boldsymbol{\theta}}_{fit}) + 2p$$

where p is the number of parameters. The second term acts as a penalty for over-parameterization of the model. The idea is that as more parameters are added the more closely the model will approximate the data. To balance this, the penalty $(2p)$ is assumed. Hence the fit is penalized with a linear function of number of parameters. Model parsimony should result in the use of such a penalized form. This is widely used for fixed effect models and is the basis of the deviance information criterion discussed below.

Another variant that is commonly used as a model choice criterion is the *Bayesian information criterion (BIC)*. This is widely used in Bayesian and hierarchical models. It asymptotically approximates a Bayes factor. It is defined as

$$-2l(\boldsymbol{y}|\widehat{\boldsymbol{\theta}}_{fit}) + p\ln m.$$

In a model with log-likelihood $l(\theta)$ the AIC or BIC value can be estimated from the output of an MCMC algorithm by

$$AIC = -2\hat{l}(\boldsymbol{\theta}) + 2p$$
$$BIC = -2\hat{l}(\boldsymbol{\theta}) + p\ln m,$$

where p is the number of parameters, m is the number of data points and

$$\hat{l}(\boldsymbol{\theta}) = \frac{1}{G}\sum_{g=1}^{G} l(\boldsymbol{y}|\boldsymbol{\theta}^g),$$

the averaged log-likelihood over G posterior samples of $\boldsymbol{\theta}$. Alternatively a posterior estimate of $\widehat{\boldsymbol{\theta}}$ (such as posterior expectation) must be first computed and then substituted into the AIC or BIC. Leonard and Hsu (1999) provide comparisons of these measures in a variety of examples. One disadvantage of the AIC or BIC is that in models with random effects, it is difficult to decide how many parameters are included within the model. For example, a unit level effect could be specified as $v_i \sim N(0, \tau)$. In this case there is one variance parameter, but there is also a separate value of v for each item. Hence, we have potentially super-saturation of the parameter space $(p > m)$ if we count the m $v_i s$ as well as τ. Should the parameterisation be 1 or $m+1$ or somewhere between these values? This quandary does not arise with random coefficient models where, for example, in a regression context we may have a p length vector of parameters, β say, that may in the simplest case have p variances.

4.1.1   Deviance Information Criterion

The *deviance information criterion (DIC)* has been proposed by Spiegelhalter et al. (2002) and is widely used in Bayesian modeling. This is defined as

$$DIC = 2E_{\boldsymbol{\theta}|\boldsymbol{y}}(D) - D[E_{\boldsymbol{\theta}|\boldsymbol{y}}(\boldsymbol{\theta})],$$

where $D(.)$ is the deviance of the model and y is the observed data. Note that the DIC is based on a comparison of the average deviance ($\overline{D} = -2\sum_{g=1}^{G} l(y|\theta^g)/G$) and the deviance of the posterior expected parameter estimates, $\hat{\theta}$ say: ($\hat{D}(\hat{\theta}) = -2l(y|\hat{\theta})$). For any sample parameter value θ^g the deviance is just $\hat{D}(\theta^g) = -2l(y|\theta^g)$. The effective number of parameters (pD) is estimated as $\widehat{pD} = \overline{D} - \hat{D}(\hat{\theta})$ and then $DIC = \overline{D} + \widehat{pD} = 2\overline{D} - \hat{D}(\hat{\theta})$. Unfortunately in some situations the $\widehat{pD}$ can be negative (as it can happen that $\hat{D}(\hat{\theta}) > \overline{D}$). Instability in pD can lead to problems in the use of this DIC. For example, mixture models, or more simply, models with multiple modes can "trick" the pD estimate because the overdispersion in such models (when the components are not correctly estimated) leads to $\hat{D}(\hat{\theta}) > \overline{D}$ (see e.g. Lunn et al., 2012). For alternative DIC formulations that can allow for mixture models see e.g. Celeux et al. (2006). It is also true that inappropriate choice of hyper-parameters for variances of parameters in hierarchical models can lead to inflation also, as can nonlinear transformations (such as changing from a Gaussian model to a log normal model). In such cases it is sometimes safer to compute the effective number of parameters from the posterior variance of the deviance. Gelman et al. (2004) on p. 182 propose the estimator $\widehat{pD} = \frac{1}{2}\frac{1}{G-1}\sum_{g=1}^{G}(\hat{D}(\theta^g) - \overline{D})^2$. This value can also be computed from sample output from a chain. (It is also available directly in R2WinBUGS.) An alternative estimator of the variance is directly available from output:

$$\widehat{var}(D) = \frac{1}{G-1}\sum_{g=1}^{G}(\hat{D}(\theta^g) - \overline{D})^2 = 2\widehat{pD}.$$ Hence a DIC based on this last

variance estimate is just $DIC = \overline{D} + \widehat{var}(D)/2$. Note that the *expected predictive deviance* (EPD: D_{pr}) is an alternative measure of model adequacy and it is based on the out-of-sample predictive ability of the fitted model. The quantity can also be approximately estimated as $\hat{D}_{pr} = 2\overline{D} - \hat{D}(\hat{\theta})$.

Recently a new measure of goodness of fit has been termed the Watanabe-Akaike Information criterion (WAIC) (see e.g. Gelman et al., 2014). This can be computed from the log pointwise predictive density (lppd), for G sampled values, as

$$lppd = \sum_{i=1}^{m} \log\{\sum_{g=1}^{G} p(y_i|\theta^g)/G\}$$

where $p(.|.)$ is the data density. The parameterisation penalty is the variance of the log pointwise predictive density summed over the data points to yield the effective number of parameters: $pWAIC = \sum_{i=1}^{m} V_{g=1}^{G}(\log p(y_i|\theta^g))$ where

$V_{g=1}^{G}(a) = \frac{1}{G-1}\sum_{g=1}^{G}(a - \bar{a})^2$. This leads to

$$WAIC = lppd - pWAIC.$$

The WAIC can be computed from sampled MCMC output and is available automatically within CARBayes and R-INLA. Spiegelhalter et al. (2014) compare the DIC and WAIC and note the potential instability of DIC and the improvement of the WAIC formulation.

4.1.2 Posterior Predictive Loss

Gelfand and Ghosh (1998) proposed a loss function based approach to model adequacy which employs the predictive distribution. The approach essentially compares the observed data to predicted data from the fitted model. Define the i th predictive data item as y_i^{pr}. Note that the predictive data can easily be obtained from a converged posterior sample. Given the current parameters at iteration $j : \boldsymbol{\theta}^{(j)}$ say, then

$$p(y_i^{pr}|\mathbf{y}) = \int p(y_i^{pr}|\boldsymbol{\theta}^{(j)})p(\boldsymbol{\theta}^{(j)}|\mathbf{y})d\boldsymbol{\theta}^{(j)}.$$

Hence the j th iteration can yield y_{ij}^{pr} from $p(y_i^{pr}|\boldsymbol{\theta}^{(j)})$. The resulting predictive value has marginal distribution $p(y_i^{pr}|\mathbf{y})$. For a Poisson distribution, this simply requires generation of counts as $y_{ij}^{pr} \leftarrow Pois(e_i\theta_i^{(j)})$.

A loss function is assumed where $L_0(\mathbf{y}, \mathbf{y}^{pr}) = f(\mathbf{y}, \mathbf{y}^{pr})$. A convenient choice of loss could be the squared error loss whereby we define the loss as: $L_0(\mathbf{y}, \mathbf{y}^{pr}) = (\mathbf{y} - \mathbf{y}^{pr})^2$.

Alternative loss functions could be proposed such as absolute error loss or more complex (quantile) forms. An overall crude measure of loss across the data is afforded by the average loss across all items: the *mean squared predictive error* (MSPE) is simply an average of the item-wise squared error loss:

$$MSPE_j = \sum_i (y_i - y_{ij}^{pr})^2/m$$

and

$$MSPE = \sum_i \sum_j (y_i - y_{ij}^{pr})^2/(G \times m),$$

where m is the number of observations and G is the sampler sample size. An alternative could be to specify an absolute error :

$$MAPE_j = \sum_i |y_i - y_{ij}^{pr}|/m$$

and

$$MAPE = \sum_i \sum_j |y_i - y_{ij}^{pr}|/(G \times m).$$

Gelfand and Ghosh (1998) proposed a more sophisticated form:

$$D_k = \frac{k}{k+1}A + B$$

$$= \frac{k}{k+1}\sum_{i=1}^{m}(y_i - \overline{y_i^{pr}})^2 + \frac{1}{m_p}\sum_{i=1}^{m}\sum_{j=1}^{m_p}(y_{ij}^{pr} - \overline{y_i^{pr}})^2$$

$$\text{where } \overline{y_i^{pr}} = \sum_{j=1}^{m_p} y_{ij}^{pr}/m_p$$

and m_p is the prediction sample size (usually $G = m_p$). Here, the k can be chosen to weight the different components. For $k = \infty$, then $D_k = A + B$. The choice of k does not usually affect the ordering of model fit. Each component measures a different feature of the fit: A represents lack of fit and B degree of smoothness. The model with lowest D_k (or $MSPE$ or $MAPE$) would be preferred.

Note that predictive data can be easily obtained from model formulae in WinBUGS: in the Poisson case with observed data y[] and predicted data ypred[], we have for the i th item

$$y[i]\tilde{}dpois(mu[i])$$

$$ypred[i]\tilde{}dpois(mu[i])$$

As the predicted values are missing they would have to be initialized.

In addition, it is possible to consider prediction-based measures as convergence diagnostics. The measures $\widehat{D}(\theta^{(j)})$, or $MSPE_j$ could be monitored using the single- or multi-chain diagnostics discussed above (Section 3.3.6). Note that for any model for which a unit likelihood contribution is available, it is possible to compute a deviance-based measure such as $\widehat{D}(\theta^{(j)})$. Hence for point process (case event), as well as count-based likelihoods, deviance measures are available whereas a residual based measure (such as MSPE) is more difficult to define for a spatial event domain. A further cross-validatory measure of goodness-of-fit called marginal predictive likelihood (MPL) or log of the MPL (LMPL) is discussed in a later in Section 4.6.

4.2   General Residuals

The analysis of residuals and summary functions of residuals forms a fundamental part of the assessment of model goodness-of-fit in any area of statistical application. In the case of disease mapping there is no exception, although full residual analysis is seldom presented in published work in the area. Often goodness-of-fit measures are aggregate functions of piecewise residuals, while measures relating to individual residuals are also available. A variety of methods are available when full residual analysis is to be undertaken. We define a piecewise residual as the standardized difference between the observed value and the fitted model value. Usually the standardisation will be based on a measure of the variability of the difference between the two values.

It is common practice to specify a residual as:

$$r_i = y_i - \widehat{y}_i \qquad\qquad (4.1)$$

or

$$r_i^s = r_i / \sqrt{\widehat{var}(r_i)}$$

where $\widehat{y}_i$ is a fitted value under a given model. When complex spatial models are considered, it is often easier to examine residuals, such as $\{r_i\}$ using Monte Carlo methods. In fact it is straightforward to implement a *parametric bootstrap* (PB) approach to residual diagnostics for likelihood models (Davison and Hinkley (1997)). The simplest case, is tract count data, where for each tract an observed count can be compared to a fitted count. In general, when Poisson likelihood models are assumed with $y_i \sim Pois\{e_i.\theta_i\}$ then it is straightforward to employ a PB by generating a set of simulated counts $\{y_{ij}^*\}$ $j = 1, ..., J$, from a Poisson distribution with mean $e_i.\widehat{\theta}_i$. In this way, a tract-wise ranking, and hence p-value, can be computed by assessing the rank of the residual within the pooled set of $J + 1$ residuals:

$$\{y_i - e_i.\widehat{\theta}_i; \{y_{ij}^* - e_i.\widehat{\theta}_i\}, j = 1, ..., J\}.$$

Denote the observed standardized residual as r_i^s, and the simulated as r_{ij}^s. Note that it is now possible to compare functions of the residuals as well as making direct comparisons. For example, in a spatial context, it may be appropriate to examine the spatial autocorrelation of the observed residuals. This may provide evidence of lack of model fit. Hence, a Monte Carlo assessment of degree of residual autocorrelation could be made by comparing Moran's I statistic for the observed residuals, say, $M(\{r_i^s\})$, to that found for the simulated count residuals $M(\{r_{ij}^s\})$, where $M(\{u\}) = \frac{u^T W u}{u^T u}$ where $u_i = r_i / \sqrt{var(r_i)}$ and $r_i = (y_i - e_i.\widehat{\theta}_i)$ and W is an adjacency matrix. Note that $E[M(\{u\})]$ may not be zero and so it would be important to allow for this fact in any assessment of residual autocorrelation.

In the above discussion the residual definition relies on an observed dependent variable (usually a count or other discrete outcome) and a model-based fitted variable value. In the situation where case event data is modeled via point process models then the domain of interest is spatial and it is more problematic to define a residual. Note that this does not apply to conditional logistic models for case event data, as a binary outcome is modeled (see Section 5.1.2). For a spatial domain it is convenient to consider a local measure of the case density and to compare it to a model fitted density. A deviance residual was proposed by Lawson (1993a), which compared a saturated density estimate with a modelled estimate. Extension to other processes has also been made (see Baddeley et al., 2005 and discussion, Baddeley et al., 2008). The measure $\widehat{D}(\boldsymbol{\theta}) = -2l(\boldsymbol{y}|\boldsymbol{\theta})$ is available for any likelihood model and so a relative comparison is possible between models with different estimated $\boldsymbol{\theta}$. Hence, $r_{d_i} = \widehat{D}_i(\boldsymbol{\theta}_s) - \overline{D}_i$ where $\overline{D}_i = -2\sum_{g=1}^{G} l(y_i|\boldsymbol{\theta}^g)/G$ and $\boldsymbol{\theta}_s$ is a saturated estimate of the parameters, or an averaged version such as $r_{d_i} = \frac{1}{G}\sum_{g=1}^{G}[\widehat{D}_i(\boldsymbol{\theta}_s) - \widehat{D}_i(\boldsymbol{\theta}^g)]$ is available. If $\boldsymbol{\theta}_s$ is fixed then these residuals are the same. A simple saturated estimate of density is $\frac{1}{A_i}$ where A_i is the area of the i th Voronoi (Dirichlet) tile. This is not a consistent estimator. Define a neighborhood set of the i th location as δ_i and the number in the set as n_{δ_i}. This set consists of all the areas that are regarded as neighbors of that point. In a tesselation the first order neighbors are usually those locations that share a common boundary with the point of interest. Hence, further averaging over tile neighbors might be useful to improve this "local" estimate. An example could be $\widehat{\lambda}_{s_i} = \frac{n_{\delta_i}+1}{A_i+\sum_{j\in\delta_i} A_j}$. In the case event situation, for a realization of cases at locations $\{s_i\}, i = 1, ..., m$, the likelihood would be a function of the intensity of the process at these locations λ_i. Hence, $\overline{D}_i = -2\sum_{g=1}^{G} l(s_i|\lambda_i^g)/G$ and $\widehat{D}_i(\boldsymbol{\theta}_s) = -2l(s_i|A_i^{-1})$ for the simple case or $\widehat{D}_i(\boldsymbol{\theta}_s) = -2l(s_i|\widehat{\lambda}_{s_i})$ where $\widehat{\lambda}_{s_i} = \frac{n_{\delta_i}+1}{A_i+\sum_{j\in\delta_i} A_j}$ for the consistent case. Hence a deviance residual can be set up for a simple non-stationary Poisson point process by computing

$$r_{d_i} = -2[l(s_i|A_i^{-1}) + \sum_{g=1}^{G} l(s_i|\lambda_i^g)/G].$$

This does not make allowance for modulation of the process however. Usually the intensity of the case event process is modulated by the "at risk" population distribution, and the intensity of the modulated process is $\lambda_i \equiv \lambda(s_i) = \lambda_0(s_i).\lambda_1(s_i;\boldsymbol{\theta})$ where $\lambda_0(s_i)$ represents this population effect and $\lambda_1(s_i;\boldsymbol{\theta})$ is the excess risk density suitably parameterized with $\boldsymbol{\theta}$. Define $\lambda_{0i} = \lambda_0(s_i)$, and $\lambda_{1i} = \lambda_1(s_i;\boldsymbol{\theta})$. Assuming that λ_{0i} is known, then the simple saturated estimate of $\lambda(s_i)$ is $\widehat{\lambda}_i = A_i^{-1}$ (as the estimate subsumes both population and excess risk effects). Hence the residual becomes

$$r_{d_i} = -2[l(s_i|A_i^{-1}) + \sum_{g=1}^{G} l(s_i|\lambda_{0i}\lambda_{1i}^g)/G].$$

The estimation of λ_{0i} may be an issue in any application, if it is not known, and this is discussed further in Section 5.1.1.

4.3   Bayesian Residuals

In a Bayesian setting it is natural to consider the appropriate version of (4.1).
Carlin and Louis (2000) describe a Bayesian residual as:

$$r_i = y_i - \frac{1}{G} \sum_{g=1}^{G} E(y_i | \theta_i^{(g)}) \tag{4.2}$$

where $E(y_i | \theta_i)$ is the expected value from the posterior predictive distribution,
and (in the context of MCMC sampling) $\{\theta_i^{(g)}\}$ is a set of parameter values
sampled from the posterior distribution.

In the tract count modeling case, with a Poisson likelihood and expectation
$e_i \theta_i$, this residual can be approximated, when a constant tract rate is assumed,
by:

$$r_i = y_i - \frac{1}{G} \sum_{g=1}^{G} e_i \theta_i^{(g)}. \tag{4.3}$$

This residual averages over the posterior sample. An alternative computa-
tional possibility is to average the $\{\theta_i^{(g)}\}$ sample, $\widehat{\theta}_i = \frac{1}{G} \sum_{g=1}^{G} \theta_i^{(g)}$ say, to
yield a posterior expected value of y_i, say $\widehat{y}_i = e_i \widehat{\theta}_i$, and to form $r_i = y_i - \widehat{y}_i$.
A further possibility is to simply form r_i at each iteration of a posterior sam-
pler and to average these over the converged sample of Spiegelhalter et al.
(1996). These residuals can provide pointwise goodness-of-fit (*GOF*) measures as well as global *GOF* measures, (such as mean squared error (MSE):
$\frac{1}{m} \sum_{i=1}^{m} r_i^2$) and can be assessed using Monte Carlo methods. For exploratory
purposes it might be useful to standardize the residuals before examination,
although this is not essential for Monte Carlo assessment. To provide for
a Monte Carlo assessment of individual unit residual behavior, a repeated
Monte Carlo simulation of independent samples from the predictive distribution would be needed. This can be achieved by taking J samples from the
converged MCMC stream with gaps of length p, where p is large enough to
ensure independence. Ranking of the residuals in the pooled set $(J + 1)$ can
then be used to provide a Monte Carlo p-value for each unit.

4.4   Predictive Residuals and Bootstrap

It is possible to disaggregate the MSPE to yield individual level residu-
als based on the predictive distribution. Define $y_i^{pr} \sim f_{pr}(\mathbf{y}^{pr} | \mathbf{y}; \boldsymbol{\theta})$ where

$f_{pr}(\mathbf{y}^{pr}|\mathbf{y};\theta) = \int f(\mathbf{y}^{pr}|\boldsymbol{\theta})p(\boldsymbol{\theta}|\mathbf{y})d\boldsymbol{\theta}$ and $f(\mathbf{y}^{pr}|\boldsymbol{\theta})$ is the likelihood of $\mathbf{y}^{pr}$ given $\boldsymbol{\theta}$. This can be approximated within a converged sample by a draw from $f(\mathbf{y}^{pr}|\boldsymbol{\theta})$. For a Poisson likelihood, at the g th iteration, with expectation $e_i\theta_i^{(g)}$, a single value $y_i^{pr(g)}$ is generated from $Pois(e_i\theta_i^{(g)})$. Hence a predictive residual can be formed from $r_i^{pr} = y_i - y_i^{pr}$. This must be averaged over the sample. This can be done in a variety of ways. For example, we could take $r_i^{pr} = \sum_{g=1}^{G}\{y_i - y_i^{pr(g)}\}/G$.

Other possibilities could be explored (see for example, Marshall and Spiegelhalter, 2003).

To further assess the distribution of residuals, it would be advantageous to be able to apply the equivalent of PB in the Bayesian setting. With convergence of a MCMC sampler, it is possible to make subsamples of the converged output. If these samples are separated by a distance (h) which will guarantee approximate independence (Robert and Casella, 2005), then a set of J such samples could be used to generate $\{y_i^{pr}\}$ $j = 1, ..., J$, with, $y_i^{pr} \leftarrow Pois(e_i\widehat{\theta}_{ij})$, and the residual computed from the data r_i can be compared to the set of J residuals computed from $y_i^{pr} - \widehat{y}_i$. In turn, these residuals can be used to assess functions of the residuals and GOF measures. The choice of J will usually be 99 or 999 depending on the level of accuracy required.

4.4.1 Conditional Predictive Ordinates (CPOs.)

It is possible to consider a different approach to inference whereby individual observations are compared to the predictive distribution with observations removed. This conditional approach has a cross-validation flavor, i.e. the value in the unit is predicted from the remaining data and compared to the observed data in the unit. The derived residual is defined as

$$r_i^{CPO} = y_i - y_{i,-i}^{rep}$$

where $y_{i,-i}^{rep}$ is the predicted value of y based on the data with the i th unit removed (Stern and Cressie, 2000). The value of $y_{i,-i}^{rep}$ is obtained from the cross-validated posterior predictive distribution:

$$p(y_{i,-i}^{rep}|\mathbf{Y}_{-i}) = \int p(y_{i,-i}^{rep}|\boldsymbol{\theta})p(\boldsymbol{\theta}|\mathbf{Y}_{-i})d\boldsymbol{\theta}$$

where $\boldsymbol{\theta}$ is a vector of model parameters. For a Poisson data likelihood, $p(y_{i,-i}^{rep}|\mathbf{Y}_{-i})$ is just a Poisson distribution with mean $e_i^*\theta_i$ where e_i^* is adjusted for the removal of the i th unit and θ_i is estimated under the cross-validated posterior distribution $p(\boldsymbol{\theta}|\mathbf{Y}_{-i})$.

As noted by Spiegelhalter et al. (1996), it is possible to make inference about such residuals within a conventional MCMC sampler via the construction of weights. Assume draws $g = 1, ..., G$ are available and define the importance weight $w_{-i}(\boldsymbol{\theta}^g) = \frac{1}{p(y_i|e_i\theta_i^g)}$. This is just the reciprocal of the Poisson probability with mean $e_i\theta_i^g$. It is then possible to compute a Monte Carlo probability for the residual via:

$$p(y_{i,-i}^{rep} \le y_i|\mathbf{Y}_{-i}) = p(y_{i,-i}^{rep} < y_i|\mathbf{Y}_{-i}) + \frac{1}{2}p(y_{i,-i}^{rep} = y_i|\mathbf{Y}_{-i})$$

$$\approx \sum_{l=0}^{y_i-1} \left\{ \sum_{g=1}^{G} p(y_{i,-i}^{rep} = g|\boldsymbol{\theta}^g)w_{-i}(\boldsymbol{\theta}^g) \right\} / w_i^T$$

$$+ \frac{1}{2} \left\{ \sum_{g=1}^{G} p(y_{i,-i}^{rep} = y_i|\boldsymbol{\theta}^g)w_{-i}(\boldsymbol{\theta}^g) \right\} / w_i^T$$

where $w_i^T = \sum_{g=1}^{G} \frac{1}{p(y_i|e_i\theta_i^g)}$. In general, a simple approach to computation of the CPO without recourse to refitting is to note that the conditional predictive ordinate for the i the unit can be obtained from

$$CPO_i^{-1} = G^{-1} \sum_{g=1}^{G} p(y_i|\boldsymbol{\theta}^g)^{-1}$$

where $p(y_i|\boldsymbol{\theta}^g)$ is the data density given the current parameters. Note that the CPO_i can be used as local measure of model fit in that $0 < CPO_i < 1$ and values closest to 1 suggest good fit while values close to zero suggest data points that are poorly fitted. An early example of the use of these diagnostics is found in Dey et al. (1997).

4.5   Interpretation of Residuals in a Bayesian Setting

Diagnostics based on residuals will be indicative of a variety of model features. What should be expected from residuals from an adequate model? In general, when a model fits well, one would expect the residuals from that model to have a number of features. First they should usually be symmetric and centered around zero. Clearly variance standardization should yield a closer approximation to this, but in general this can only be approximate. Second they should not show any particular structure and should appear to be reasonably random. However, as to distribution, it is not clear that residuals should have a zero mean Gaussian form (as suggested by the use of normal quantile plots in, e.g. Carlin and Louis (2000)). In the following example, I examine a dataset for congenital anomaly mortality counts for the 46 counties of South Carolina for the year 1990. The standardized mortality ratio for these data using the statewide standard population rate is shown in Section 1.1. Figure 4.1 displays Bayesian residuals from a converged Poisson-gamma model fitted to congenital anomaly mortality for South Carolina counties 1990. These residuals seem to have some structure although their QQ plot is relatively straight.

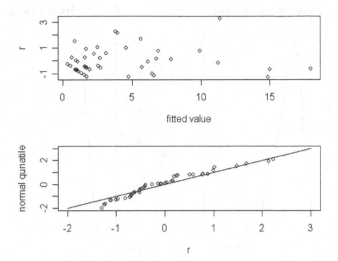

FIGURE 4.1: Residual plots of Bayesian residuals for 46 counties of South Carolina for congenital anomaly mortality data 1990. Gamma-Poisson model fit with standardised residual versus fitted value (top) and QQ plot of standardised residual versus standard normal quantile (bottom).

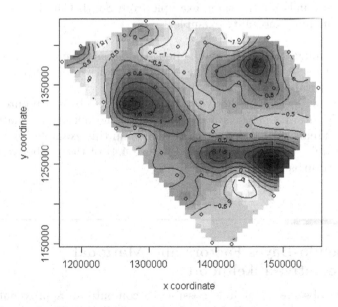

FIGURE 4.2: Bayes residual map for the South Carolina congenital anomaly mortality data. The county centroids are marked as ◇.

FIGURE 4.3: Residual plots using density estimation. Bayes residuals and predictive residuals for the same example from South Carolina county-level congenital anomaly mortality example.

Figure 4.2 displays the corresponding image and contour map of the smoothed residual surface. It is noticeable that the high positive residuals are grouped in relatively rural areas.

As a comparison, the predictive residuals have also been computed for this example. Figure 4.3 displays a comparison of the density estimates of the different residuals. These appear quite similar in this example, with a longer right hand tail. The associated map (Figure 4.4) of the predictive residuals shows a rural-urban difference also.

4.6   Pseudo-Bayes Factors and Marginal Predictive Likelihood

It has been observed that it is possible to compute an approximation to a Bayes factor based on CPO_i values (Gelfand and Dey, 1994; Dey et al., 1997). For two different models $M1$ and $M2$ the pseudo-Bayes factor for $M1$ versus $M2$ can be written as $PBF = \prod_i CPO_i(M1) / \prod_i CPO_i(M2)$, and for a given

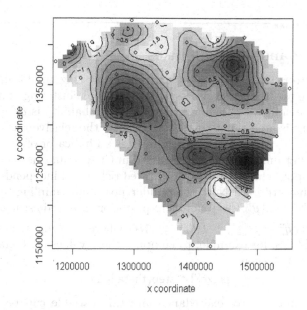

FIGURE 4.4: Predictive residual map for the South Carolina congenital anomaly mortality example.

model ($M1$ say) then $PL(M1) = \sum_i \log(CPO_i(M1))$ is an estimate of the log marginal predictive likelihood (LMPL) for that model. A comparison can be made between any competing models based on values of $PL(.)$. In general, the less negative the value of $PL(.)$ the higher the probability of the model. MPL, DIC, WAIC and MSPE are all commonly used as overall model goodness-of-fit measures.

4.7   Other Diagnostics

Another approach to evaluating model behavior is to consider the relation of prior probability model(s), and neighborhoods and their local effect on the posterior distribution and hence parameter estimates. Scheel et al. (2011) have proposed the use of the *local critique plot* which is constructed by examining a plot of the cumulative distribution of the local prior and the cumulative normalized likelihood. In addition a cumulative local marginal posterior distribution can be examined. In this way unusual dependencies can be detected in nodes of the hierarchical model.

4.8   Exceedance Probabilities

Exceedance probabilities are important when assessing the localized spatial behavior of the model and the assessment of unusual clustering or aggregation of disease. The simplest case of an exceedance probability is $q_i^c = \Pr(\theta_i > c)$. The probability is an estimate of how frequently the relative risk exceeds the null risk value ($\theta_i = 1$) and can be regarded as an indicator of "how unusual" the risk is in that unit. As will be discussed in Chapter 6, this leads to assessment of "hot spot" clusters: areas of elevated risk found independently of any cluster grouping criteria. Note that under posterior sampling, a converged sample of $\left\{ \theta^{m+1}, ..., \theta^{m+m_p} \right\}$ can yield posterior expected estimates of these probabilities as $\hat{q}_i^c = \sum_{g=m+1}^{m+m_p} I(\theta_i^{(g)} > c)/G$ where $G = m_p$. It is straightforward on WinBUGS, for example, to compute these values. If theta[i] is set to store the current θ_i then

$$\text{prexc[i]} < \text{-step(theta[i]-1)}$$

will store the indicator of exceedance and this variable can be monitored. The posterior average of the variable will yield the posterior estimate of the probability. Large values of $\hat{q}_i^c$ would be suggestive of unusual areas of risk. In fact this measure has been proposed as a method for detecting clusters (Richardson et al., 2004; Abellan et al., 2008; Hossain and Lawson, 2006, 2010). While it certainly can be used to examine maps for individual hot spots of risk, the measure itself does not directly measure clustering of risk in terms of spatial aggregation. In fact, the measure can be applied to any underlying model when a relative risk parameter is estimated and hence may be model dependent. This is discussed more fully in the chapter on disease cluster modeling (Chapter 6).

There are a number of issues associated with the evaluation of $\hat{q}_i^c$. First the value of c must be specified. Second the level of probability that is regarded as "*unusual*" (a say) must also be fixed. However there is a trade-off between these measures. Not only will there be different features found if you change the threshold (c), but there will be different interpretations if the threshold is changed from say $a_1 = 0.95$ to $a_2 = 0.99$ or lowered to $a_0 = 0.90$. Hence a simple rule such as, *classify as unusual any region where* $\hat{q}_i^1 > a_1$, might be equivalent to $\hat{q}_i^2 > a_0$. Whereas the null level of risk may be $c = 1$ there is no reason to assume as a threshold $\hat{q}_i^1 > 0.95$. Of course usually a or c is fixed. It should be noted that $\hat{q}_i^c$ is a function of a model and so is not necessarily going to yield the same information as, say a residual r_i. While both depend on model elements, a residual usually also contains extra (at least) uncorrelated noise and should, if the model fits well, not contain any further structure. On the other hand, posterior estimates of relative risk will include modeled components of risk (such as trend or correlation) and should be relatively free of extra noise.

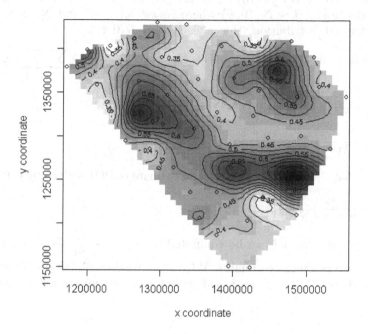

FIGURE 4.5: The exceedance probability contour map of $\hat{q}_i^1$ (*i.e.* $c = 1$) for the South Carolina county-level congenital anomaly mortality for 1990.

In the example of the South Carolina congenital anomaly mortality for 1990, the standardized residuals and the $\hat{q}_i^1$ seem to reflect areas of relatively unusual risk. Both reflect rural areas where incidence is marked. The $\hat{q}_i^1$ map displayed in Figure 4.5 shows these areas well. None of the areas exceed 0.95 in this case. A comparison with Figure 4.2 suggests that even though the model does reflect the elevated risk areas, in these areas there is excess risk unaccounted for.

4.9 Exercises

1) The Bernoulli model with outcome y_i is assumed for the binary label for a set of m case and n control events: $S : \{ s_1, ..., s_m, s_{m+1}, ..., s_{m+n} \}$. Let us, denote the outcome as $y_i \equiv y(s_i)$. The probability, conditional on a site s_i, of being a case, is $p_i \equiv p(s_i)$. Assume a logistic link to a covariate

model: $logit\ p_i = \eta_i = x_i^T \beta$ with an intercept and single covariate x_{1i}. The parameter vector is $\beta : [\beta_0, \beta_1]'$. The parameters have zero mean Gaussian prior distributions: $\beta \sim N_2(0, \Gamma)$ where $\Gamma = diag(\tau_0, \tau_1)$ and N_2 denotes a bivariate normal distribution. Hence, each $\beta_* \sim N(0, \tau_*)$.

a) Show that the DIC for this model is given by $DIC = -2\frac{1}{G} \sum\limits_{g=1}^{G} \sum\limits_{i=1}^{m+n} \log[p_i^g] +$

$$\frac{1}{2}\frac{1}{G-1} \sum\limits_{g=1}^{G} \left[2(\frac{1}{G} \sum\limits_{g=1}^{G} \sum\limits_{i=1}^{m+n} \log[p_i^g] - \sum\limits_{i=1}^{m+n} \log[p_i^g]) \right]^2$$

where $p_i^g = \frac{[\exp(\beta_0^g + \beta_1^g x_{1i})]^{y_i}}{1 + \exp(\beta_0^g + \beta_1^g x_{1i})}$.

b) Show that the conditional predictive ordinate (CPO) is given by $CPO_i =$

$$\left\{ \frac{1}{G} \sum\limits_{g=1}^{G} \left[\frac{[\exp(\beta_0^g + \beta_1^g x_{1i})]^{y_i}}{1 + \exp(\beta_0^g + \beta_1^g x_{1i})} \right]^{-1} \right\}^{-1}.$$

c) A Bayesian residual can be computed from $y_i - \frac{\exp(\widehat{\beta}_0 + \widehat{\beta}_1 x_{1i})}{1 + \exp(\widehat{\beta}_0 + \widehat{\beta}_1 x_{1i})}$, where $\widehat{\beta}_0, \widehat{\beta}_1$ are posterior mean estimates. Why is this residual difficult to interpret? What remedy could be suggested to allow a more meaningful residual analysis?

2) A case event realization within a window of area T is defined as $\{s_i\}$, $i = 1, ..., m$. A modulated Poisson process model is assumed with first order intensity, conditional on a parameter vector θ, governed by $\lambda(s) = \lambda_0(s).\lambda_1(s|\theta)$. Assume that $\lambda_1(s|\theta) = 1 + \alpha \exp\{-\beta\|s - s_0\|\}$ where s_0 is a fixed spatial location. Assume also that $\lambda_0(s)$ is fixed and known. The prior distributions for the parameters are $\alpha_t = \log \alpha \sim N(0, \tau_\alpha)$ and $\beta \sim N(0, \tau_\beta)$. The log likelihood is given by

$$l(s|\alpha, \beta) = \sum\limits_{i=1}^{m} \ln(1 + \alpha \exp\{-\beta d_i\}) - \Lambda(\alpha, \beta)$$

where $d_i = \|s_i - s_0\|$ and $\Lambda(\alpha, \beta) = \int\limits_{T} \lambda_0(u).[1 + \alpha \exp\{-\beta\|u - s_0\|\}]du$.

a) Show that, for this model under MCMC iterative sampling with a converged sample of size G, the $DIC = \overline{D} + \widehat{pD}$ where $\overline{D} = -2\sum\limits_{g=1}^{G}[\sum\limits_{i=1}^{m} \ln(1 +$

$\alpha^g \exp\{-\beta^g d_i\}) - \Lambda(\alpha^g, \beta^g)]/G$ and $\widehat{pD} = \frac{1}{2}\frac{1}{G-1}\sum\limits_{g=1}^{G}([-2\sum\limits_{i=1}^{m} \ln(1 +$

$\alpha^g \exp\{-\beta^g d_i\}) - \Lambda(\alpha^g, \beta^g)] - \overline{D})^2$.

b) Show that a deviance residual can be computed from r_{d_i} = $-2[l(s_i|A_i^{-1})+\sum_{g=1}^{G}[\ln(1+\alpha^g \exp\{-\beta^g d_i\})-\Lambda^*(\alpha^g,\beta^g)])/G]$ where $\Lambda_i^*(\alpha^g,\beta^g)$ = $w_i\lambda_0(s_i).[1+\alpha\exp\{-\beta||s_i-s_0||\}]$ and w_i is a Berman-Turner integration weight (see Chapter 5) for the i th unit.

c) Suggest a total model discrepancy measure based on this residual.

3) For the model in 2) consider an exceedance probability for the resulting estimated intensity function. The focus of interest is the relative risk $\lambda_1(s|\boldsymbol{\theta})$. Under posterior sampling the estimated risk is $\lambda_1(s|\boldsymbol{\theta}) = [1+\alpha^g \exp\{-\beta^g d_i\}]$. However usually it is assumed that $\alpha > 0$, and usually, but not necessarily, $\beta > 0$ (as the estimated risk could increase with distance from the fixed point). Hence we cannot have $\lambda_1(s|\boldsymbol{\theta}) < 1$, as will be seen later in Chapter 8. This risk is a natural form for the function of distance. However a simpler form that is often assumed is a purely multiplicative one:

$$\lambda_1(s|\boldsymbol{\theta}) = \rho.\exp\{-\beta d_i\}$$

where $\rho = \exp(\alpha)$. Show that an exceedance probability $(\widehat{g}_i)$ can be computed from:

$$\widehat{g}_i = \widehat{\Pr}(\lambda_1(s_i|\boldsymbol{\theta}) > 1) = \sum_{g=1}^{G} I(\exp\{\alpha^g - \beta^g d_i\}/G).$$

Part II

Themes

5

Disease Map Reconstruction and Relative Risk Estimation

5.1 Introduction to Case Event and Count Likelihoods

5.1.1 Poisson Process Model

Define a study area as T and within that area m events of disease occur. These events usually address locations of the cases. A case could be an incident or prevalent case or could be a death certificate address. We assume that the cases are geo-coded down to a point (with respect to the scale of the total study region). Hence they form a point process in space. Define $\{s_i\}, i = 1, ..., m$ as the set of all cases within T. This is called a *realization* of the disease process, in that we assume that all cases within the study area are recorded. This is a common form of data available from government agencies. Sub-samples of the spatial domain, where incomplete realizations are taken are not considered at this point.

The basic point process model assumed for such data within disease mapping is the heterogeneous Poisson process with first order intensity $\lambda(s)$. The basic assumptions of this model are that points (case events) are independently spatially-distributed and governed by the first order intensity. Due to the independence assumption, we can derive a likelihood for a realization of a set of events within a spatial region. For the study region defined above the unconditional likelihood of m events is just

$$L(\{\mathbf{s}\}|\psi) = \frac{1}{m!} \prod_{i=1}^{m} \lambda(s_i|\psi) \exp\{-\Lambda_T\} \tag{5.1}$$

$$\text{where } \Lambda_T = \int_T \lambda(u|\psi)du.$$

The function Λ_T is the integral of the intensity over the study region, ψ is a parameter vector, and $\lambda(s_i|\psi)$ is the first order intensity evaluated at the case event location s_i. Denote this likelihood as $PP[\{\mathbf{s}\}|\psi]$. This likelihood can be maximized with respect to the parameters in ψ and likelihood-based inference could be pursued. The only difficulty in its evaluation is the estimation of the spatial integral. However a variety of approaches can be used for numerical integration of this function and with suitable weighting schemes this likelihood can be evaluated even with conventional linear modeling functions within software packages (such as glm in R) (see e.g. Berman and Turner, 1992, Lawson, 2006b, App. C). An example of such a weighted log-likelihood approximation is:

$$l(\{\mathbf{s}\}|\psi) = \sum_{i=1}^{m} \ln \lambda(s_i|\psi) - \Lambda_T, \qquad (5.2)$$

where $\Lambda_T \approx \sum_{i=1}^{m} w_i \lambda(s_i|\psi)$ and w_i is an integration weight. This scheme per se is not accurate and more weights are needed. In the more general scheme of Berman and Turner a set of additional mesh points (of size m_{aug}) is added to the data. The augmented set $(N = m + m_{aug})$ is used in the likelihood with indicator function, I_k:

$$l(\{\mathbf{s}\}|\psi) = \sum_{k=1}^{N} w_k \{\frac{I_k}{w_k} \ln \lambda(s_k|\psi) - \lambda(s_k|\psi)\},$$

$$\text{where } \int_T \lambda(u|\psi)du = \sum_{k=1}^{N} w_k \lambda(s_k|\psi).$$

This has the form of a weighted Poisson likelihood, with $I_k = 1$ for a case and 0 otherwise. Diggle (1990) gives an example of the use of a likelihood such as (5.1) in a spatial health data problem.

In disease mapping applications, it is usual to parameterize $\lambda(s|\psi)$ as a function of two components. The first component makes allowance for the underlying population in the study region, and the second component is usually specified with the modelled components (i.e. those components describing the "excess" risk within the study area).

A typical specification would be

$$\lambda(s|\psi) = \lambda_0(s|\psi_0).\lambda_1(s|\psi_1). \qquad (5.3)$$

Here it is assumed that $\lambda_0(s|\psi_0)$ is a spatially-varying function of the population "at risk" of the disease in question. It is parameterized by ψ_0. The second function, $\lambda_1(s|\psi_1)$, is parameterized by ψ_1 and includes any linear or non-linear predictors involving covariates or other descriptive modeling terms thought appropriate in the application. Often we assume, for positivity, that $\lambda_1(s_i|\psi_1) = \exp\{\eta_i\}$ where η_i is a parameterized linear predictor allowing a link to covariates measured at the individual level. The covariates could include spatially-referenced functions as well as case-specific measures. Note that $\psi : \{\psi_0, \psi_1\}$. The function $\lambda_0(s|\psi_0)$ is a nuisance function which must be allowed for but which is not usually of interest from a modeling perspective.

5.1.2   Conditional Logistic Model

When a bivariate realization of cases and controls is available it is possible to make conditional inference on this joint realization. Define the case events as $s_i : i = 1, ..., m$ and the control events as $s_i : i = m + 1, ..., N$ where $N = m + n$ is the total number of events. Associated with each location is a binary variable (y_i) which labels the event either as a case ($y_i = 1$) or a control ($y_i = 0$). Assume also that the point process model governing each event type (case or control) is a heterogeneous Poisson process with intensity $\lambda(s|\psi)$ for cases and $\lambda_0(s|\psi_0)$ for controls. The superposition of the two processes is also a heterogeneous Poisson process with intensity $\lambda_0(s|\psi_0) + \lambda(s|\psi) = \lambda_0(s|\psi_0)[1 + \lambda_1(s|\psi_1)]$. Conditioned on the joint realization of these processes, it is straightforward to derive the conditional probability of a case at any location as

$$\Pr(y_i = 1) = \frac{\lambda_0(s_i|\psi_0).\lambda_1(s_i|\psi_1)}{\lambda_0(s_i|\psi_0)[1 + \lambda_1(s_i|\psi_1)]}$$

$$= \frac{\lambda_1(s_i|\psi_1)}{1 + \lambda_1(s_i|\psi_1)} = p_i \qquad (5.4)$$

and

$$\Pr(y_i = 0) = \frac{1}{1 + \lambda_1(s_i|\psi_1)} = 1 - p_i. \qquad (5.5)$$

The important implication of this result is that the background nuisance function $\lambda_0(s_i|\psi_0)$ drops out of the formulation and, further, this formulation leads to a standard logistic regression if a linear predictor is assumed within $\lambda_1(s_i|\psi_1)$. For example, a log linear formulation for $\lambda_1(s_i|\psi_1)$ leads to a logit link to p_i, i.e.

$$p_i = \frac{\exp(\eta_i)}{1 + \exp(\eta_i)},$$

where $\eta_i = x_i'\beta$ and x_i' is the i th row of the design matrix of covariates and β is the corresponding p-length parameter vector. Note that slightly different formulations can lead to non-standard forms (see e.g. Diggle and Rowlingson (1994)). In some applications, non-linear links to certain covariates may be appropriate (see Section 8).

Further, if the probability model in (5.4) applies, then the likelihood of the realization of cases and controls is simply

$$L(\psi_1|\mathbf{s}) = \prod_{i \in cases} p_i \prod_{i \in controls} 1 - p_i$$

$$= \prod_{i=1}^{N} \left[\frac{\{\exp(\eta_i)\}^{y_i}}{1 + \exp(\eta_i)} \right]. \qquad (5.6)$$

Hence, in this case, the analysis reduces to that of a logistic likelihood, and this has the advantage that the "at risk" population nuisance function does not have to be estimated. This model is ideally suited to situations where it is natural to have a control and case realization, where conditioning on the spatial pattern is reasonable. A fuller review of case event models in this context is found in Lawson (2012) and Onicescu and Lawson (2016).

5.1.3 Binomial Model for Count Data

In the case where we examine arbitrary small areas (such as census tracts, counties, postal zones, municipalities, health districts), usually a count of disease is observed within each spatial unit. Define this count as y_i and assume that there are m small areas. We also consider that there is a finite population within each small area out of which the count of disease has arisen. Denote this as $n_i \; \forall_i$. In this situation, we can consider a binomial model for the count data conditional on the observed population in the areas. Hence we can assume that given the probability of a case is p_i, then y_i is distributed independently as

$$y_i \sim bin(p_i, n_i)$$

and that the likelihood is given by

$$L(y_i|p_i, n_i) = \prod_{i=1}^{m} \binom{n_i}{y_i} p_i^{y_i} (1 - p_i)^{(n_i - y_i)}. \tag{5.7}$$

It is usual for a suitable link function for the probability p_i to a linear predictor to be chosen. The commonest would be a logit link so that

$$p_i = \frac{\exp(\eta_i)}{1 + \exp(\eta_i)}.$$

Here, we envisage the model specification within η_i, to include spatial and non-spatial components. Two applications which are well suited to this approach are the analysis of sex ratios of births, and the analysis of birth outcomes (e.g. birth abnormalities) compared to total births. Sex ratios are often derived from the number of female (or male) births compared to the total birth population count in an area. A ratio is often formed, though this is not necessary in our modeling context. In this case the p_i will often be close to 0.5 and spatially-localized deviations in p_i may suggest adverse environmental risk (Williams et al., 1992). The count of abnormal births (these could include any abnormality found at birth) can be related to total births in an area. Variations in abnormal birth count could relate to environmental as well as health service variability (over time and space)(Morgan et al., 2004).

5.1.4 Poisson Model for Count Data

Perhaps the most commonly encountered model for small area count data is the Poisson model. This model is appropriate when there is a relatively low count of disease and the population is relatively large in each small area. Often the disease count y_i is assumed to have a mean μ_i and is independently distributed as

$$y_i \sim Poisson(\mu_i). \tag{5.8}$$

The likelihood is given by

$$L(\mathbf{y}|\boldsymbol{\mu}) = \prod_{i=1}^{m} \mu_i^{y_i} \exp(-\mu_i)/y_i!. \tag{5.9}$$

The mean function is usually considered to consist of two components: *i)* a component representing the background population effect, and *ii)* a component representing the excess risk within an area. This second component is often termed the *relative risk*. The first component is commonly estimated or computed by comparison to rates of the disease in a standard population and a local expected rate is obtained. This is often termed *standardisation* (Inskip et al., 1983). Hence, we would usually assume that the data is independently distributed with expectation

$$E(y_i) = \mu_i = e_i\theta_i$$

where e_i is the expected rate for the i th area and θ_i is the relative risk for the i th area. As we will be developing Bayesian hierarchical models we will consider $\{y_i\}$ to be conditionally independent given knowledge of $\{\theta_i\}$. The expected rate is usually assumed to be fixed for the time period considered in the spatial example, although there is literature on the estimation of small area rates that suggests this may be naive (Ghosh and Rao, 1994, Rao, 2003).

5.1.4.1 Standardisation

The expected rates or counts that are used in Poisson models are usually assumed to be fixed quantities. They are a product of *indirect* standardisation. (The alternative to indirect standardisation is direct standardisation which converts the outcome counts to rates thereby changing the data model.) For indirect standardisation, which is the commonest form of adjustment, a standard population is chosen to which the disease outcome is to be compared. In the standard population the overall disease rate is computed and then applied to each local area based on its population size. For example,

$$R = \frac{\sum y_i}{\sum p_i} \text{ and } e_i = p_i R$$

where p_i is the local area population and the sum is over all relevant areas. Essentially this rate proportionately allocates the disease risk based on local population. The choice of standard population, over which to calculate the rate R, is crucial for the computation of the expected rate or count. Often for spatial studies the whole study area is used as reference, as the internal relative changes in risk are of importance. However, larger areas could also be used or indeed sometimes an area external to the study region could be chosen. For example, a study of risk within New Jersey could have as standard population the whole of the northeast US. An external reference population could be useful when two study areas are to be compared. For example, if a comparison of New Jersey to Pennsylvania is to be made then New England could be the reference, say. Alternatively the combined states could act as references for each state. The choice of reference standard population is crucial and different choices could lead to different conclusions concerning risk. In most studies the expected rate or count is assumed to be fixed once computed. Note also that more sophisticated standardisation can be used whereby sub-strata in the population can be adjusted for. For instance age and gender distribution can be included as part of the adjustment of the expected rate. This could be important when there is a strong spatial differentiation in sub-strata distribution across a study area.

5.1.4.2   Relative Risk

Usually the focus of interest will be the modeling of the relative risk. The most common approach is to assume a logarithmic link to a linear predictor model:

$$\log \theta_i = \eta_i.$$

This form of model has seen widespread use in the analysis of small area count data in a range of applications (see e.g. Leyland and Goldstein, 2001, Chapter 10, Stevenson et al., 2005, Waller and Gotway, 2004, Chapter 9).

5.2   Specification of Predictor in Case Event and Count Models

In all the above models a predictor function (η_i) was specified to relate to the mean of the random outcome variable, via a suitable link function. Often the predictor function is assumed to be linear and a function of fixed covariates and also possibly random effects. We define this in a general form, for p covariates, here as

$$\eta_i = x_i'\beta + z_i'\xi,$$
$$= \beta_1 x_{1i}........\beta_p x_{pi} + \xi_1 z_{1i}......\xi_q z_{qi}$$

where x_i' is the i th row of a covariate design matrix $\mathbf{x}$ of dimension $m \times p$, β is a $(p \times 1)$ vector of regression parameters, ξ is a $(q \times 1)$ unit vector and z_i' is a row vector of individual level random effects, of which there are q. In this formulation the unknown parameters are β and $\mathbf{z}$ the $(m \times q)$ matrix of random effects for each unit. Note that in any given application it is possible to specify subsets of these covariates or random effects. Covariates for case event data could include different types of specific level measures such as an individual's age, gender, smoking status, health provider, etc., or could be environmental covariates which may have been interpolated to the address location of the individual (such as soil chemical measures or air pollution levels). For count data in small areas, it is likely that covariates will be obtained at the small area level. For example, for census tracts, there are likely to be socioeconomic variables such as poverty (percent of population below an income level), car ownership, median income level, available from the census. In addition, some variates could be included as supra-area variables such as health district in which the tract lies. Environmental covariates could also be interpolated to be used at the census tract level. For example air pollution measures could be averaged over the tract.

In some special applications non-linear link functions are used, and in others, mixtures of link functions are used. One special application area where this is found is the analysis of putative hazards (see Chapter 8), where specific distance- and/or direction-based covariates are used to assess evidence for a relation between disease risk and a fixed (putative) source of health hazard. For example, one simple example of this is the conditional logistic modeling of disease cases around a fixed source. Let distance and direction from the source to the i th location be d_i and ϕ_i respectively, then a mixed linear and non-linear link model is commonly assumed where

$$\eta_i = \{1 + \beta_1 \exp(-\beta_2 d_i)\}. \exp\{\beta_0 + \beta_3 \cos(\phi_i) + \beta_4 \sin(\phi_i)\}.$$

Here the distance effect link is non-linear, while the overall rate (β_0) and directional components are log-linear. The explanation and justification for this formulation is deferred to Chapter 8.

Fixed covariate models can be used to make simple descriptions of disease variation. In particular it is possible to use the spatial coordinates of case events (or in the case of count data, centroids of small areas) as covariates. These can be used to model the long range variation of risk: *spatial trend*. For example, let's assume that the i th unit x-y coordinates are (x_{si}, y_{si}). We could define a polynomial trend model such as:

$$\eta_i = \beta_0 + \sum_{l=1}^{L} \beta_{xl} x_{si}^l + \sum_{l=1}^{L} \beta_{yl} y_{si}^l + \sum_{k=1}^{K} \sum_{l=1}^{L} \beta_{lk} x_{si}^l y_{si}^k.$$

This form of model can describe a range of smoothly varying non-linear surface forms. However, except for very simple models, these forms are not parsimonious and also cannot capture the extra random variation that often exists in disease incidence data.

5.2.1   Bayesian Linear Model

In the Bayesian paradigm all parameters are stochastic, if not fixed, and are therefore assumed to have prior distributions. Hence in the covariate model

$$\eta_i = x_i'\beta,$$

the β parameters are assumed to have prior distributions. Hence this can be formulated as

$$P(\beta, \tau_\beta | data) \propto L(data|\beta, \tau_\beta) f(\beta|\tau_\beta)$$

where $f(\beta|\tau_\beta)$ is the joint distribution of the covariate parameters conditional on the hyperparameter vector τ_β. Often we regard these parameters as independent and so

$$f(\beta|\tau_\beta) = \prod_{j=1}^{p} f_j(\beta_j|\tau_{\beta_j}).$$

More generally it is commonly assumed that the covariate parameters can be described by a Gaussian distribution and if the parameters are allowed to be correlated then we could have the multivariate Gaussian specification:

$$f(\beta|\tau_\beta) = \mathbf{N}_p(\mathbf{0}, \mathbf{\Sigma}_\beta),$$

where under this prior assumption, $E(\beta|\tau_\beta) = \mathbf{0}$ and $\mathbf{\Sigma}_\beta$ is the conditional covariance of the parameters. The most common specification assumes prior independence and is:

$$f(\beta|\tau_\beta) = \prod_{j=1}^{p} N(0, \tau_{\beta_j}),$$

where $N(0, \tau_{\beta_j})$ is a zero mean single variable Gaussian distribution with variance τ_{β_j}. At this point an assumption about variation in the hyperparameters is usually made. At the next level of the hierarchy hyperprior distributions are assumed for τ_β. The definition of these distributions could be important in defining the model behavior. For example if a vague hyperprior is assumed for τ_{β_j} this may lead to extra variation when limited learning is available from the data. This can affect computation of DIC and convergence diagnostics. While uniform hyperpriors (on a large positive range) can lead to improper posterior distributions, it has been found that a uniform distribution for the standard deviation can be useful (Gelman (2006)), i.e. $\sqrt{\tau_{\beta_j}} \sim U(0, A)$ where A has a large positive value.

Alternative suggestions are usually in the form of gamma or inverse gamma distributions with large variances. For example, Kelsall and Wakefield (2002) proposed the use of gamma (0.2, 0.0001) with expectation 2000 and variance

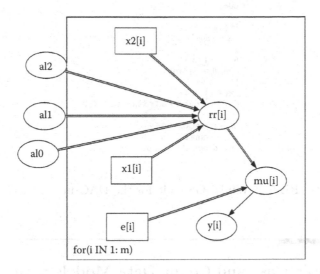

FIGURE 5.1: A directed acyclic graph (in WinBUGS Doodle format) for a simple Poisson Bayesian regression with log linear relative risk and two covariates.

20,000,000, whereas Banerjee et al. (2004) examine various alternative specifications including gamma (0.001, 0.001). One common specification (Thomas et al., 2004) is gamma (0.5, 0.0005) which has expectation 1000 and variance 2,000,000. A default specification for precision of a Gaussian distribution on R-INLA is gamma (1, 0.00005) which also has a large variance. Weakly informative priors have also been suggested, such as gamma (2, 1) or gamma (2, 1/A) where A is the reasonable span of the parameter (Simpson et al., 2015).

While these prior specifications lead to relative uninformativeness, their use has been criticized by Gelman (2006) (see also Lambert et al., 2005), in favor of half-Cauchy and uniform prior distributions on the standard deviation. See also https://github.com/stan-dev/stan/wiki/prior-choice-recommendations for other discussion.

To summarize the hierarchy for such covariate models, Figure 5.1 displays a directed acyclic graph for a simple Bayesian hierarchical covariate model with two covariates (x_1, x_2) and relative risk defined as $\theta_i = \exp(\beta_0 + \beta_1 x_{1i} + \beta_2 x_{2i})$ for data $\{y_i, e_i\}$ for m regions. The regression parameters are assumed to have independent zero mean Gaussian prior distributions with fixed precisions. Figure 5.2 displays the corresponding WinBUGS code.

```
model;
{
  for( i in 1 : m ) {
    y[i] ~ dpois(mu[i])
  }
  for( i in 1 : m ) {
    mu[i] <- e[i] * rr[i]
  }
  for( i in 1 : m ) {
    rr[i] <- exp(al0 + al1 * x1[i] + al2 * x2[i])
  }
  al0 ~ dnorm( 0.0,1.0E-6)
  al1 ~ dnorm( 0.0,1.0E-6)
  al2 ~ dnorm( 0.0,1.0E-6)
}
```

FIGURE 5.2: WinBUGS code for the DAG in Figure 5.1.

5.3 Simple Case and Count Data Models with Uncorrelated Random Effects

In the previous section, some simple models were developed. These consisted of functions of fixed observed covariates. In a Bayesian model formulation all parameters are stochastic and so the extension to the addition of random effects is relatively straightforward. In fact the term "mixed" model (linear mixed model: LMM, normal linear mixed model: NLMM, or generalized linear mixed model: GLMM) is strictly inappropriate as there are no fixed effects in a Bayesian model.

The simple regression models described above often do not capture the extent of variation present in count data. Overdispersion or spatial correlation due to unobserved confounders will usually not be captured by simple covariate models and often it is appropriate to include some additional term or terms in a model which can capture such effects.

Initially, overdispersion or extra-variation can be accommodated by either a) inclusion of a prior distribution for the relative risk, (such as a Poisson-gamma model) or b) by extension of the linear or non-linear predictor term to include an extra random effect (log-normal model).. In both cases the model addresses uncorrelated heterogeneity (UH) in the outcome.

5.3.1 Gamma and Beta Models

5.3.1.1 Gamma Models

The simplest extension to the likelihood model that accommodates extra variation is one in which the parameter of interest in the likelihood is given a prior distribution. One case event example would be where the intensity is specified at the i th location as $\lambda_0(s_i|\psi_0).\lambda_1(s|\psi_1) \equiv \lambda_{0i}.\lambda_{1i}$, suppressing the parameter dependence for simplicity. Note that λ_{1i} plays the role of a relative risk parameter. This parameter can be assigned a prior distribution, such as a gamma distribution to model extra-variation. For most applications where count data are commonly found a Poisson likelihood is assumed. We will focus on these models in the remainder of this section. The Poisson parameter θ_i could be assigned a *gamma*(a, b) prior distribution. In this case, the prior expectation and variance would be respectively a/b and a/b^2. This could allow for extra variation or overdispersion. Here we assume that parameters a, b are fixed and known. This formulation is attractive as it leads to a closed form for the posterior distribution of $\{\theta_i\}$, i.e.

$$[\theta_i|y_i, e_i, a, b] \sim gamma(a^*, b^*) \tag{5.10}$$

where $a^* = y_i + a$, $b^* = e_i + b$.

Hence, conjugacy leads to a gamma posterior distribution with posterior mean and variance given by $\frac{y_i+a}{e_i+b}$ and $\frac{y_i+a}{(e_i+b)^2}$ respectively. Note that a variety of θ_i estimates are found depending on the values of a and b. Lawson and Williams (2001) pp. 78-79 demonstrate the effect of different values of a and b on relative risk maps. Samples from this posterior distribution are straightforwardly obtained (e.g. using the rgamma function on R). The prior predictive distribution of $\mathbf{y}^*$ is also relevant in this case as it leads to a distribution often used for overdispersed count data: the negative binomial. Here we have the joint distribution as:

$$[\mathbf{y}^*|\mathbf{y}, a, b] = \int f(\mathbf{y}^*|\theta) f(\theta|a, b) d\theta \tag{5.11}$$

$$= \prod_{i=1}^{m} \left[\frac{b^a}{\Gamma(a)} \frac{\Gamma(y_i^* + a)}{(e_i + b)^{(y_i^* + a)}} \right].$$

5.3.1.1.1 Hyperprior Distributions One extension to the above model is to consider a set of hyperprior distributions for the parameters of the gamma prior (a and b). Often these are assumed to also have prior distributions on the positive real line such as gamma (a', b') with $a' > 0$, $b' = 1$ or $b' > 1$. Note that only if hyperprior distributions are assumed for a and b can they be estimated using the posterior distribution.

5.3.1.1.2 Linear Parameterization One approach to incorporating more sophisticated model components into the relative risk model is to model the parameters of the gamma prior distribution. For example, gamma linear models can be specified where

$$[\theta_i|y_i, e_i, a, b_i] \sim gamma(a, b_i)$$

where $b_i = a/\mu_i$ *and*

$$\mu_i = \eta_i.$$

In this formulation the prior expectation is μ_i and the prior variance is μ_i^2/a. While this formulation could be used for modeling, often the direct linkage between the variance and mean could be seen as a disadvantage. As will be seen later, a log-normal parameterization is often favored for such models.

5.3.1.2   Beta Models

When Bernoulli or binomial likelihood models are assumed (such as (5.6) or (5.7)) then one may need to consider prior distributions for the probability parameter p_i. Commonly a beta prior distribution is assumed:

$$[p_i|\alpha,\beta] \sim beta(\alpha,\beta).$$

Here the prior expectation and variance would be $\frac{\alpha}{\alpha+\beta}$ and $\frac{\alpha\beta}{(\alpha+\beta)^2(\alpha+\beta+1)}$. This distribution can flexibly specify a range of forms from peaked ($\alpha = \beta, \beta > 1$) to uniform ($\alpha = \beta = 1$) and U-shaped ($\alpha = \beta = 0.5$) to skewed or monotonically decreasing or increasing. In the case of the binomial distribution this prior distribution, with α, β fixed, leads to a beta posterior distribution, i.e.

$$[\mathbf{p}|\mathbf{y},\mathbf{n},\alpha,\beta] = B(\alpha,\beta)^{-m} \prod_{i=1}^{m}[\binom{n_i}{y_i}p_i^{y_i}(1-p_i)^{(n_i-y_i)}.p_i^{\alpha-1}(1-p_i)^{\beta-1}]$$

$$= B(\alpha,\beta)^{-m} \prod_{i=1}^{m}[\binom{n_i}{y_i}p_i^{y_i+\alpha-1}(1-p_{i^i})^{(n_i-y_i+\beta-1)}].$$

This is the product of m independent beta distributions with parameters $y_i+\alpha$, $n_i-y_i+\beta$. Hence the beta posterior distribution for p_i has expectation $\frac{y_i+\alpha}{n_i+\beta+\alpha}$ and variance $\frac{(y_i+\alpha)(n_i-y_i+\beta)}{(n_i+\beta+\alpha)^2(n_i+\beta+\alpha+1)}$.

5.3.1.2.1   Hyperprior Distributions   The parameters α, and β are strictly positive and these could also have hyperprior distributions. However, unless these parameters are restricted to the unit interval, then distributions such as the gamma, exponential or inverse gamma or inverse exponential would have to be assumed as hyperprior distributions.

5.3.1.2.2   Linear Parameterization   An alternative specification for modeling covariate effects is to specify a linear or non-linear predictor with a link to a parameter or parameters. For example, it is possible to consider a parameterization such as $\alpha_i = \exp(\eta_i)$ and $\beta_i = \psi\alpha_i$ where ψ is a linkage parameter with prior mean given by $\frac{1}{1+\psi}$. When $\psi = 1$, then the distribution is symmetric. The disadvantage with this formulation is that a single parameter is assigned to the linear predictor and a dependence is specified between α_i and β_i. One possible alternative is to model the prior mean as $\text{logit}(\frac{\alpha_i}{\alpha_i+\beta_i}) = \eta_i$. However this also forces a dependence between α_i and β_i.

```
model
{
for (i in 1:m){
y[i]~dpois(mu[i])
mu[i]<-e[i]*theta[i]
smr[i]<-y[i]/e[i]
log(theta[i])<-al0+v[i]
v[i]~dnorm(0,tau.v)
PP[i]<-step(theta[i]-1)
}
#other prior distributions

al0~dnorm(0,tau.0)
tau.v~dgamma(1,0.5)
tau.0~dgamma(1,0.5)
}
```

FIGURE 5.3: Poisson-log-normal model for the Georgia county-level oral cancer mortality data. The model assumes a zero mean Gaussian prior distribution for the uncorrelated heterogeneity random effect . The posterior expected exceedance probability ($\widehat{\Pr}(\theta > 1)$) is computed as PP[].

5.3.2   Log-Normal and Logistic-Normal Models

One simple device that is very popular in disease mapping applications is to assume a direct linkage between a linear or non-linear predictor (η_i) and the parameter of interest (such as θ_i or p_i). This offers a convenient method of introducing a range of covariate effects and unobserved random effects within a simple formulation. The general structure of this formulation is $\eta_i = x_i'\beta + z_i'\gamma$. The simplest form involving uncorrelated heterogeneity would be

$$\eta_i = z_{1i}$$

where z_{1i} is an uncorrelated random effect.

An example of the application of this model is given in Figure 5.3 where the WinBUGS code is presented and in Figure 5.4 where the posterior expected relative risk estimates under the Poisson-log-normal model are displayed. In this example, the counties of Georgia are modelled and the outcome of interest is the count of oral cancer deaths in these counties for a given year (2004). In this example the county-wise expected rate was computed from the state-wide oral cancer rate for 2004. The likelihood model assumed for this example is Poisson with $y_i \sim Poisson(e_i\theta_i)$ and $\eta_i = \alpha_0 + z_{1i}$.

Here the extra-variation is modeled as uncorrelated heterogeneity (UH) with a zero mean Gaussian prior distribution, i.e. $z_{1i} \sim N(0, \tau_{z_1})$.

FIGURE 5.4: Georgia county level mortality counts for oral cancer, 2004. Uncorrelated heterogeneity random effect model. County-wise expected rate computed from the state-wide oral cancer rate. Row-wise from top left: standardised mortality ratio; posterior expected relative risk estimates; posterior expected exceedance probability $(\widehat{Pr}(\theta_i > 1))$.

5.4   Correlated Heterogeneity Models

Uncorrelated heterogeneity (UH) models with gamma or beta prior distributions for the relative risk are useful but have a number of drawbacks. First, as noted above, a gamma distribution does not easily provide for extensions into covariate adjustment or modeling, and, second, there is no simple and adaptable generalization of the gamma distribution with spatially correlated parameters. Wolpert and Ickstadt (1998) provided an example of using correlated gamma field models, but these models have been shown to have poor performance under simulated evaluation (Best et al., 2005). The advantages of incorporating a Gaussian specification are many. First, a random effect which is log-Gaussian behaves in a similar way to a gamma variate, but the Gaussian model can include a correlation structure. Hence, for the case where it is suspected that random effects are correlated, it is simpler to specify a log Gaussian form for *any* extra variation present. The simplest extension is to consider additive components describing different aspects of the variation thought to exist in the data.

For a spatial Gaussian process, (Ripley, 1981, p. 10), any finite realization has a multivariate normal distribution with mean and covariance inherited from the process itself, i.e. $\mathbf{x} \sim MVN(\mu, K)$, where μ is an m length mean vector and K is an $m \times m$ positive definite covariance matrix. Note that this is not the only possible specification of a prior structure to model correlated heterogeniety (CH) (see also Møller et al., 1998).

There are many ways of incorporating such heterogeneity in models, and some of these are reviewed here. First, it is often important to include a variety of random effects in a model. For example, both CH and UH might be included (see Section 5.5). One flexible method for the inclusion of such terms is to include a log-linear term with additive random effects. Besag et al. (1991) first suggested, for tract count effects, a rate parametrization of the form,

$$\exp\{x_i'\beta + u_i + v_i\},$$

where $x_i'\beta$ is a trend or fixed covariate component, u_i and v_i are correlated and uncorrelated heterogeneity, respectively. These components then have separate prior distributions. Often the specification of the correlated component is considered to have either an intrinsic Gaussian (CAR) prior distribution or a fully specified multivariate normal prior distribution.

5.4.1 Conditional Autoregressive (CAR) Models

5.4.1.1 Improper CAR (ICAR) Models

The intrinsic autoregression improper difference prior distribution, developed from the lattice models of Kunsch (1987), uses the definition of spatial distribution in terms of differences and allows the use of a singular normal joint distribution. This was first proposed by Besag et al. (1991). Hence, the prior for $\{u\}$ is defined as

$$p(\mathbf{u}|r) \propto \frac{1}{r^{m/2}} \exp\left\{-\frac{1}{2r} \sum_i \sum_{j \in \delta_i} (u_i - u_j)^2\right\}, \qquad (5.12)$$

where δ_i is a neighborhood of the i th tract. The neighborhood δ_i was assumed to be defined for the first neighbor only. Hence, this is an example of a Markov random field model (see e.g. Rue and Held (2005)). More general weighting schemes could be used. For example neighborhoods could consist of first and second neighbors (defined by common boundary) or by a distance cut-off (for example, a region is a neighbor if the centroid is within a certain distance of the region in question). The uncorrelated heterogeneity (v_i) was defined by Besag et al. (1991) to have a conventional zero-mean Gaussian prior distribution:

$$p(v) \propto \sigma^{-m/2} \exp\left\{-\frac{1}{2\sigma} \sum_{i=1}^{m} v_i^2\right\}. \tag{5.13}$$

Both r and σ were assumed by Besag et al. (1991) to have improper inverse exponential hyperpriors:

$$\text{prior}(r, \sigma) \propto e^{-\epsilon/2r} e^{-\epsilon/2\sigma}, \qquad \sigma, r > 0, \tag{5.14}$$

where ϵ was taken as 0.001. These prior distributions penalize the absorbing state at zero, but provide considerable indifference over a large range. Alternative hyperpriors for these parameters which are now commonly used are in the gamma and inverse gamma family, which can be defined to penalize at zero but yield considerable uniformity over a wide range. In addition, these types of hyperpriors can also provide peaked distributions if required.

The full posterior distribution for the original formulation where a Poisson likelihood is assumed for the tract counts is given by

$$P(u, v, r, \sigma | y_i) =$$

$$\prod_{i=1}^{m} \{\exp(-e_i\theta_i)(e_i\theta_i)^{y_i}/y_i!\}$$

$$\times \frac{1}{r^{m/2}} \exp\left\{-\frac{1}{2r} \sum_{i} \sum_{j \in \delta_i} (u_i - u_j)^2\right\}$$

$$\times \sigma^{-m/2} \exp\left\{-\frac{1}{2\sigma} \sum_{i=1}^{m} v_i^2\right\} \times \text{prior}(r, \sigma).$$

This posterior distribution can be sampled using MCMC algorithms such as the Gibbs or metropolis–Hastings samplers. A Gibbs sampler was used in the original example, as conditional distributions for the parameters were available in that formulation.

An advantage of the intrinsic Gaussian formulation is that the conditional moments are defined as simple functions of the neighboring values and number of neighbors ($n_{\delta i}$).

$$E(u_i| \ldots) = \overline{u_i} \text{ and}$$
$$var(u_i| \ldots) = r/n_{\delta i},$$

and the conditional distribution is defined as:

$$[u_i| \ldots] \sim N(\overline{u_i}, r/n_{\delta i}),$$

where $\overline{u_i} = \sum_{j \in \delta_i} u_j/n_{\delta i}$, the average over the neighborhood of the i th region.

5.4.1.2   Proper CAR (PCAR) Models

While the intrinsic CAR model introduced above is useful in defining a correlated heterogeneity prior distribution, this is not the only specification of a Gaussian Markov random field (GMRF) model available. In fact, the improper CAR is a special case of a more general formulation where neighborhood dependence is admitted but which allows an additional correlation parameter (Stern and Cressie, 1999). Define the spatially-referenced vector of interest as $\{u_i\}$. One specification of the proper CAR formulation yields:

$$[u_i|\ldots] \sim N(\mu_i, r/n_{\delta i}) \tag{5.15}$$

$$\mu_i = t_i + \phi \sum_{j \in \delta_i} (u_j - t_j)/n_{\delta_i} \tag{5.16}$$

where t_i is the trend ($=x_i'\beta$), r is the variance, and ϕ is a correlation parameter. It can be shown that to ensure definiteness of the covariance matrix, ϕ must lie on a predefined range which is a function of the eigenvalues of a matrix. In detail, the range includes the smallest and largest eigenvalues ($\phi_{\min} = \eta_1^{-1}, \phi_{\max} = \eta_m^{-1}$) of $diag\{n_{\delta_i}^{-1/2}\}.C.diag\{n_{\delta_i}^{1/2}\}$ where $C_{ij} = c_{ij}$, i.e. ($\phi_{\min} < \phi < \phi_{\max}$) and

$$c_{ij} = \begin{cases} \frac{1}{n_{\delta_i}} & \text{if } i \sim j \\ 0 & \text{otherwise} \end{cases}.$$

Of course, $\phi_{\min}$ and $\phi_{\max}$ can be precomputed before using the proper CAR as a prior distribution. It could simply be assumed that a (hyper) prior distribution for ϕ is $U(\phi_{\min}, \phi_{\max})$. As noted by Stern and Cressie (1999) this specification does lead to a simple form for the partial correlation between different sites. Note that in the simple case of no trend ($t_i = 0$) then the model reduces to

$$[u_i|\ldots] \sim N(\mu_i, r/n_{\delta i}) \tag{5.17}$$

$$\mu_i = \phi \bar{u}_i. \tag{5.18}$$

The main advantage of this model formulation is that it more closely mimics fully specified Gaussian covariance models, as it has a variance and correlation parameter specified, does not require matrix inversion within sampling algorithms, and can also be used as a data likelihood. In Appendix D, Table D.1, a further range of CAR specifications is provided with different assumptions concerning neighborhood relations.

5.4.1.3   Gaussian Process Convolution (PC) and GCM Models

Higdon (2002) and Calder et al. (2002), (see also Calder, 2007, 2008 and Cressie and Wikle, 2011, Chapter 4) suggested the use of kernel convolution models to discretely approximate continuous spatial Gaussian fields. The basic idea is that at location s the field is given by

$$u(s) = \sum_{j=1}^{M} k(w_j - s)x(w_j)$$

where $k(w_j - s)$ is a distance kernel, $\{w_j\}$ a set of evaluation points, and independent noise $x(w_j) \overset{iid}{\sim} N(0, \tau_u)$. This is a smoothing of white noise which leads to a correlated effect. Usually the kernel is assumed to be symmetric bivariate Gaussian and it can be precomputed and so the main parameter controlling the field is τ_u, and the evaluation mesh. Higdon (2002) noted that for the continuous analogue, the given kernel has a one-to-one relation to the covariance of the resulting Gaussian field. This allows for great flexibility and the idea of *reduced rank* models has led to predictive process models (Banerjee et al., 2008).

A recent application of the PC approach to parsimonious modeling of correlated spatial noise is found in Onicescu et al. (2017a). A closely related model is the Gaussian component mixture (GCM) which uses neighborhood adjacencies instead of a distance kernel:

$$u_i = \frac{1}{w_i} \sum_{j=1}^{n_{\delta_i}} w_{ij} x_j$$

$$\text{with} \quad x_j \overset{iid}{\sim} N(0, \tau_u)$$

and $w_{ij} = 1$ within the i th neighborhood and $w_i = \sum_{j=1}^{n_{\delta_i}} w_{ij}$.

Hence, for these weights, $u_i = \frac{1}{n_{\delta_i}} \sum_{j=1}^{n_{\delta_i}} x_j$, the neighborhood average. The resulting GCM does seem to approximate ICAR effects well, as demonstrated by Moraga and Lawson (2012). Note that the spatial model assumed in multilevel modeling closely resembles the GCM (see e.g. Leyland and Goldstein (2001), pp. 143-157) but includes $n_{\delta_i} + 1$ white noise terms.

5.4.1.4 Case Event Models

For case event data, where a point process model is appropriate, it is still possible to consider a form of log Gaussian Cox process where the intensity of the process is governed by a spatial Gaussian process and conditional on the intensity the case distribution is a Poisson process. As an approximation to the Gaussian process a CAR prior distribution can be proposed. For example, define the first order intensity of the case events as

$$\lambda(s) = \lambda_0(s) \exp\{\beta + S(s)\}$$

where $S(s)$ is the Gaussian process component. At a given case location, s_i, this will yield a likelihood contribution

$$\lambda(s_i) = \lambda_0(s_i) \exp\{\beta + S(s_i)\}.$$

By considering an intrinsic Gaussian specification for $S(s_i)$ we can proceed by assuming that the prior distribution for $\{S(s_i)\}$ is a conditional autoregressive specification, i.e. for short, define $S_i \equiv S(s_i)$, and hence

$$[S_i | \ldots] \sim N(\overline{S}_{\delta_i}, r/n_{\delta i})$$

where $\overline{S}_{\delta_i}$ is the mean of the S values in the neighborhood of S_i. This idea relies on the definition of a neighborhood. For case events, the definition of a neighborhood is problematic. Unlike polygonal regions there is not a simple definition of neighboring points. One possibility, would be to define a circular radius around each event and to include all points as neighbors if they fell within the radius. Of course, the size of the radius is arbitrary and hence the neighborhoods could be arbitrary also. An alternative would be to define a neighborhood by tesselation. Tesselation is a form of tiling which uniquely divides the space of points into areas. Usually, these tilings describe the area closest to the point only. In that sense they define "territories". Defining the neighborhood in this way leads to a Voronoi/Dirichlet tesselation. This tiling leads to sets of neighbors defined by the adjoining edges of the tile. An adjoining neighbor is known as a *natural neighbor*. The tiling defines *natural neighborhoods*. It is possible to define first (or greater) order neighbors in this way. On R the package deldir can be modified for this task. The dual of the Voronoi tesselation is the Delauney triangulation formed by the bisectors of the tile edges. Each point will also have a number of Delauney neighbors. These are often the same as the Voronoi neighbors but they can also differ. One advantage of the Delauney neighbors is that the triangulation always forms a convex hull in the points, and to some degree suffers less from edge effects than the Voronoi tesselation. Figure 5.5 displays six points with associated Voronoi tesselation (short dash) and Delauney triangulation (long dash).

Hence a hierarchical model can be specified with the i th likelihood contribution $\lambda(s_i) = \lambda_0(s_i) \exp\{\eta_i + v_i + S_i\}$ where $\eta_i = x_i'\beta$ is a linear predictor with fixed covariate vector x_i', and

$$[S_i | \ldots] \sim N(\overline{S}_{\delta_i}, r/n_{\delta i})$$
$$v_i \sim N(0, \kappa_v)$$
$$\beta \sim \mathbf{N}(\mathbf{0}, \mathbf{\Gamma}_\beta).$$

Of course, in this formulation, $\lambda_0(s_i)$ must be estimated, and also the integral of the intensity must be computed. A Berman-Turner approximation scheme (Berman and Turner, 1992) could be used for this purpose. An example of this type of analysis is given in Hossain and Lawson (2008). The conditional

FIGURE 5.5: Plot of six arbitrary points with associated Voronoi tesselation (short dash) and Delauney triangulation (long dash).

logistic likelihood model (5.6) can be fitted if a control disease is available, and this obviates the necessity of estimating $\lambda_0(s_i)$ (see e.g. Lawson, 2006b, Chapter 8.4, Appendix C). However the specification of the spatial structure is different as the joint distribution of cases and controls is considered under the conditional model.

5.4.2   Fully Specified Covariance Models

An alternative specification involves only one random effect for both CH and UH. This can be achieved by specifying a prior distribution having two parameters governing these effects. For example, the covariance matrix of an MVN prior distribution can be parametrically modelled with such terms (Diggle et al., 1998; Wikle, 2002). This approach is akin to universal kriging (Wackernagel, 2003; Cressie, 1993), which employs covariance models including variance and covariance range parameters. It has been dubbed "*generalized linear spatial modeling*." A software library is available in R (geoRglm). Usually, these parameters define a multiplicative relation between CH and UH. For the full Bayesian analysis of this model, use is often made of posterior sampling algorithms.

In the parametric approach of Diggle et al. (1998), which was originally specified for point process models, the first-order intensity of the process was specified as

$$\lambda(s) = \lambda_0(s) \exp\{\beta + S(s)\}, \tag{5.19}$$

where β is a non-zero mean level of the process, and $S(s)$ is a zero mean Gaussian process with, for example, a powered exponential correlation function defined for the distance d_{ij} between the i th and j th locations as $\rho(d_{ij}) = \exp\{-(d_{ij}/\phi)^\kappa\}$ and variance σ^2. Other forms of covariance function can be specified. One popular example is the Matérn class defined for the distance (d_{ij}) as

$$\rho(d_{ij}) = (d_{ij}/\phi)^\kappa K_\kappa(d_{ij}/\phi)/[2^{\kappa-1}\Gamma(\kappa)] \tag{5.20}$$

where $K_\kappa(.)$ is a modified Bessel function of the third kind. In this case, the parameter vector $\theta = (\beta, \sigma, \phi, \kappa)$ is updated via a metropolis–Hastings-like (Langevin-Hastings) step, followed by pointwise updating of the S surface. Conditional simulation of S surface values at arbitrary spatial locations (non-data locations) can be achieved by inclusion of an additional step once the sampler has converged. Covariates can be included in this formulation in a variety of ways. For count data, the equivalent Poisson mean specification could be

$$\mu_i = e_i \exp\{\beta + S_i\},$$

where $\mathbf{S} \sim \mathbf{MVN}(\mathbf{0}, \mathbf{\Gamma})$ and $\mathbf{\Gamma}$ is a spatial covariance matrix (Kelsall and Wakefield, 2002). In comparisons of CAR and fully specified covariance models there appear to be different conclusions about which are more useful in recovering relative risk in disease maps (Best et al., 2005, Henderson et al., 2002). A comparison of software for these models is given in Chapter 15. Note that when the above formulation is applied to a point process, a log Gaussian Cox process (LGCP) results. There is now a range of software available for fitting such models: the R package lgcp provides MCMC implementation and the stochastic partial differential equation (SPDE) approach on a finite element mesh is available in INLA (Illian et al., 2012). Comparison of these approaches is given in Taylor and Diggle (2014) and Taylor et al. (2015).

5.5 Convolution Models

Often it is important to employ both CH and UH random effects within the specification of η_i. The rationale for this lies in the basic assumption that unobserved effects within a study area could take on a variety of forms. It is always prudent to include a UH effect to allow for uncorrelated extra variation. However, without prior knowledge of the unobserved confounding, there is no reason to exclude either effect from the analysis and it is simple to include both effects within an additive model formulation such as $\eta_i = v_i + u_i$. In general,

```
model
{
for (i in 1:m){
y[i]~dpois(mu[i])
mu[i]<-e[i]*theta[i]
smr[i]<-y[i]/e[i]
log(theta[i])<-al0+v[i]+u[i]
v[i]~dnorm(0,tau.v)
PP[i]<-step(theta[i]-1)
}
#ICAR prior distribution
u[1:m]~car.normal(adj[],weights[],num[],tau.u)
#weights
for(k in 1:sumNumNeigh){weights[k]<-1}

#other prior distributions

al0~dnorm(0,tau.0)
tau.u~dgamma(1,0.5)
tau.v~dgamma(1,0.5)
tau.0~dgamma(1,0.5)
}
```

FIGURE 5.6: Code for a convolution model with both UH and CH components. CAR model used for the CH component. The posterior expected exceedance probability ($\widehat{\Pr}(\theta > 1)$) is computed in PP[]. The SMR is also computed.

these two random effects are not well identified, although the assumption of a correlated prior distribution for u_i should provide a degree of identification between UH and CH. Given that we are usually interested in the total effect of unobserved confounding, the sum of the effects is the important component and that is well identified. Discussion of these identifiability issues is given in Eberley and Carlin (2000). Occasionally, the computation of the relative variance contribution (intraclass correlation) can be useful. If the variance of the UH and CH component are, respectively, κ_v and κ_u, then the result is given by $\frac{\kappa_v}{\kappa_v+\kappa_u}$. Of course if these components are not identified then this computation will not be useful. In the following figures, (Figures 5.6 and 5.7) the ODC and some selected output from a convolution model for the Georgia oral cancer data are presented.

Convolution models, or as they are also known: BYM models, with ICAR CH random effects, have been shown to be robust under simulation to a wide range of underlying true risk models (Lawson et al., 2000) and so are widely used for the analysis of relative risk in disease mapping.

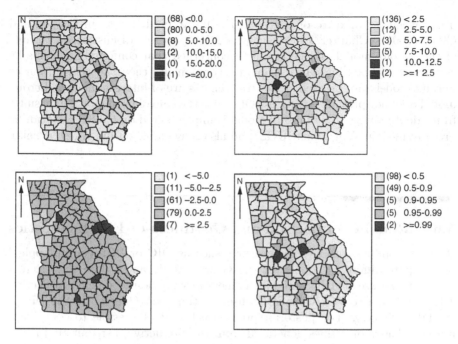

FIGURE 5.7: Georgia county level mortality counts for oral cancer, 2004. Convolution model with UH and CH effects. County-wise expected rate computed from the state-wide oral cancer rate. Row-wise from top left: standardised mortality ratio; posterior expected relative risk estimates; correlated random effect (CH) u_i ; posterior expected exceedance probability $(\widehat{Pr}(\theta_i > 1))$.

5.5.1 Leroux Prior Specification

The BYM convolution model is defined by $\exp\{\alpha_0 + u_i + v_i\}$,with suitable CH prior distribution for u_i and UH prior distribution for v_i and an intercept α_0.The lack of indentification in this setup has led to an alternative proposal that has now been implemented in various packages: the Leroux model. Leroux et al. (2000) first proposed a model whereby the covariance structure of the model is a mixture between uncorrelated and correlated effects. More formally

$$\mathbf{u} \sim \mathbf{N}(\mathbf{0}, \tau^2 Q(W, \rho)^{-1}),$$

where the precision matrix is defined as $Q(W, \rho) = \rho[diag(W1) - W] + (1 - \rho)I$ and W is a neighborhood weight matrix. In this formulation the ρ parameter controls the overall degree to which the effect is correlated: $\rho = 1$ gives an ICAR model whereas $\rho = 0$ gives an uncorrelated (UH) effect. Hence the model allows the degree of smoothing or clustering to be estimated, without having to use a convolution with two components. An alternative idea is to use a weighted mixture of effects in the linear predictor as in Lawson and Clark (2002). Chapter 4 of Congdon (2010) provides discussion of variants of the Leroux model. If one is simply concerned that confounding effects be estimated without interest in separate components then the convolution or Leroux models can be used. If separate effects are of interest then the Leroux model does not provide estimates. Of course the convolution model may suffer from identifiability. The Leroux model is implemented on CARBayes, can be programmed on Win/OpenBUGS, and also is available on INLA (leroux.inla).

5.6   Model Comparison and Goodness-of-Fit Diagnostics

Relative goodness-of-fit (gof) measures such as DIC or PPL can be applied to compare different models for the Georgia oral cancer data set. Here I focus on the use of DIC and MSPE measures previously defined in Section 4.1, in relation to Win/OpenBUGS use. Within Win/OpenBUGS the DIC is available directly. For posterior predictive loss it is possible to compute a measure based on values generated from the predictive distribution $\{y_i^{pred}\}$ and compare these to the observed data via a suitable loss function. For binary data an absolute value loss is useful as it measures the proportionate misclassification under the model. For positive outcomes (such as Poisson or binomial data) often a squared error loss is used, although the absolute value loss may also be useful. In general the computation involves the averaging of the loss (f) over the data and posterior sample (of size G):

$$MSPE = \sum_{j}^{m} \sum_{i} f(y_i - y_{ij}^{pred})/(m \times G). \qquad (5.21)$$

In the case of a Poisson likelihood, the following lines in an ODC, which produce squared error loss for each observation, compute a point-wise PPL:

$$ypred[i] \sim dpois(mu[i])$$
$$PPL[i] < -pow(ypred[i] - y[i], 2).$$

Both the individual values of

$$PPL_i = \sum_{j} f(y_i - y_{ij}^{pred})/G \qquad (5.22)$$

and the average over all the data (5.21) can be useful in diagnosing local and global lack-of-fit. As a comparison with the BYM convolution, Leroux, GCM convolution and UH models we have computed the DIC, MAPE (absolute error), and MSPE (squared error) for each model. Table 5.1 displays the

Measure	GCM Convolution	Leroux Model	BYM Convolution	UH Model
DIC	417.6	446.3	445.0	383.30
pD	37.1	1.088	1.210	40.70
MAPE	0.995	1.161	1.166	0.994
MSPE	2.777	4.092	4.102	2.75

Table 5.1: Comparison of GCM convolution, BYM convolution, Leroux, and uncorrelated heterogeneity (UH) models for the Georgia oral cancer data set

results. Overall the UH model yields the lower DIC and is lower on both the absolute and squared error loss. Of the correlation models, the GCM has lowest DIC, and MSPE and MAPE are close to the UH model, but yield higher pD than other correlation models. Note that the DIC criterion measures how well the model fits the observed data, allowing for parameterization, while the posterior predictive loss (PPL) criteria compare the predictive ability of the models.

Figure 5.8 displays the point-wise PPL for squared error and absolute error loss for the Georgia oral cancer mortality data set under the BYM convolution and UH models. The maps suggest a marked concentration of loss in a few regions in the northwest of the state (Fulton, Cobb, Dekalb, and Gwinett Counties). Fulton County contains the largest urban area in the state (Atlanta), and shows the highest PPL under both squared error and absolute error. Note that under the better-fitting UH model the loss in most areas is lower than under the convolution model.

5.6.1 Residual Spatial Autocorrelation

While models for disease maps can be assessed for global fit and also at the individual unit level via residual diagnostics, there remains the question of whether any residual spatial structure has been left within the data after a model fit. One approach to this is to consider that a good model fit should leave residuals from the fit with little or no spatial correlation. Hence a test for spatial correlation in the residuals from a model fit would be a useful guide to whether the model has managed to account for the spatial variation adequately. It is possible within a posterior sampling algorithm to compute a measure of spatial autocorrelation and to average this in the final sample. This will provide an estimate of the correlation and the sample will also provide a credible interval or standard deviation of the estimate. In this way it is possible to avoid the need to consider the sampling distribution of the computed statistic, by using functionals of the posterior distribution via posterior sampling.

Various statistics could be used to measure autocorrelation but probably the most common spatial autocorrelation statistic is Moran's I (see e.g. Cliff

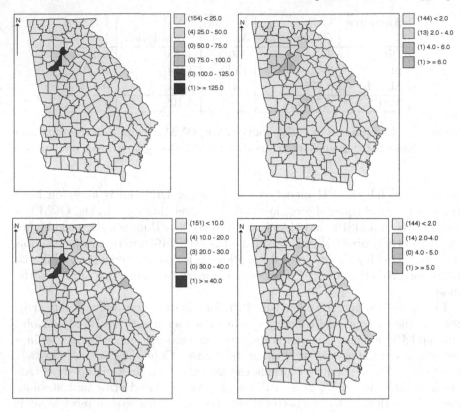

FIGURE 5.8: Georgia county-level oral cancer mortality, 2004. Top row: CII model, bottom row: UII model. Point-wise posterior predictive loss for squared error loss (left panel)and absolute error loss (right panel), averaged over a converged sample of size 10000.

and Ord, 1981; Cressie, 1993, Section 6.7). This is usually defined as a ratio of quadratic forms:

$$I = \mathbf{e}'W\mathbf{e}/\mathbf{e}'\mathbf{e}$$

where $\mathbf{e} = \{e_i, ..., e_m\}$ and $e_i = (y_i - \widehat{y}_i)/\sqrt{var(\widehat{y}_i)}$ and W is the 0/1 $m \times m$ adjacency matrix for the regions with elements w_{ij}. For a Poisson data model this residual could be defined as $e_i = (y_i - \widehat{\mu}_i)/\sqrt{\widehat{\mu}_i}$. Congdon et al. (2007) following Fotheringham et al. (2002), noted that given $\{e_i\}$ and a set of 0/1 adjacencies $\{w_{ij}\}$ a regression of e_i on e_i^* where $e_i^* = \sum_j w_{ij}e_j$ will yield a slope parameter that is an estimate of I. That is, fitting the linear model $e_i = a_0 + \rho e_i^* + \epsilon_i$ will yield the posterior average of ρ as an estimate of I. The WinBUGS code for this is given below. In this code the cumulative number of neighbors up to the ith data point is defined as cum[] with the convention that cum[1]=0. Values of cum[] are used as indices to select neighboring residuals.

In the following, y[] is the outcome and mu[] is the mean (assuming a Poisson data model), e[] is the residual and estar[] is the sum of the neighbouring residuals, adj[] is the adjacency list for the regions, and sumNN is the sum of the number of neighbours:

```
for (i in 1:m){

y[i]~dpois(mu[i])

e[i]<-(y[i]-mu[i])/sqrt(mu[i])
estar[i]<-sum(we[cum[i]+1:cum[i+1]])
de[i]<-e[i]-mean(e[])
d.estar[i]<-estar[i]-mean(estar[])
dt[i]<-de[i]*d.estar[i]
db[i]<-pow(d.estar[i],2)}

for (j in 1: sumNN) {we[j,i]<-e[adj[i]]}

rho<-sum(dt[])/sum(db[])
```

For the 2004 Georgia oral cancer mortality data with a fitted convolution model in Section 5.5 (see e.g. Figure 5.7), I have computed the posterior average Moran's I. In this case, the result is $\hat{\rho} = 0.0018$ (standard deviation: 0.0334) with 95% credible interval : (-0.0692, 0.0613), which suggests limited autocorrelation left after fitting the convolution model.

5.7 Alternative Risk Models

In Sections 5.4 and 5.4.2 models for risk variation which are most commonly found in applications were discussed. However there are a number of alternative modeling approaches that are less commonly applied but might be of considerable use in certain applications. The first of these utilizes a likelihood approximation (pseudolikelihood) that was first suggested by Besag (1975). The second is a relaxation of the parametric assumptions inherent in the second level prior specifications for the covariance and convolution models.

5.7.1 Autologistic Models

Besag (1975) first suggested the possibility of modeling spatially-distributed continuous variables via a conditioning on neighborhoods. Instead of considering the full conditional distributions within a likelihood function, a likelihood that is simply a product of the local conditioning within a neighborhood was proposed. Hence, for a count data example with $\{y_i\}$ observed within m arbitrary small areas we examine the $\Pr(y_i|\{y_j\}_{j\in\delta_i}|\boldsymbol{\theta})$ where $j \neq i$ and δ_i is a neighborhood of the i th small area. The pseudolikelihood is then simply assumed to be given by

$$L_p(\{y_i\}|\boldsymbol{\theta}) = \prod_{i=1}^{m} \Pr(y_i|\{y_j\}_{j\in\delta_i}|\boldsymbol{\theta}).$$

This likelihood can take a range of simple forms when specific parameterizations are assumed for $\Pr(y_i|\{y_j\}_{j\in\delta_i}|\boldsymbol{\theta})$. For example, the most common and, possibly, most widely used example, is where the outcome in the small areas is binary, so that y_i takes the value 1/0. In this case,

$$\Pr(y_i|\{y_j\}_{j\in\delta_i},\boldsymbol{\theta}) = \theta_i^{y_i}(1-\theta_i)^{1-y_i}$$
$$\text{and } \theta_i = \frac{\exp\{\alpha_i + \lambda S_{\delta_i}\}}{1 + \exp\{\alpha_i + \lambda S_{\delta_i}\}}$$
$$\text{and } S_{\delta_i} = \sum_{j\in\delta_i} y_j.$$

Hence the likelihood is simply:

$$L_p(\{y_i\}|\boldsymbol{\theta}) = \prod_{i=1}^{m} \left\{ \frac{[\exp\{\alpha_i + \lambda S_{\delta_i}\}]^{y_i}}{1 + \exp\{\alpha_i + \lambda S_{\delta_i}\}} \right\}.$$

Notice that this form simply conditions on the sum of neighboring values and as this sum can be precomputed before analysis, this model could be fitted using conventional logistic regression software. One simple extension to this model is to include covariates via a linear predictor. Hence, $\theta_i = \frac{\exp\{\lambda S_{\delta_i}+x_i'\beta\}}{1+\exp\{\lambda S_{\delta_i}+x_i'\beta\}}$ could be specified, where $x_i'\beta$ is a linear predictor with row vector of i th covariate values x_i' and parameter vector β. Alternatively, covariate representation which is centered could be advantageous (Caragea and Kaiser, 2006). Similarly the model could be extended to include random effects. For example, we could include uncorrelated unit level random effects (ν_i) within the logit term:

$$\theta_i = \frac{\exp\{\lambda S_{\delta_i} + x_i'\beta + \nu_i\}}{1 + \exp\{\lambda S_{\delta_i} + x_i'\beta + v_i\}}. \tag{5.23}$$

In the Bayesian context, logit $\theta_i = \lambda S_{\delta_i} + x_i'\beta + \nu_i$ can be treated as other spatial models in that both covariates and random effects can be present. Both the parameter vector β, λ and random effect ν_i (in Example (5.23)) will have prior distributions at the second level of the hierarchy. Further hyperprior specifications would also be available.

Note that this likelihood includes spatial correlation at the first level of the hierarchy. Hence it does not require the inclusion of any spatially-structured

Model	pD	DIC	MSPE
Autologistic	1.97	112.18	0.4268(0.0503)
Autologistic+UH	25.16	109.04	0.3142 (0.0525)
Convolution (UH+CH)	27.69	103.98	0.2844 (0.0515)

Table 5.2: DICs for three models: autologistic with no random effects; autologistic with UH component; convolution model with UH and CH

random effect at the second level of the hierarchy. Modifications to the model where further spatial components describing second or higher order dependence could be made. For example, $S_{\delta_i} = \omega \sum_{j \in \delta_i} y_j + (1 - \omega) \sum_{l \in \delta_{2i}} y_l$ where δ_{2i} is a second order neighborhood (such as neighbors of first order neighbors, once counted) and ω is a weighting parameter with $0 < \omega < 1$. This would allow for some measure of distance-related dependence in the model. Note that S_{δ_i} cannot be precomputed unless ω is fixed, but the sums: $\sum_{j \in \delta_i} y_j$, and $\sum_{l \in \delta_{2i}} y_l$, can be.

An example of the application of an autologistic model and comparison with a conventional random effect model is demonstrated in Figures 5.9 and 5.10 in which the Ohio county level respiratory cancer mortality data set (1968) is examined. The expected rates were computed from the statewide age-gender stratified rate applied to count population strata and summed. The county level standardised mortality ratios (smr_i) of respiratory cancer for the year 1968 are available in the form of an exceedance indicator. The standardisation was performed using the state-wide rate for 18 age-gender groups for that year. The data was made available as a dichotomized variate via the criteria:

$$y_i = \begin{cases} 1 \text{ if } smr_i > 2 \\ 0 \text{ otherwise} \end{cases}.$$

Interest focuses on the analysis of high risk status and hence on this binary outcome at the county level. Of course, the arbitrary dichotomisation may lead to concerns about use of artificial screening levels, and this indeed should be a consideration in any analysis where thresholding has taken place. Of course, when only dichotomised data is available, it is appropriate to seek to model it directly. In Figure 5.9, the contrast between the posterior expected estimates under the simple autologistic model (top row), which are relatively uniform, and the probability of risk exceedance and UH component (v_i) (bottom row) demonstrate four regions of elevated risk. When the convolution model with UH and CH components is fitted, as shown in Figure 5.10, it is clear that the areas of elevated risk are further highlighted and indeed the CH component seems to isolate the southeastern corner of the state as a particularly elevated region.

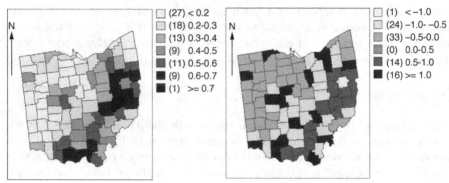

FIGURE 5.9: Autologistic model with a single binary covariate and a first order dependence on the sum of neighbouring values. Top row: posterior expected probability of SMR exceedance (>2); bottom row: same model with a county level heterogeneity random effect (v_i): UH. Left: posterior expected probability of SMR exceedance. Right: posterior expected UH component.

The difference in DIC is less marked between the different models in this case. Table 5.2 displays the DIC results. Both the autologistic model with UH and the standard convolution model appear to yield better models than the simple autologistic. Interestingly, the convolution model yields a lower DIC than both autologistic models. However, the autologistic model with uncorrelated heterogeneity is much closer to the DIC of the convolution model and is more parsimonious.

For this example, models with both first order and second order neighborhoods were examined, but overall the best model remained the first order neighborhood based on DIC.

Note that the difference between the autologistic model and convolution model lies mainly in the fact that the spatial correlation is at different levels of the hierarchy and that the likelihood used for the autologistic model is

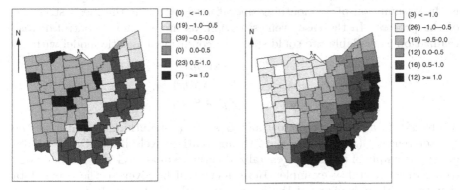

FIGURE 5.10: Standard binary convolution model with UH and CH CAR components. Top row: posterior expected probability of SMR exceedance; Bottom row left: UH component; right: CH component.

a pseudolikelihood. Pseudolikelihoods are known to be reasonable approximations where spatial correlation is not strong. In space-time, autologistic models can also be considered (see Chapter 12).

5.7.1.1 Other Auto Models

Besides the autologistic model, it is possible to consider other auto models based on Poisson, binomial, negative binomial, or other exponential family distributions. Usually pseudolikelihood must be used for estimation as normalizing constants are intractable. In addition, in some cases, the parameterization of the model must be constrained. In the auto Poisson model, the most general form of the intensity is defined, for the i th region, as $\lambda_i = \exp(\alpha_i + \sum_{j=1}^{m} \eta_{ij} y_j)$ and so constraints must be put on the $\eta_{ij}s$ to ensure negativity. In the case of the auto binomial, the logit of the probability of a case in the i th area is an un-normalized function of the surrounding case totals (rather than proportions): logit $p_i = \alpha_i + \sum_{j=1}^{m} \eta_{ij} y_j$. Of the range of auto models available it is clear that the autologistic is the most popular in applications and likely the simplest to implement and interpret.

5.7.2   Spline-Based Models

As an alternative to strictly parametric models, it is possible to assume a semi-parametric approach to the modeling of the spatially-structured component of a disease risk model. The use of spline models is of course not limited to employment in spatial smoothing, but I will concentrate on that aspect here. The basic idea behind a semi-parametric representation of a spatial model is the assumption of a smoothing operator to represent the mean structure of the process. In the case event situation, assuming that a control disease realization is available, we could specify a conditional logistic model with

$$p_i = \frac{\exp(\eta_i)}{1 + \exp(\eta_i)}$$
$$\text{with } \eta_i = x_i'\beta + S(s_i)$$

where $x_i'\beta$ is a linear predictor, and $S(s_i)$ is a smoothing operator at the geo-reference s_i (location) of the i th observation. Kelsall and Diggle (1998) give an example of using this generalized additive model (GAM) methodology to a cancer mortality example. Here, focus will be given to the count data situation although many of the issues found there also apply to case event data.

In the count data situation, make the usual assumption of observed data $\{y_i\}$, $i = 1, ..., m$ and $y_i \sim Poisson(\mu_i)$. Further define the geo-reference for i th observation as $s_i : (x_{i1}, x_{i2})$. This could be a centroid of the small area or other associated point reference. Here it is assumed that $\log \mu_i = S(s_i)$ where $S(.)$ will be defined as smoothing operator. A variety of choices are available for $S(.)$. Here, I focus on spline models which are attractive in applications and have strong links to Gaussian process models (see e.g. French and Wand, 2004). Define the mean level as

$$\log \mu_i = \alpha_0 + \sum_{j=1}^{2} \alpha_j x_{ij} + \sum_{j=1}^{n_\kappa} \psi_j C\{||s_i - \kappa_j||\}$$
$$= \mathbf{x}_i'\alpha + z_i'\psi,$$

where $\{\kappa_j\}$,$j = 1, ..., n_\kappa$ is a set of knots (fixed locations in space), $\{\psi_j\}$ is a Gaussian random effect , $z_i' = \{z_1, ..., z_{n_\kappa}\}$ and

$$\mathbf{z} = [C\{||s_i - \kappa_j||/\rho\}]_{1 \le i \le m, \, 1 \le j \le n_\kappa},$$

the covariance function defined here as

$$C\{a\} = (1 + |a|)e^{-|a|}.$$

Define the square matrix

$$\omega = [C\{||\kappa_i - \kappa_j||/\rho\}]_{1 \leq i,j \leq n_\kappa}$$

and the joint random effect prior distribution as

$$\psi \sim \mathbf{N}(\mathbf{0}, \tau\omega^{-1}).$$

A reparameterization of $\mathbf{z}_* = \mathbf{z}\omega^{-1/2}$, $\psi_* = \omega^{1/2}\psi$ yields a linear mixed model with $cov(\psi_*) = \tau\mathbf{I}$, and then $\log \mu_i = \mathbf{x}'_i\alpha + \mathbf{z}_{*i}\psi_*$. In French and Wand (2004), the value of ρ is fixed in advance. This allows the precomputing of the covariance matrix and reparameterization so that standard software can be used. This type of spline modeling is termed *low rank kriging*. In general, it would be useful to estimate ρ as this controls the degree of smoothing. Figures 5.11 and 5.12 display the resulting posterior expected (PE) estimate maps for the two models. In Figure 5.11 the PE relative risk and CH component are shown. The model fitted to the log relative risk included a planar trend in the x, y centroids as well as additive UH and CH components. For the spline model for comparability the log relative risk was also a function of planar trend in centroid locations but an additive spline term was also included. The covariance was assumed to be defined by $\omega = [C\{||\kappa_i - \kappa_j||/\rho\}]_{1 \leq i,j \leq n_\kappa}$ where ρ was fixed at $\hat{\rho} = \max(||s_i - s_j||) \; \forall_{i,j}$. This tends to produce a very smooth surface effect as can be seen in the Figure 5.12 where the top panel shows the PE relative risk with a much reduced range. The bottom panel displays the spline effect which includes both spatial and non-spatial effects (beyond the trend component). Overall the relative risk pattern is mostly similar between the two models. However, in this case the spline model did not provide a good model based on DIC. The DIC for the spline model was 761.52. Whereas for the convolution model it was 623.15. Of course, if ρ were to be estimated then it is possible that a much improved fit could be achieved. Alternative spline-based approaches, have been proposed by Zhang et al. (2006) to spatio-temporal multivariate modeling, and by Macnab (2007) in a comparison of spline methods for temporal components of spatial maps.

5.7.3   Zip Regression Models

If a disease is rare then there will be considerable sparsity in the data. The implication of this is that few cases are observed within the study area, or, for small areas, zero counts are common. In this case, the spatial distribution of cases will often form isolated clusters. A good example of this is childhood leukemia which is a rare disease but is known to cluster. The major question that is posed by this situation is whether the standard models for disease mapping hold when such sparsity of data arises. A priori the conventional log relative risk model where the log of the risk is modeled with Gaussian effects may be simply inadequate to deal with a situation where the rate is close to

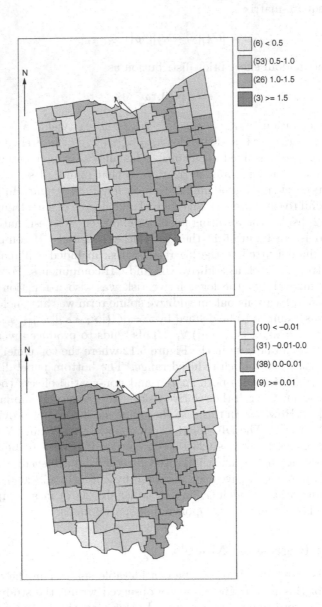

FIGURE 5.11: CAR model fit for Ohio county level respiratory cancer mortality for 1979. Model includes a planar trend in centroids and both UH and CH CAR components. Top row: posterior expected relative risk; bottom row: CH component.

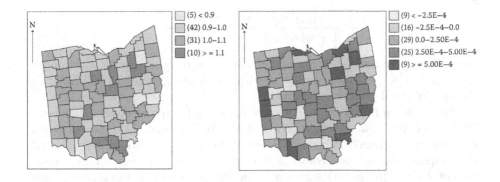

FIGURE 5.12: Spline model fit based on the low rank kriging model as a linear mixed model with a fixed covariance parameter ρ ($\widehat{\rho} = \max(||x_i - x_j||)$ $\forall ij$) where x_i is the i th centroid.

the boundary of its space (i.e. $\lambda_0(s) \simeq 0$, or the expected rate $e_i \simeq 0$). Singular information methods may be useful here (see e.g. Bottai et al. (2007)).

Two alternatives can be immediately envisioned. First it may be possible to directly model the locations of the disease clusters via object models (Lawson and Denison, 2002). These models do not make simple global assumptions about surface form, but rather seek to estimate locations of objects (in this case clusters). As it turns out these models can recover risk surfaces reasonably well. Examples of this are found in Lawson (2006b), Section 6.5. An alternative is to consider the marginal distribution of the concentration of cases. In the sense that any arbitrary area or mesh area will yield a local concentration of cases, it might be noted that under sparsity, many areas will have zero cases and a few will have small positive numbers. For count data in arbitrary regions this could lead to an overdispersed distribution and even multi-modality in the marginal distribution. Note that this effect in count data may not be adequately modelled by an overdispersed distribution such as the negative binomial.

One solution is to consider a mixture of processes so that the low intensity is separately modelled from the peaks. For case event data we could assume

$$\lambda(s) = \lambda_0(s)[w(s) + (1 - w(s))\lambda_1(s)]$$

where $w(s)$ is spatially dependent weight which controls which process is dominant locally. This can lead to a logistic model when a control disease realization is present in that the probability of a given location s_i being a case is:

$$\frac{w(s_i) + (1 - w(s_i))\lambda_1(s_i)}{1 + w(s_i) + (1 - w(s_i))\lambda_1(s_i)}$$

Hence it may be interesting to include covariates or other effects within both $w(s_i)$ and $\lambda_1(s_i)$. I do not pursue this approach here. Instead I will focus on the area of small area count data.

There is much literature on mixture modeling for sparse counts (Lambert, 1992, Boehning et al., 1999, Agarwal et al., 2002, Dagne, 2004, Ghosh et al., 2006 amongst others). When a mixture of Poisson distributions is considered the simplest case is a two component mixture where zero counts have a component $1 - p + p\exp(-\mu)$ where μ is the Poisson mean and non-zero counts have component $p\exp(-\mu).\mu^y/y!$. In general the distribution is given by

$$f(y; p|\mu) = (1 - p)P_0(y, 0) + pP_0(y, \mu)$$

where $P_0(y, \mu)$ is the Poisson distribution with mean μ. The inclusion of covariates can proceed as usual via link to the mean (e.g. $\mu = \exp(\mathbf{x}'\beta)$). In addition, covariates can be included in the mixture weight (p). For example we could have

$$p = \frac{\exp(\mathbf{w}\prime\gamma)}{1 + \exp(\mathbf{w}\prime\gamma)}.$$

where $\mathbf{w}\prime\gamma$ is a predictor with additional covariates and effects. In general, for observed data $\{y_i\}, i = 1, ..., m$ and expected counts $\{e_i\}$ the model is specified

$$[y_i|e_i, \theta_i] \sim (1 - p)Pois(0) + pPois(e_i.\theta_i). \tag{5.24}$$

A further modification clarifies the role of the components. For example, we might consider that this problem is one where an unobserved classification variable treats the zeroes as structural $(z = 1)$ or usual Poisson $(z = 0)$. In this case z is unobserved and must be estimated. This can be done within a data augmentation loop. In that case, the incomplete data likelihood is

$$[y_i|e_i, \theta_i, z_i = 0] \sim Pois(e_i.\theta_i) \tag{5.25}$$
$$[y_i|e_i, \theta_i, z_i = 1] \sim Pois(e_i.\theta_i^*).$$

Then the second stage would be to generate the allocation variables from $[z_i|y_i, e_i, \{\theta_i, \theta_i^*\}]$. Usually $[z_i|y_i, e_i, \{\theta_i, \theta_i^*\}] \sim Bern(p_i)$ for two components. The complete data likelihood (Marin and Robert, 2007) used to estimate the parameters would be

$$L(\{y_i\}, z) = \prod_{i=1}^{m} p_{z_i} Pois(y_i; e_i.\theta_{z_i}).$$

Figure 5.13 displays a ZIP regression analysis for the Georgia county level asthma mortality counts for the year 2000. The posterior average relative risk estimates for high and low risk counties 1 and 144 (crude SMR = 3.85,

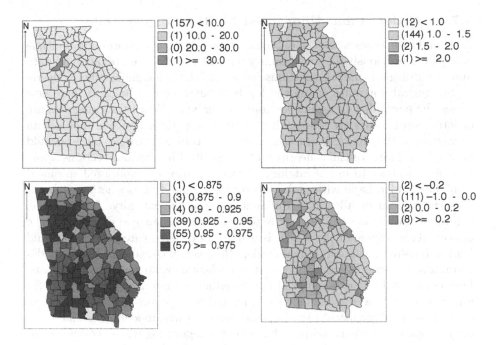

FIGURE 5.13: ZIP Bayesian model with two components applied to the Georgia asthma mortality data for 2000. Row-wise from top left: posterior expected Poisson mean, relative risk, component probability, and uncorrelated heterogeneity (UH).

0.0) and the posterior average relative risk estimates are 1.322, and 0.573 under a converged convolution model, whereas under a two component ZIP model (5.24), with no spatially-correlated component, the posterior expected estimates of relative risk were 1.818, 1.094. Although these estimates seem to have similar ranges, the latter model has shifted both estimates away from zero. Interestingly, in both models the Atlanta area appears to yield a very high posterior average relative risk estimate, and also a high probability of membership in the full Poisson model. The ZIP model did not include a CH component unlike the convolution model. Some residual structure remains as evidenced by the posterior expected UH map.

Finally, it should be apparent that the idea of mixtures of components can be generalized to a wide variety of situations. The primary area of application may be the incorporation of (unobserved) multiple scales of aggregation within one analysis when it is believed that different components represent these different scales. This is discussed more extensively in Chapter 9.

5.7.4   Ordered and Unordered Multi-Category Data

A special case arises when the outcome of interest is in the form of a multiple category. I previously discussed binary data as a special case of binomial data and in autologistic models (Sections 5.1.3 and 5.7.1). Extending these criteria to categorical outcomes where the levels of outcome can be > 2, there now arises the possibility of ordinal or nominal analysis. When the categories are ordered, such as disease stages, then ordinal (logistic) regression models can be assumed. If ordering is not apparent then nominal (logistic) models could be applied. Many of the concerns and issues cited in previous sections apply here with respect to use of random effects and prior distributional specification. One added issue with multi-category outcome data is whether different structures could be allowed at different levels of the category. For example, commonly available within cancer registry data is the stage at diagnosis of the cancer. This staging of the cancer is usually an ordered category. This would lead to consideration of a ordinal model. However, unstaged cancers are indeterminate in terms of stage, and so it is unclear at what level they would be best considered. This might lead one to either assume a nominal model for all the staged data or to exclude the unstaged from an ordinal analysis. Zhou et al. (2008a) demonstrate the application of ordinal models: baseline category logits, proportional odds and adjacent category logits, to geo-referenced cancer registry data in South Carolina. They found that baseline category logits model fitted best in terms of DIC, and that a model with a spatially-correlated random effect for the regional stage and an uncorrelated random effect for the distant stage was best fitting in this case.

5.7.5   Latent Structure Models

An alternative to conventional modeling of the mean level of risk is to consider that the risk is composed of a combination of unobserved risk levels. These risk levels are latent and so there is no directly observed data concerning their form. These types of models really fall between areas. On the one hand they are used to provide overall estimates of relative risk (and so are relative risk models). On the other hand, they are also used to isolate underlying patterns of risk and so the latent risk levels may be of importance. Some of the models discussed in Chapter 6 could be regarded as latent structure models also. For example the hidden process or object models can be used to provide relative risk estimates, besides estimates of cluster locations (see e.g. Lawson, 2006b, Section 6.5.3). The hidden Markov models of Green and Richardson (2002) provide estimates of relative risk as do the mixture component models of Fernandez and Green (2002). Partition-based models can also be considered in this way. Here we provide a brief summary of spatial component models, both latent and known. In Section 12.3.3 a brief review of space-time latent models is given.

5.7.5.1 Mixture Models

A number of examples of mixture-based models have been proposed for relative risk estimation. Often these have been applied to count data and so the following discussion focuses on that data form. First of all, fixed component mixtures have been proposed. A simple example of these would in fact be ZIP regression. More generally, in a Bayesian context, one can consider a fixed component model that consists of sums of random terms within the mean predictor. These models could be termed mean mixture models. The convolution model of Section 5.5 is a mean mixture with two fixed components. An extension of this type of model was proposed by Lawson and Clark (2002) where a mean mixture of a CAR component and an L1 norm component was used to preserve discontinuities and boundary effects (see also Congdon, 2005, Chapter 8). The mixing parameter was allowed to vary spatially and so a posterior expected mixing field was estimated. The basic model specification was

$$[y_i|e_i\theta_i] \sim Poiss(e_i\theta_i)$$
$$\log(\theta_i) = v_i + w_i u_{1,i} + (1 - w_i)u_{2,i}$$
$$w_i \sim beta(\alpha, \alpha).$$

Note that $u_{1,i}$ and $u_{2,i}$ are CAR and $L1$ norm spatial prior distributions respectively and $v_i \sim N(0, \tau_v)$. The mixing parameter $\{w_i\}$ was allowed to vary spatially, albeit with a common exchangeable distribution, and so a posterior expected mixing field could be estimated. Figure 5.14 displays the posterior expected components of the three component mixture fitted to North Carolina sudden infant death syndrome (SIDS) data. The L1 norm field appears quite different from the CAR component field (which seems to display a west-east trend). Each field provides unique information concerning the different components of the model supported in the data.

In general an extended mixture of fixed random components with different prior assumptions could be imagined.

An early example of hidden mixture modeling, albeit in an empirical Bayes context, was proposed by Schlattman and Böhning (1993). Their approach assumed that the distribution governing the observed data is a mixture of Poisson distributions:

$$f(y_i|\mathbf{p}, e_i, \boldsymbol{\theta}) = \sum_{k=1}^{K} p_k Pois(y_i|e_i\theta_k) \qquad (5.26)$$

with mixing probabilities $\{p_k\}$ and $\sum_k p_k = 1$. Both $\mathbf{p}$ and θ are unknown. Suitable prior distributions for the components have to be specified. Besides estimation of $\mathbf{p}$ and θ, it is possible to estimate the risk in each area from

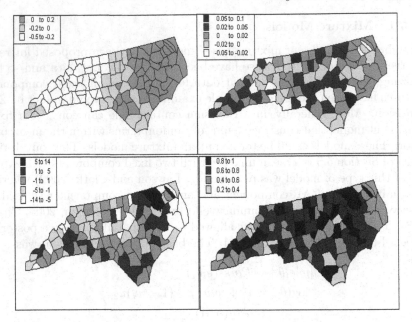

FIGURE 5.14: Three-component mixture model for SIDS in North Carolina (as reported in Lawson and Clark (2002)). Row-wise from top left: $u_{1,i}$ component, v_i component, $u_{2,i}$ component, and w_i component. Posterior averages reported.

posterior sampling based on

$$\widehat{\theta}_i = \frac{1}{G} \sum_{g=1}^{G} \sum_{k=1}^{K} \theta_k^g p_k^g Pois(y_i|e_i\theta_k^g) / \sum_{k=1}^{K} p_k^g Pois(y_i|e_i\theta_k^g) \qquad (5.27)$$

for the case of fixed K. When K is not fixed then a prior distribution would have to be specified for K. For that case, (5.27) could be used with K replaced by K_g. The choice of prior specification under could be various. Clearly for the probabilities one could use a Dirichlet distribution:

$$\mathbf{p} \sim Dir(\alpha)$$

where $\{\alpha_k\}$, $k = 1, ..., K$, while for the $\{\alpha_k\}$ gamma prior distributions could be specified. In addition, prior specification for the $\{\theta_k\}$ could be based on gamma distributions. For example,

$$\theta_k \sim Ga(c_k a, a)$$

would yield a prior mean of c_k. Suitable hyperprior distributions can be assumed for the positive parameters c_k.

An ordering constraint on the components may be required if K is not fixed and the components are to be identified. Posterior sampling for the fixed K case is straightforward. For the non-fixed case, then a prior distribution must be assumed for K. This is often a Poisson with fixed rate, i.e. $K \sim Poiss(d)$, or a uniform distribution up to a fixed maximum: $K \sim U(1, K_{\max})$.

When spatial dependence is to be included, one approach has been to assume that, instead of a Dirichlet distribution for the weights, the weights have a spatial dependence structure. Fernandez and Green (2002) suggest a variety of models. One proposal, the logistic normal model, specifies that

$$f(y_i|\mathbf{p}, e_i, \boldsymbol{\theta}) = \sum_{k=1}^{K} p_{ik} Pois(y_i|e_i\theta_k)$$

$$p_{ik} = \eta_{ik}(\phi)/\sum_{l=1}^{L} \eta_{il}(\phi)$$

$$where \ \eta_{ik}(\phi) = \exp\{x_{ik}/\phi\}$$

where $\{x_{ik}\}$ is a set of spatially-correlated random field components indexed by the i th area, and ϕ is a spatial correlation parameter. The fields are given proper CAR prior distributions to ensure propriety. The relative risk estimates from posterior sampling are obtained via allocation of components. A more extensive review of mixture modeling can be found in Hossain and Lawson (2016).

Major alternatives to these mixture type models are those that posit factorial decomposition of the risk in each area. For multiple diseases only, where y_{ij} is the observed count in the i th area and the j th disease, Wang and Wall (2003) first proposed a model where a spatial factor underlay the risk:

$$y_{ij} \sim Poiss(e_{ij}\theta_{ij})$$

$$\log(\theta_{ij}) = \log(e_{ij}) + \lambda_j f_i$$

where $\lambda_j f_i = \log(\theta_{ij}/e_{ij})$ and f_i is the spatially-referenced common risk factor. It is further assumed that

$$\mathbf{f} \sim \mathbf{N}(\mathbf{0}, \boldsymbol{\Sigma})$$

with unit variance, $\Sigma_{ij} = \exp(-d_{ij}/\phi)$, and $\sum f_i = 0$ for identifiability. Subsequently, Liu et al. (2005) extended the proposal to structural equation models. Of course, these approaches are not univariate, and there are a wide range of dimension reduction possibilities when multivariate outcomes or multiple predictors are included within models. I do not pursue this here. In Section 12.3.3, I examine the possibility of space-time latent modeling.

Another potentially useful development is the use of Dirichlet process (DP) mixing models to provide more flexible spatial structures (Ishwaran and James, 2002, 2001; Gelfand et al., 2005; Griffin and Steel, 2006; Kim et al.,

2006; Duan et al., 2007; Cai and Dunson, 2008; Rodriguez et al., 2009; Hossain et al., 2012; Kottas, 2016). In addition, there is a possibility that DP mixtures could provide a flexible approach to variable dimension modeling within clustering or variable selection scenarios (Kim et al., 2006). In a later section, (Chapter 12) we give an example of a DP process mixture used to examine latent structure in spatio-temporal data.

5.7.6 Quantile Regression

Commonly, in disease mapping applications, the mean of the outcome is the default focus of modeling. However, it is possible to consider other quantiles such as the median: median Poisson regression could be considered. More generally, it is possible to consider any quantile as a focus and to set up a fitting process that provides estimated parameters for that outcome quantile. Quantile regression (Koenker and Bassett, 1978; Koenker, 2005) sets up a functional approach whereby a specific quantile is modelled via a loss fucntion. Yu and Moyeed (2001) provide the Bayesian extension of quantile regression for continuous outcomes. Usually a quantile loss function is incorporated into a likelihood, and the asymmetric Laplace (ASL) distribution is often used for this:

$$L_\gamma(y|\boldsymbol{\theta}, \mathbf{x}) = \frac{\gamma(1-\gamma)}{\lambda} \exp[-\rho_\gamma\{y - g(\mathbf{x}^T\beta)\}/\lambda] \qquad (5.28)$$

where $\lambda > 0$ is a scale parameter, $g(.)$ is a link function, γ is the quantile, and $\rho_\gamma\{.\}$ is the loss function for the γ quantile. Note that the likelihood is specific to the quantile chosen and so also different β_γ could be assumed. Lee and Neocleous (2010) provide an example of applying an ASL to log transformed and jittered disease count data, where $g(\mathbf{x}^T\beta) = \mathbf{x}^T\beta_\gamma + \log(e_i)$. Presumably the predictor can be extended to include random effects as well as fixed covariates. Alternative discrete approaches are given by Machado and Silva (2005).

Inferential issues arises with quantile methods. The fact that there is an infinity of quantiles means the choice of which to fit to is not always clear. Additionally, the question of whether the model should fit well to all the quantiles, or only chosen ones, has to be made. Should the 'best fitting' quantile be sought? If so then γ would have to be estimated. These issues are largely unresolved. A recent review of quantile regression in epidemiological applications is given by Reich (2016).

5.8 Edge Effects

The importance of the assessment of edge effects in any spatial statistical application cannot be underestimated. Edge effects play a larger role in spatial problems than in, say, time-series. Specifically, we define edge effects as "any effect upon the analysis of the observed data brought about by the proximity of the study area boundary." The effects of the edges of a study area are largely the result of the effects of *spatial censoring*. That is, the fact that observations outside the window are not considered and therefore cannot contribute to analysis within the window. This mirrors the effects of temporal censoring in say, survival analysis, where, for example, the outcome for some subjects may not be observed because the observation period has stopped before the outcome appears.

Of course, all censoring depends on the idea that observations are dependent in *some* way. That is, the occurrence of observations outside the window of observation relies on observations within the window. In the spatial case, it is easily possible for individual disease response to relate to "missing" observations outside the window. For example, it may be that an environmental health hazard is located outside e.g. in the case of viral etiology, an infected person or carrier is located outside. For diseases which have uncertain etiology, it could be possible that factors underlying the incidence of the disease have a distribution that is spatially dependent and hence the disease incidence reflects this structure even when individual responses are independent. If, in addition, some unknown genetic etiology underpinned the disease incidence and has spatial expression, the incidence of disease could relate to unobserved genetically linked subjects outside the observation region.

In addition, such spatial censoring can affect estimation procedures, even when no explicit spatial dependence is proposed. For example, spatial smoothing methods, including geostatistical methods (kriging), splines or convolution random effect models, use data from different regions of the observed window in the estimation of risk at a location. Hence, if no correction is pursued for this effect at the edges, some edge distortion will result. In other cases parametric estimation may require the computation of averages of values in neighborhoods of a chosen point. Hence, close to edges there could be considerable distortion induced by missing neighbors. This edge problem can not only induce bias in estimation, but also tends to lead to considerable increases in estimator variance at such locations, and hence to low reliability of estimation. An example of the effect can be seen immediately when a CAR distribution is assumed. In that model, the conditional variance for the i th area is defined, in the notation of (5.18), as r/n_{δ_i}. This dependence on the number of neighbors (n_{δ_i}) implies that, for a given r, a reduction of neighbor number will increase the variance.

A number of methods have been proposed to deal with such edge effects. These methods have been in part developed within stochastic geometry, where it is often assumed that the process under study is first- and second-order stationary and isotropic (Ripley, 1988). These methods include (1) correction methods applied to smoothers or other estimators, for example, using weights relating to the proximity of the external boundary, (2) employing guard areas to provide external information to allow better boundary area estimation within the window, and (3) simulation of missing data outside the window and iterative re-estimation or model fitting. (The use of toroidal correction is not usually appropriate in the analysis of disease incidence data, as it is not usually appropriate to make the appropriate stationarity assumptions.) This final method has significant advantages if used within iterative simulation methods such as data augmentation (Gilks et al., 1996; Tanner, 1996; Robert and Casella, 2005) or general MCMC algorithms, as the external data can be treated as parameters in the estimation sequence.

An example of the degree to which edge effects could affect the application of convolution models was examined under simulation by Vidal-Rodiero and Lawson (2005). In that study, counties within a large multi-state region of the United States were examined and external county hulls were peeled from the observation window to examine the effects of different neighborhood "depths" on estimation. Figure 5.15 displays the effect of stripping out a sequence of hulls of small areas around a central area. In the simulation study a large number of states within central USA were amalgamated and the counties gathered into one study area. Successive hulls of counties were then stripped and the effect of this stripping was noted. The effect of stripping on four different models (convolution (BYM), Poisson-gamma (PG), Poisson log normal (PLN) and fixed SMR (C)) was assessed. The six sets (6 through 11) indicate internal regions at different depths where the relative risk was estimated. The outer sets of counties (sets 1 through 5) were successively stripped. Four models were fitted and the different set results are given in Figure 5.15. It is clear that sets close to the sets close to the boundary (e.g. sets 6 and 7) that is stripped show bigger differences in average relative risk.

5.8.1   Edge Weighting Schemes and MCMC Methods

The two basic methods of dealing with edge effects are (1) the use of weighting or correction systems, which usually apply different weights to observations depending on their proximity to the study boundary, and (2) the use of guard areas, which are outside our study region.

5.8.1.0.1   Weighting Systems   Usually, it is appropriate to set up weights which relate the position of the event or tract to the external boundary. These weights, $\{w_i\}$ say, can be included in subsequent estimation and inference. Often the form will be $w_i = f(d_i)$ where d_i is the distance to the boundary, from a fixed point in a small area or in the case event situation from the case event itself. Another alternative for small area data would be to use the length of boundary of the small area in common with the study area boundary. In that case, one could propose $w_i = f(l_i)$ where l_i is the common boundary length. For example, the proportion of the total boundary length of the small area common with the study boundary might be a useful measure. The weight for an observation is usually intended to act as a surrogate for the

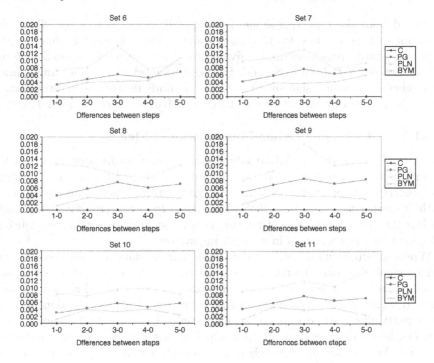

FIGURE 5.15: Model effects of hull stripping: six sets of internal regions (sets 6 through 11)) with four different models (convolution (BYM), Poisson-gamma (PG), Poisson log normal (PLN) and fixed SMR (C)).

degree of missing information at that location and so may differ depending on the nature and purpose of the analysis. Some sensitivity to the specification of these weights will inevitably occur and should be assessed in any case study. More detail on suitable weights can be found in Lawson (2006b), Chapter 5.

Defining an indicator for closeness to the boundary for each area, when in the tract count case, some external standardised rates are available, it is possible to structure an expectation-dependent weight for a particular tract, e.g. based on the ratio of the sum of all adjacent area expectations to the sum of all such expectations within the study window. Other suitable weighting schemes could be based on the proportion of the number of observed neighbors.

Guard areas An alternative approach is to employ guard areas. These areas are external to the main study window of interest. These areas could be boundary tracts of the study window or could be added to the window to provide a guard area, in the case of tract counts. In the case event situation, the guard area could be some fixed distance from the external boundary (Ripley, 1988). The areas are used in the estimation process but they are

excluded from the reporting stage, as they will be prone to edge effects. If
boundary tracts are used for this, then some loss of information must result.
External guard areas have many advantages. First, they can be used *with* or
without their related data. Second, they can be used within data augmentation schemes in a Bayesian setting. These methods regard the external areas
as a missing data problem (see e.g. Little and Rubin, 2002, Chapter 10).

5.8.1.1   MCMC and Other Computational Methods

It is usually straightforward to adapt conventional estimation methods to
accommodate edge-weighted data. In addition, if guard areas are selected
and observations are available within the guard area, it is possible to proceed
with inference by using the whole data but selectively reporting areas not
within the guard area. Note that this is not the same as setting $w_i = 0$ for
all guard area observations in a weighting system.

When external guard areas are available but no data are observed, resort
must usually be made to missing data methods. An intermediate situation
arises when in the tract count case some external standardised rates are available. In that case it is possible to structure an expectation-dependent weight
for a particular tract, e.g. based on the ratio of the sum of all adjacent area
expectations to the sum of all such expectations within the study window.
This can be used as an edge weight within such a weighting system. An example of a study of different edge remedies for count data can be found in
Lawson et al. (1999).

5.8.2   Discussion and Extension to Space-Time

In the situation in which case events are studied where censoring is present
and could be important (i.e. when there is clustering or other correlated heterogeneity), it is advisable to use an internal guard area or an external guard
area with augmentation via MCMC. In cases where only a small proportion of
the study window is close to the boundaries and only general (overall) parameter estimation is concerned, then it may suffice to use edge weighting schemes.
If residuals are to be weighted, then it may suffice to label the residuals only
for exploratory purposes.

In the situation where counts are examined, then it is also advisable to use
an internal guard area or external area with augmentation via MCMC. In
some cases, an external guard area of *real* data may also be available. This
may often be the case when routinely collected data are being examined. In
this case, analysis can proceed using the external area *only to correct internal
estimates*. Edge weighting can be used also, and the simplest approach would
be to use the proportion of the region *not* on the external boundary. Residuals
can be labelled for exploratory purposes. The assumptions underlined in any
correction method are that the model be correctly specified and that it could
be extended to the areas not observed. In particular, whether an adjustment

can really be obtained when ignoring the information on the outer areas is questionable. Edge effect bias should be less prominent when an unstructured exchangeable model is chosen. Since each area relative risk would be regressed toward a grand mean, the information lacking for the unobserved external areas is very small compared to information from the observed areas. Of course, such a simple model where common expectation is found is highly unlikely to be a good model in this area.

Extending the edge effect problem to consideration of space-time data, the situation is more complex as spatial edge effects can interact with temporal edge effects. The use of sequential weighting, based on distance from time and space boundaries, may be appropriate (Lawson and Viel, 1995). For tract counts observed in distinct time periods only, the most appropriate method is likely to be based on distance from time and space boundaries, although it may be possible to provide an external spatial and/or temporal guard area either with real data or via augmentation and MCMC methods.

The use of augmentation methods can also be fruitfully employed in this context. If the external areas are known, but information concerning the disease of interest is not available in these external areas, it is possible to regard such missing or censored data as parameters which can be estimated within an iterative sampling algorithm, such as an MCMC algorithm. In addition, if partial information were known (for example the standardised rates in the external areas), then we could condition these missing data count estimates on the known information.

5.9 Exercises

5.9.1 Maximum Likelihood

To provide a backdrop for the Bayesian analysis we present some basic results for likelihoods from simple mapping models.

1) A state in the US has m counties. Within these counties, total births and births with abnormalities are recorded. The births with abnormalities (Ba) are a subset of all births.

The probability that a birth in the i th region is a Ba is θ_i. Each birth has an independent risk of being Ba. We observe $\{y_i\}$, $i = 1, ..., m$ Ba events in the m counties and the total births in the m counties are denoted $\{n_i\}$.

a) A likelihood model for these data could be a binomial with probability θ_i, as in (5.7) above. Explain why this is appropriate.

b) If we assume there is a common probability across all regions, show that the maximum likelihood estimator of θ_i is given by, $\widehat{\theta} = \sum_{i=1}^{m} y_i / \sum_{i=1}^{m} n_i$.

c) A logistic linear model results if we assume a logistic link between θ_i and a linear predictor. For the model $\theta_i = \exp(\beta_0)/\{1 + \exp(\beta_0)\}$, show that the maximum likelihood estimator of β_0, either directly or by invariance, is given by

$$\widehat{\beta}_0 = \log\left\{ \frac{S_y}{S_n} \Big/ \left(1 - \frac{S_y}{S_n}\right) \right\},$$

where $S_y = \sum_{i=1}^{m} y_i$, $S_n = \sum_{i=1}^{m} n_i$.

d) Show that the large sample standard error of $\widehat{\beta}_0$ is given by

$$se(\widehat{\beta}_0) = \left\{ S_y - S_y^2/S_n \right\}^{-\frac{1}{2}}.$$

2) Case event data is observed within a study area W. There are m events in W and their locations are denoted by $\{s_i\}$, $i = 1, ..., m$. A realization of control events is also available in the same window: $\{s_j\}$, $j = m+1, ..., m+n$. The conditional log-likelihood for these data can be written as:

$$l = \sum_{i=1}^{m+n} y_i \eta_i - \sum_{i=1}^{m+n} \log[1 + \exp(\eta_i)],$$

where y_i is now an indicator variable taking the value 1 for a case and 0 for a control (see (5.6) above), and $\exp(\eta_i) = \rho f(s_i; \alpha) = \exp(\alpha_0 - \alpha d_i)$ where d_i is the distance from a fixed point to the i th location and $f(s_i; \alpha) = \exp(-\alpha d_i)$. Assume we want to test for a distance effect between the case locations and a fixed point.

a) Show that under the null hypothesis H_0: $\alpha = 0$, the maximum likelihood estimator of ρ is just m/n.

b) If you substitute this estimator into the likelihood above, a possible test statistic to find whether distance is significant is based on the first derivative of the likelihood WRT α. This is known as a score test statistic.

Show that under H_0: $\alpha = 0$, the test statistic is given by:

$$\frac{m}{m+n}\left[\sum_{i=1}^{m} d_i + \sum_{j=m+1}^{m+n} d_j \right] - \sum_{i=1}^{m} d_i.$$

5.9.2   Poisson-Gamma Model: Posterior and Predictive Inference

A random sample of size m from a Poisson distribution with parameter θ is denoted $x_1, ..., x_m$. The parameter has a prior distribution :

$$g(\theta) = \begin{cases} \lambda e^{-\lambda\theta} & \lambda > 0 \\ 0 & \text{elsewhere} \end{cases}.$$

The posterior distribution of θ is given by:

$$P(\theta|\{x_i\}) = \frac{\beta^{s+1}}{\Gamma(s+1)}\theta^s e^{-\theta\beta}$$

where $\beta = m + \lambda$, and $\Gamma()$ is the gamma function and $s = \sum x_i$, assuming λ is fixed.

Derive the prior predictive distribution $Pr(x|x_1, ..., x_m)$ and hence find $Pr(x > 2|x_1, ..., x_m)$ when $\lambda = 2$.

5.9.3   Poisson-Gamma Model: Empirical Bayes

For the Poisson likelihood model with gamma prior distribution defined in (5.10), the unconditional distribution of y_i given a, b is negative binomial. This is also the prior predictive distribution of y_i. The marginalized log-likelihood is given by

$$L(a,b) = \sum_i \left[\log \frac{\Gamma(y_i + a)}{\Gamma(a)} + b\log(a) - (y_i + a)\log(e_i + a)\right].$$

This likelihood is free of $\{\theta_i\}$ and can be maximized to yield *empirical Bayes* estimates of a, and b (Clayton and Kaldor, 1987). Show that this leads to normal equations, which can be solved for $\widehat{a}$, and $\widehat{b}$:

$$\frac{\widehat{b}}{\widehat{a}} = \frac{1}{m}\sum_i \frac{(y_i + \widehat{a})}{(e_i + \widehat{b})}$$

$$\sum_i^m \sum_{j=0}^{y_i-1} \frac{1}{\widehat{a}+j} + m\log(\widehat{b}) - \sum_i^m \log(e_i + \widehat{b}) = 0.$$

6

Disease Cluster Detection

In the study of disease spatial distribution it is often appropriate to ask questions related to the local properties of the relative risk surface rather than models of relative risk per se. Local properties of the surface could include peaks of risk, sharp boundaries between areas of risk, or local heterogeneities in risk. These different features relate to surface properties but not directly to a value at a specific location. Relative risk estimation (or disease mapping; Chapter 5) concerns the "global" smoothing of risk and estimation of true underlying risk level (height of the risk surface), whereas cluster detection is focused on local features of the risk surface where elevations of risk or depressions of risk occur. Hence it is clear that cluster detection is fundamentally different from relative risk estimation in its focus. However the difference can become blurred, as methods that are used for risk estimation can be extended to allow certain types of cluster detection. This will be discussed more fully in later sections.

6.1 Cluster Definitions

Before discussing cluster detection and estimation methods it is important to define the nature of the clusters and/or clustering to be studied.

There are a variety of definitions of clusters and clustering. Different definitions of clusters or clustering will lead to differences in the ability of detection methods. First it should be noted that sometimes the correlated heterogeneity term in relative risk models is called a clustering term (see e.g. Clayton and Bernardinelli, 1992). This implies that the term captures aggregation in the risk and indeed this does lead to an effect where neighboring areas have similar risk levels. This is a global feature of the risk however, and also induces a smoothing of risk. This begs the question of how we define clusters and clustering: should they be global features or local in nature?

Global clustering basically assumes that the risk surface is clustered or has areas of like elevated (reduced) risk. An uncorrelated surface, on the other hand, should display random changes in risk with changes in location and so should be much more variable in risk level and have few contiguous areas of like risk.

FIGURE 6.1: Simulated examples of correlated (A) and uncorrelated surfaces. Simulation using the R function GaussRF with mean 1.0.

Figure 6.1 displays a comparison between an uncorrelated and correlated risk surface. Figure 6.1 A shows areas of elevated risk that may qualify as clusters (by some definition). However, modeling the overall clustering does not address cluster locations specifically. Hence this form of clustering does not address localized behavior or the locations of clusters per se. This is often termed general clustering (Besag and Newell (1991)).

A general definition of a (spatial) cluster is: *Any spatially-bounded area of significantly elevated (reduced) risk.* This is clearly very general and requires further definition. By *spatially-bounded* I mean that the cluster must have some spatial integrity. This could be a neighborhood criterion such as *areas must be adjoining* or *at least two adjoining areas must meet a criterion,* or could be defined to have a certain type of external boundary (e.g. risk differences around the cluster must meet a criterion). A simple criterion that is often assumed is known as *hot spot clustering.* In hot spot clustering, any area or region can be regarded as a cluster. This is due to the assumption of a zero neighborhood criterion, i.e. no insistence on adjacency of regions within clusters. This is a convenient and simple criterion and is often assumed to be the only criterion. It is commonly used in epidemiology (see e.g. Richardson et al., 2004). Without prior knowledge of the behavior of the disease, this criterion is appealing. It could be useful for preliminary screening of data, for example.

However, this hot spot definition ignores any contiguity that may be thought to be inherent in relevant clusters. For example, it might be important that clusters of a given threshold size be investigated. This threshold size could be defined as a minimum number of contiguous areas. Hence, only groups of contiguous regions of "unusual" risk could qualify as clusters. On the other hand, in the case of infectious diseases, it may be that a certain shape and size of cluster are important in understanding disease spread.

In this chapter I will mainly consider three different scenarios for clustering:

(a) Single region hot spot relative risk detection

(b) Clusters as objects or groupings

(c) Clusters defined as residuals

6.1.1   Hot Spot Clustering

Hot spot clustering is often the most intuitive form of clustering and may be that which most public health professionals consider as their definition. In hot spot clustering, any area or region can be regarded as a cluster. This is due to the assumption of a zero neighborhood criterion, i.e. no insistence on adjacency of regions within clusters. Simply any area displaying "excess" or "unusual" risk by some criterion, is a hot spot. This is a relatively nonparametric definition.

6.1.2   Clusters as Objects or Groupings

Clustering might be considered to be apparent in a data set when a specific form of grouping is apparent. This grouping would usually be predefined. Usually the criterion would also have a neighborhood or proximity condition. That is, only neighboring or proximal areas (which meet other criteria) can be considered to be in a cluster. Hence some parametric conditions must be met under this definition.

6.1.3   Clusters Defined as Residuals

Often it is convenient to consider clusters as a residual feature of data. For example, let's assume that y_i is the count of disease within the i th census tract within a study area. Let's also assume that our basic model for the average count μ_i (i.e. $E(y_i) = \mu_i$) is

$$\log \mu_i = a_i + e_i.$$

Here a_i could consist of a linear or non-linear predictor as a function of covariates and could also consist of random effects of different kinds. To simplify the idea we assume that a_i is the smooth part of the model and e_i is the rough or residual part. The basic idea is that if we model a_i to include all relevant non-clustering confounder effects then the residual component must contain residual clustering information. Hence if we examine the estimated value of e_i then this will contain information about any clusters unaccounted for in a_i. Of course this does not account for any pure noise that might also be found in e_i. This means that an estimate of e_i could have at least two components: clustered and unclustered (or frailty). There could, be additional components depending on whether the confounding in a_i was adequately specified or estimated.

There are a number of approaches to isolating the residual clustering. First, it is possible to include a pure noise term within a_i and to consider e_i as a cluster term. For example we could assume that $a_i = f(v_i; covariates)$ where $f(.)$ is a function of a uncorrelated noise at the observation level (v_i : frailty or random effect term) and a function of covariates. Second, a smoothed version of e_i, $s(e_i)$ say, could be examined in the hope that the pure noise is smoothed out. Of course this begs the question of which component should include the clustering: should it be a model component or a residual component? If the clustering is likely to be irregular and we can be assured that no clustering confounding effects are to be found in the model component, then a residual or smoothed residual might be useful. On the other hand, if there is any prior knowledge of the form of clustering to be expected, then it may be more important to include some of that information within the model itself. The real underlying issue is the ability of models and estimation procedures to differentiate spatial scales of clustering.

6.2 Cluster Detection Using Residuals

First, assume that we observe disease outcome data within a spatial window.

6.2.1 Case Event Data

6.2.1.1 Unconditional Analysis

For the case event scenario we have $\{s_i\}, i = 1, ..., m$ events observed within the window T. Modeling here focuses on the first order intensity and its parameterization. Assume that $\lambda(s|\psi) = \lambda_0(s|\psi_0).\lambda_1(s|\psi_1)$ as defined in Chapter 5. We focus first on the specification of a residual for a point process governed by $\lambda(s|\psi)$. First, in the spirit of classical residual analysis, it is clear that we want to compare fitted values to observed values. This is not simple as we have locations as observed data. One way to circumvent this problem is to consider a function of the observed data which can be compared with an intensity estimate at location s_i, $\lambda(s_i|\widehat{\psi})$ say. One such function could be a saturated or nonparametric intensity estimate $(\widehat{\lambda}_{loc}(s_i))$, say, where *loc* denotes a local estimator). Essentially this gives a slight aggregation of the data, but it allows for a direct comparison of model to data. Hence we can define a residual as:

$$r_i^{loc} = \widehat{\lambda}_{loc}(s_i) - \lambda(s_i|\widehat{\psi})$$

or in the case of a saturated estimate (Lawson, 1993a)

$$r_i^{sat} = \widehat{\lambda}_{sat}(s_i) - \lambda(s_i|\widehat{\psi}).$$

Baddeley et al. (2005) discuss more general cases applied to a range of processes. An example of a local estimate of intensity could be derived from a suitably edge-weighted density estimate (Diggle, 1985). An example of the use of the saturated estimator is as follows.

First assume that an estimator is available for the background intensity $\lambda_0(s_i|\psi_0)$, $\lambda_0(s_i|\widehat{\psi}_0) \equiv \lambda_{0i}$ say. Also assume that it can be used as a plug-in estimator within $\lambda(s|\psi)$. If this is the case, then we can compute $r_i^{sat} = \lambda_{0i}[\widehat{\lambda_{1sat}}(s_i) - \lambda_1(s_i|\widehat{\psi})]$. For a simple heterogeneous Poisson process model with intensity $\lambda_0(s|\psi_0).\lambda_1(s|\psi_1)$ and using an integral weighting scheme (as described in Chapter 5), the saturate estimate of the intensity at s_i is $1/(w_i\lambda_{0i})$. A simple weight (which provides a crude estimator of the local intensity) is $w_i = A_i$ where A_i is the Dirichlet tile area surrounding s_i, based on a tesselation of the case events. Hence a simple residual could be based on

$$r_i^{sat} = \lambda_{0i}[(w_i\lambda_{0i})^{-1} - \lambda_1(s_i|\widehat{\psi})]$$
$$= w_i^{-1} - \lambda_{0i}\lambda_1(s_i|\widehat{\psi}).$$

The use of such tile areas must be carefully considered as edge effect distortion can occur with tesselation and so boundary regions of the study window should be treated with caution. Of course the error in estimation of the background intensity is ignored here and a crude approximation to the saturated intensity is assumed. Note that r_i^{sat} or r_i^{loc} can be computed within a posterior sampler and so a posterior expectation of the residuals can be estimated.

Figure 6.2 displays an example of the use of posterior expectation of r_i^{sat} for a model for the well known larynx cancer data set from Lancashire, UK, 1974 through 1983. This data set has been analyzed many times and consists of the residential address locations of cases of larynx cancer with the residential addresses of cases of respiratory cancer as a control disease (see e.g. Diggle, 1990; Lawson, 2006b, Chapter 1), with distance decline component (variable d_i) around the fixed point (3.545, 4.140), an incinerator. The motivation for this type of analysis relates to assessment of health hazards around putative sources (putative source analysis). This is discussed more fully in Chapter 8. The model for the first order intensity is defined to depend on this distance: $\lambda_1(s_i|\theta) = \beta_0[1 + \exp(-\beta_1 d_i)]$. Appendix B contains the WinBUGS code for this example. The map displays the contours for the posterior sample estimate of $\Pr(r_i^{sat} > 0)$, the residual exceedance probability.

FIGURE 6.2: Map of Lancashire larynx cancer case distribution with super-imposed contour map of exceedance probability (0.7, 0.8, 0.9) for the residual (r_i^{sat}) from a Bayesian model assuming Berman-Turner Dirichlet tile integration weights and non-parametric density estimate of background risk computed from the respiratory cancer control distribution.

To allow for extra unobserved variation in this map an uncorrelated random effect term can also be included in the model. Appendix A.6 displays the code used for this model. Figure 6.3 displays the resulting posterior average residual exceedance probability map for the model with $\lambda_1(s_i|\theta) = \beta_0[1 + \exp(-\beta_1 d_i)]\exp(v_i)$ where $v_i \sim N(0, \tau_v)$ and $\beta_* \sim N(0, \tau_{\beta_*})$. The hyperparameter specifications are given in the appendix. Both figures suggest that there is slight evidence for an excess of aggregation in the north of the study region (where there is a large area where $\Pr(r_i^{sat} > 0) > 0.9$). There is also weaker evidence of an excess in the area to the west of the putative source (3.545, 4.140), where $\Pr(r_i^{sat} > 0) > 0.8$ on average. There are also marked edge effects close to the study region corners due to the distortion of the tesselation suspension algorithm. Figure 6.4 displays a similar picture after removal of extra noise.

6.2.1.2   Conditional Logistic Analysis

An alternative approach to the analysis of case event data is to consider the joint realization of cases and controls and to model the conditional probability of a case if an event has occurred at a location. This approach was discussed in Chapter 5 and has the advantage that the background effect factors out

FIGURE 6.3: Version of Figure 6.2 in which the model has included a random uncorrelated effect (v_i) to allow for extra variation in the risk: $\lambda_1(s_i|\theta) = \beta_0[1 + \exp(-\beta_1 d_i)]\exp(v_i)$.

of the likelihood. Define the joint realization of m cases and n controls as $s_i : i = 1, ..., N$ with $N = m + n$. Also define a binary label variable $\{y_i\}$ which labels the event either as a case $(y_i = 1)$ or a control $(y_i = 0)$. The resulting conditional likelihood has a logistic form:

$$L(\psi_1|\mathbf{s}) = \prod_{i \in cases} p_i \prod_{i \in controls} 1 - p_i$$

$$= \prod_{i=1}^{N} \left[\frac{\{\exp(\eta_i)\}^{y_i}}{1 + \exp(\eta_i)} \right]$$

where $p_i = \frac{\exp(\eta_i)}{1+\exp(\eta_i)}$ and $\eta_i = x_i'\beta$ and x_i' is the i th row of the design matrix of covariates and β is the corresponding p-length parameter vector. Hence in this form a Bernoulli likelihood can be assumed for the data and a hierarchical model can be established for the linear predictor $\eta_i = x_i'\beta$. In general, it is straightforward to extend this formulation to the inclusion of random effects in a generalized linear mixed form. Bayesian residuals such as $r_i = y_i - \widehat{p}_i/\widehat{se}(y_i - \widehat{p}_i)$ (or directly standardised version: $r_i = (y_i - \widehat{p}_i)/\sqrt{\widehat{p}_i(1 - \widehat{p}_i)}$) are available, where $\widehat{p}_i$ is the average value of p_i from the posterior sample. Residuals from binary data models are often difficult to interpret due to the limited variation in the dependent variable (0/1), and the usual recommendation for their examination is to group or aggregate the results. A wide variety of

FIGURE 6.4: Contour map of the standardised Bayesian residual for the logistic case-control spatial model applied to the larynx cancer data from Lancashire. The display shows the posterior average residual for a sample size of 5000 after burn-in.

aggregation methods could be used. Spatial aggregation methods might be considered here. Figure 6.4 displays the mapped surface of the standardised Bayesian residual using $r_i = (y_i - \widehat{p}_i)/\sqrt{\widehat{p}_i(1 - \widehat{p}_i)}$ where $\widehat{p}_i$ is computed from the converged posterior sample. Appendix A.6 displays the code for this model. The result was obtained using R2WinBUGS. The model assumed for this example also has an additive distance effect and is specified by

$$p_i = \frac{\lambda_i}{1 + \lambda_i}$$
$$\lambda_i = \exp\{\alpha_0 + v_i\}.\{1 + \exp(-\alpha_1 d_i)\}.$$

Figure 6.5 displays the thresholded mapped surface of the $\Pr(r_i > 2)$ for values (0.05, 0.1, 0.2). This suggests some evidence of clustering or unusual aggregation in the north and also in the vicinity of the putative location in the south.

   Note that all of these models assume negligible clustering under the model and that any residual effects will include the clustering. These models do not explicitly model clustering, but model only long range and uncorrelated variation. Hence we make the tacit assumption that any remaining aggregation of cases will be found in the residual component. Of course other effects which were excluded from the model could be present in the residuals.

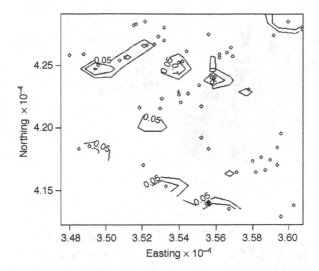

FIGURE 6.5: Map of the contoured surface of $\Pr(r_i > 2)$ estimated from the converged posterior sample for the standardised Bayesian residual in Figure 6.4.

6.2.2 Count Data

For count data, it is assumed that either a Poisson data likelihood or a binomial likelihood is relevant. Note that an autologistic model could also be specified.

6.2.2.1 Poisson Likelihood

In the case of a Poisson likelihood, assume that y_i, $i = 1, ..., m$ are counts of cases of disease and e_i $i = 1, ..., m$ are expected rates of the disease in m small areas, and so $y_i \sim Poiss(e_i\theta_i)$ given θ_i. The log relative risk is usually modeled and so $\log \theta_i$ is the modeling focus. Bayesian residuals for this likelihood are easily computed in standardised form as $r_i = (y_i - e_i\widehat{\theta}_i)/\sqrt{e_i\widehat{\theta}_i}$ where $\widehat{\theta}_i$ is the average value of the θ_i possibly from the converged posterior sample. In this case, the Georgia oral cancer data was examined with a Poisson data likelihood and model $\log \theta_i = \alpha_0 + v_i$ where

$$\alpha_0 \sim U(a, b)$$
$$v_i \sim N(0, \tau_v)$$

with τ_v set large and (a, b) a large negative to positive range. Appendix A.6 has details of the WinBUGS code used. No correlated random effect is included here as it is assumed that clustering is to be found in residuals.

FIGURE 6.6: Georgia county maps of Bayesian residuals from a converged posterior sampler with uncorrelated random effect term. Top to bottom: $Pr(r_i > 2)$ and $Pr(r_i > 3)$.

Figure 6.6 displays the results from a converged sampler based on 10,000 burn-in and sample size of 2000. The display shows the average estimate of $Pr(r_i > 2)$ and $Pr(r_i > 3)$ for r_i given above. The most extreme region appears to be the far west of Georgia. It should be noted that there is considerable noise in these residuals, particularly for $Pr(r_i > 2)$.

6.2.2.2 Binomial Likelihood

In the case of a binomial likelihood assume m small areas, and that in the ith area there is a finite population n_i out of which y_i disease cases occur. The probability of a case is p_i. The data model is thus $y_i \sim bin(p_i, n_i)$ given p_i, and the usual assumption is made that logit $(p_i) = f(\eta_i)$, where η_i is a linear or non-linear predictor. Of course, various ingredients can be specified for $f(\eta_i)$, including the addition of random effects to yield a binomial generalized linear mixed model. A Bayesian residual for this model is given in standardised form as $r_i = (y_i - n_i \widehat{p}_i) / \sqrt{n_i \widehat{p}_i (1 - \widehat{p}_i)}$ where $\widehat{p}_i$ is the average of p_i values found in the converged posterior sample.

While the above discussion has focused on simple residual diagnostics, albeit from posterior samples, there is also the possibility of examining *predictive* residuals for any given model. A predictive residual can be computed for each observation unit as

$$r_i^{pr} = y_i - y_i^{pr}$$

where $y_i^{pr} = \frac{1}{G} \sum_{g=1}^{G} f(y_i | \theta^g)$, and $f(y_i | \theta^g)$ is the likelihood given the current value of θ^g. Of course this will usually be small compared to the standard Bayesian residual. Note that for a given data model, y_i^{pred} can be easily generated on WinBUGS. For the binomial example above the code could be:

y[i]~dbin(p[i],n[i])

ypred[i]~dbin(p[i],n[i])

rpred[i]<-y[i]-ypred[i]

An alternative approach to residual analysis could be based in the construction of a residual envelope, based on the comparison of the Bayesian residual: $r_i = y_i - \widehat{y}_i$ with $r_i^* = y_i^{pred} - \widehat{y}_i$. Unusual residuals could be analysed by assessing the ranking of r_i among the a series of B simulated $\{r_{ib}^*\}$ $b = 1, ..., B$, as discussed in Section 4.4. Further, a p-value surface can be computed from a tally of exceedances:

$$P_{v_i} = \Pr(|r_i| > |r_i^*|) = \frac{1}{B} \sum_{b=1}^{B} I(|r_i| > |r_{ib}^*|).$$

The mapped surface of P_{v_i} could be examined for areas of unusually elevated values and hence provide a tool for hot spot detection.

6.3 Cluster Detection Using Posterior Measures

Another approach to cluster detection is to consider measures of quantities monitored in the posterior that may contain clustering information. One such measure is related to estimates of first order intensity (case event data) or relative risk (Poisson count data) or case probability (binomial count data). If we have captured the clustering tendency within our estimate of any of these

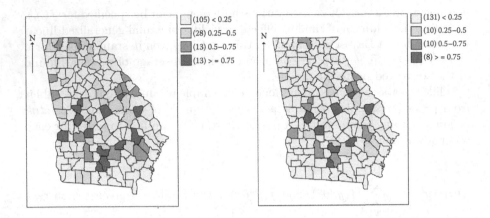

FIGURE 6.7: Georgia oral cancer: maps of $\widehat{\Pr}(\theta > c)$ for $c = 2$ and $c = 3$ for a model with uncorrelated random effect (UH).

quantities then we could examine their posterior sample behavior. Perhaps the most commonly used example of this is the use of exceedance probability in relation to relative risk estimates for individual areas for count data (see e.g. Richardson et al., 2004; Abellan et al., 2008; Lawson and Rotejanaprasert, 2014). Define the exceedance probability as the probability that the relative risk θ exceeds some threshold level (c): $\Pr(\theta > c)$. This is often estimated from posterior sample values $\{\theta_i^g\}_{g=1,\dots,G}$ via

$$\widehat{\Pr}(\theta_i > c) = \sum_{g=1}^{G} I(\theta_i^g > c)/G$$

$$\text{where } I(a) = \begin{cases} 1 \text{ if } a \text{ true} \\ 0 \text{ otherwise} \end{cases}.$$

Of course, there are two choices that must be made when evaluating $\widehat{\Pr}(\theta_i > c)$. First, the value of c must be chosen. Second, the threshold for the probability must also be chosen, i.e. $\widehat{\Pr}(\theta_i > c) > b$ where b might be set to some conventional level such as 0.95, 0.975, 0.99, etc. In fact, there is a tradeoff between these two quantities and usually one must be fixed before considering the value of the other. Figure 6.7 displays the posterior expected exceedance probability maps: $\widehat{\Pr}(\theta_i > c)$ for $c = 2$, and $c = 3$ for the Georgia oral cancer data when a relative risk model with a UH component was fitted (see Section 5.3.2).

One major concern with the use of exceedance probability for single regions is that it is designed only to detect *hot spot* clusters (i.e. single region signalling) and does not consider any other information concerning possible forms of clusters or even neighborhood information. Some attempt has been made to enhance this post hoc measure by inclusion of neighborhoods by Hossain and Lawson (2006). For the neighborhood of the i th area defined as δ_i and the number of neighbors as n_i:

$$\overline{q_i} = \sum_{j=0}^{n_i} q_{ij}/(n_i + 1)$$

$$\text{where } q_{ij} = \Pr(\theta_j > c) \ \forall \ j \in \delta_i$$

$$\text{and } q_{i0} = \Pr(\theta_i > c).$$

This measures $\overline{q_i}$ and q_{i0} can be used to detect different forms of clustering. Other more sophisticated measures have also been proposed (see e.g. Hossain and Lawson, 2006, 2010 for details).

A second concern with the use of exceedance probabilities is of course that the usefulness of the measure depends on the model that has been fitted to the data. It is conceivable that a poorly fitting model will not demonstrate any exceedances related to clustering and may leave the clustering of interest in the residual noise. An extreme example of this is displayed in Figure 6.8. In that figure the same data set is examined with completely different models. The data set covers South Carolina county level congenital anomaly deaths for 1990 (see also Lawson et al., 2003, Chapter 8). The expected rates were computed for an 8-year period. In the left panel a Poisson log linear trend model was assumed and in the right panel shows a convolution model. The trend model was $\log \theta_i = \alpha_0 + \alpha_1 x_i + \alpha_2 y_i$, (where the predictors are the coordinates of the centroids of the regions), and zero mean Gaussian prior distributions for the regression parameters, whereas the right panel was $\log \theta_i = \alpha_0 + u_i + v_i$ where the u_i, v_i are correlated and uncorrelated heterogeneity terms with the usual CAR and zero mean Gaussian prior distributions. Without examination of the goodness of fit of these models it is clear that there could be considerable latitude for misinterpretation if exceedance probabilities are used in isolation to assess (hot spot) clustering. In fact in this example the DIC for the trend model was 171.81 (pD: 2.85) whereas the DIC for the convolution model was 174.46 (pD: 11.57). If the trend model were preferred then the apparent clustering is in fact trended, whereas if the convolution model is favored then the clustering appears much more isolated.

As in the count data situation we can also examine exceedances for other data types and models. For example, in the case event example, intensity exceedance could be examined as $\Pr(\hat{\lambda}_1(s_i) > 1)$, whereas for the binary or binomial data the exceedance of the case probability could be used: $\Pr(\hat{p}_i > 0.5)$. These can also be mapped of course. However the rider concerning the goodness-of-fit of the model as highlighted by Figure 6.8 also applies here. Lawson and Rotejanaprasert (2014) provide a recent example of applying exceedance measures to childhood brain cancer in Florida.

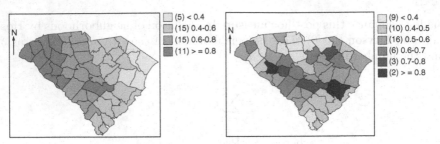

FIGURE 6.8: Display of exceedance probabilities for two models. Left panel: simple first order trend; right panel: convolution model with UH and CH only and no trend for the same data set. South Carolina county level congenital mortality, 1990.

6.4   Cluster Models

It is also possible to design models which explicitly describe the clustering behavior of the data. In this way parameters and functions can be defined that summarize this behavior. It should be noted that clustering behavior is often regarded as a second order feature of the data. By second order I mean, "relating to the mutual covariation of the data." Hence it is often assumed that covariance modeling will capture clustering in data. This is often termed *general* clustering. However, as noted in Section 6.1, while general covariance modeling can capture the overall mutual covariation (as in Figure 6.1 A) it does *not* lead to identification or detection of clusters per se. In the following section I focus on the detection of clusters, rather than general clustering.

6.4.1   Case Event Data

In the analysis of point processes (PPs) there is a set of models designed to describe clustering. For an introductory overview, which focuses mainly on general cluster testing, see Diggle (2003), Chapter 9. Basic models often assumed for PPs, which allow clustering, are the Poisson cluster process and the Cox process. In the Poisson (Neyman-Scott) cluster process (PcP) an underlying process of parents (unobserved cluster centers) is assumed and offspring (observed points) are generated randomly in number and location. This generation is controlled by distributions. Clearly this formulation is most appropriate in examples where parent generation such as seed dispersal in ecology occurs.

An alternative to a PcP is found in the Cox process where a non-negative stochastic process ($\Lambda(s)$) governs the intensity of a heterogeneous Poisson process (hPP). Conditional on the realization of the stochastic process the events follow a hPP. In this case

$$\lambda(s) = E[\Lambda(s)]$$

where the expectation is with respect to the process. Note that this formulation allows the inclusion of spatial correlation via a specification such as $\Lambda(s) = \exp\{S(s)\}$ where $S(s)$ is a spatial Gaussian process. This is sometimes known as a log-Gaussian Cox process (LGCP) (see e.g. Møller et al., 1998, Liang et al., 2009, Taylor et al., 2015). Note also that an intensity process of the form

$$\Lambda(s) = \mu \sum_{j=1}^{\infty} h(s - c_j) \tag{6.1}$$

can be assumed, where $h(s - c_j)$ is a bivariate pdf, and c_j are cluster centers. If the centers are assumed to have a homogeneous PP then this is also a PcP. Of course these models were derived mainly for ecological examples and not for disease case events. However we can take as a starting point a model for case events that includes population modulation in the first order intensity, and that also allows clustering via unobserved process of centers.

6.4.1.1 Object Models

Define the first order intensity as

$$\lambda(s|\psi) = \lambda_0(s|\psi_0).\lambda_1(s|\psi_1).$$

Assume that the case events form a hPP conditional on parameters in ψ_1. In the basic hPP likelihood, dependence on ψ_0 would also have to be considered. Often $\lambda_0(s|\psi_0)$ is estimated nonparametrically and a profile likelihood is assumed. Alternatively ψ_0 could be estimated within a posterior sampler. Here focus is made on the specification of $\lambda_1(s|\psi_1)$. Following from the basic definitions of PcPs and Cox processes it is possible to formulate a Bayesian cluster model that relies on underlying unobserved cluster center locations, but is not restricted to the restrictive assumptions of the classical PcP. Define the excess intensity at s_i as

$$\lambda_1(s_i|\psi_1) = \mu_0 \sum_{j=1}^{K} h(s_i - c_j; \tau) \tag{6.2}$$

where a finite number of centers is considered inside (or close to) the study window. For practical purposes, K is assumed to be relatively small (usually in the range of $1 - 20$). The parameter τ controls the scale of the distribution. Note that in this formulation we do not insist that $\{c_j\}$ follow a homogeneous PP, nor is the cluster distribution function $h(s_i - c_j; \tau)$ restricted to a pdf, although it must be non-negative. A simple extension of this allows for individual level covariates within a predictor (η_i):

$$\lambda_1(s_i|\psi_1) = \exp(\rho_0 + \eta_i). \sum_{j=1}^{K} h(s_i - c_j; \tau) \qquad (6.3)$$

where $\mu_0 = \exp(\rho_0)$.

In the following, intensity (6.2) will be examined. In general, intensity (6.2) can be regarded as a mixture intensity with unknown numbers of components and component values (cluster center locations). A general Bayesian model formulation can be

$$[\{s_i\}|\psi_0, \mu_0, \tau, K, \mathbf{c}] \sim \prod_{i=1}^{m} \lambda(s_i|\psi). \exp\left\{-\int_T \lambda(u|\psi)du\right\}$$

where $\psi \equiv \{\psi_0, \mu_0, K, \mathbf{c}\}$, with $\psi_1 \equiv \{\rho_0, \tau, K, \mathbf{c}\}$

$$\lambda(s_i|\psi_1) = \lambda_0(s_i|\psi_0)\lambda_1(s_i|\psi_1)$$

$$\lambda_1(s_i|\psi_1) = \exp(\rho_0). \sum_{j=1}^{K} h(s_i - c_j; \tau)$$

$$\rho_0 \sim Ga(a, b)$$
$$K \sim Pois(\gamma)$$
$$\{c_j\} \sim U(A_T)$$
$$\tau \sim Ga(c, d).$$

Here, the prior distributions reflect our beliefs concerning the nature of the parameter variation. As ρ_0 is the case event rate we assume a positive distribution (in this case a Gamma distribution). The parameter γ essentially controls the parent rate (center rate) and in this case the prior for the number of centers (K) is Poisson with rate γ. Other alternatives can be assumed for this distribution. A uniform distribution on a small positive range would be possible. Another possibility is to assume that the centers are mutually inhibited and to assume a distribution that will provide this inhibition. Such a distribution could be a Markov process form such as a Strauss distribution (Móller and Waagpetersen, 2004, Chapter 6). The τ parameter is assumed to appear as a precision term in the cluster distribution function: $h(s_i - c_j; \tau)$. A typical symmetric specification for this distribution is distance based:

$$h(s_i - c_j; \tau) = \frac{\tau}{2\pi} \exp\left\{-\tau d_{ij}^2/2\right\}$$
$$\text{where } d_{ij} = \|s_i - c_j\|.$$

Other forms are of course possible including allowing the precision to vary with location and asymmetry of the directional form. Many examples exist where variants of these specifications have been applied to cluster detection problems (e.g. Lawson, 1995; Lawson and Clark, 1999b; Lawson, 2000; Cressie and Lawson, 2000; Clark and Lawson, 2002). One variant that has been assumed commonly is to change the link between the cluster term and the background risk. For example, there is some justification to assume that areas of maps could be little affected by clustering if far from a parent location. In these areas the background rate $(\lambda_0(s_i|\psi_0))$ should remain. The multiplicative link, assumed in $\lambda_1(s_i|\psi_1)$, may be improved by assuming an additive-multiplicative link as well as the introduction of linkage parameters (a, b):

$$\lambda_1(s_i|\psi_1) = \exp(\rho_0). \left\{ a + b \sum_{j=1}^{K} h(s_i - c_j; \tau) \right\}. \tag{6.4}$$

6.4.1.2 Estimation Issues

The full posterior distribution for this model is proportional to

$$[\{s_i\}|\psi_0, \mu_0, a, b, \tau, K, \mathbf{c}].P_1(\psi_0, \mu_0, a, b, \tau).P_2(K, \mathbf{c})$$

where $P_*(.)$ denotes the joint prior distribution. Given the mixture form of the likelihood, it is not straightforward to develop a simple posterior sampling algorithm. Both the number of centers (K) and their locations $(\mathbf{c})$ are unknown. Hence it is not possible to use straightforward Gibbs sampling. In addition we don't require assignment of data to centers and so no allocation variables are used, unlike other mixture problems (Marin and Robert, 2007, Chapter 6). One simple approximate approach is to evaluate a range of fixed component models with different fixed K. The model with the highest marginal posterior probability is chosen (K^*) and the sampler is rerun with fixed K^*. This two stage method is not efficient however. Another alternative would be to use the fixed dimension metropolized Carlin-Chib algorithm (Godsill, 2001; Kuo and Mallick, 1998). Instead, for variable dimension problems such as this, resort can be made to reversible jump MCMC (Green, 1995). A special form of this algorithm called a spatial birth-death MCMC can be used. In this algorithm centers at different iterations are added, deleted or moved based on proposal and acceptance criteria. In this way the location and the number of centers can be sampled jointly. Details of these algorithms are given in van Lieshout and Baddeley (2002) and Lawson (2001), Appendix C.

Figure 6.9 displays one part of the posterior output from a birth-death MCMC sampler run on the Lancashire larynx cancer example. For this case the prior distributions assumed were Strauss for the joint distribution of centers and number of centers (with fixed inhibition parameter), additive-multiplicative link was used with $a = 1$, $b = 1$, and a symmetric Gaussian cluster distribution was used with precision parameter κ^{-1}. The population background was estimated via a density estimation but the smoothing parameter was sampled in the posterior distribution. The model was given an

FIGURE 6.9: Lancashire larynx cancer: MCMC birth-death output. The posterior expected probability density surface of the cluster center locations obtained by overlay of center realisations from different K values.

$InvGa(1, 100)$ prior distribution. Additional random effect terms were also included in this model. For further details of this example see Lawson (2000)). Both the number of centers and location vary over iterations in this example. Hence summarization of the posterior output is not straightforward: different distributions of parameters will be associated with different numbers of centers. One gross summary of the cluster center distribution is available whereby the density estimate surface of the centers overlain from different realizations is presented. This is simply an average over different K values. Of course this can be criticized as it ignores the possibility that markedly different spatial realizations could occur with different K values. In fact this is a general problem with mixture models. It is interesting to note that an area of elevated probability density appears close to a putative source (incinerator at location: 35450, 41400).

How do these models perform and are they realistic for disease cluster detection? In general, the simplistic assumptions made by point process models are really inadequate to describe clustering in spatial disease data. First, clustering tends to occur not as a common spatial field but often as isolated areas. Even when multiple clusters occur it is unlikely they will be of similar size or shape. In addition, clusters do not form regular shapes and any spatial time cross-section may show different stages of cluster development. For instance, there may be an infectious agent which differentially affects different areas at different times. A time-slice spatial map will then show different cluster forms in different areas. Another factor is that scales of clustering can appear on spatial maps. This is not considered in simple cluster PP models.

Given the possibility that unobserved confounders are present then the resulting clustering will be: a) unlikely to be summarized by a common model with global clustering components and b) cluster distribution functions with regular forms may not fit the irregular variation found.

The use of birth-death MCMC with cluster models is not as limited as it may at first seem. The disadvantages of this form of modeling are a) tuning of reversible jump MCMC is often needed and so the method is not readily available, b) interpretation of output is more difficult due to the sampling over a joint distribution of centers and number of centers, and c) possible rigidity of the model specification.

However there are a number of advantages. First, the model can easily be modified to include variants such as spatially dependent cluster variances (thereby allowing different sizes of clusters in different areas) and even a semiparametric definition of $h(s_i - c_j; \tau)$ which would allow some adaptation to local conditions. Second, it is also important to realize that by posterior sampling and averaging over posterior samples it is possible to gain flexibility; even with a rigid symmetric form such as $\frac{\tau}{2\pi} \exp\left\{-\tau d_{ij}^2/2\right\}$ it is easy to see that the resulting cluster density map does not reflect a common global form (Figure 6.9) and indeed highlights the irregularity in the data. This of course is quite unlike the rigidity found in commonly-used cluster testing methods like SatScan (http://www.satscan.org/, Jung, 2016). In addition there is a wealth of information provided from a posterior sampler that can even include additional clustering information. For instance, the posterior marginal distribution of number of centers can yield information about multiple scales of clustering (even when these are not included in the model specification). Figure 6.10 displays a histogram of the posterior marginal center rate parameter for a different data example. In that example there appears to be a major peak at 6 to 7 centers whereas subsidiary peaks appear at 10 to 11 and also at 13. This may suggest different scales of processes operating in the study window.

Finally, it is also possible to increase the flexibility of the model by introduction of extra noise in the cluster sum. For example, the introduction of a random effect parameter for each of the centers: $\sum_{j=1}^{K} \exp(\psi_j).h(s_i - c_j; \tau)$

with $\psi_j \sim N(0, \tau_\psi)$, can lead to improved estimation of the overall intensity of the process. Another option that could be exploited which allows the sampling of mixtures more nonparametrically is the use of Dirichlet process prior distributions for mixtures (Ishwaran and James, 2001; Ishwaran and James, 2002; Kim et al., 2006; Kottas et al., 2008; Kottas, 2016).

FIGURE 6.10: Posterior expected distribution of number of centers from a converged sampler.

6.4.1.3 Data-Dependent Models

Another possible approach to modeling is to consider models that do not assume a hidden process of centers but model the data interdependence directly. Such data-dependent models have various forms depending on assumptions.

6.4.1.3.1 Partition Models and Regression Trees

Partition models attempt to divide up the space of the point process into segments or partitions. Each partition has a parameter or parameters associated with it. The partitions are usually disjoint and provide complete coverage of the study domain (T). An example of disjoint partition (or tiling) is the Dirichlet tesselation which is constructed around each point of the process. Each tile consists of allocations closer to the associated point than to any other. Figure 6.11 displays such a tesselation of the Lancashire larynx cancer data set. It is clear from the display that small tiles (small tile area) are associated with aggregations of cases. The area in the south of the study region is particularly marked. The formal statistical properties of such a tesselation are known (Barndorff-Nielsen et al., 1999) for most processes (such as the marginal distribution of tile areas).

However in partition modeling, the tesselation is used in a different manner. Byers and Raftery (2002) describe an approach where a Dirichlet tesselation is used to group events together. Hence a tiling consisting of K tiles with areas a_k is superimposed on the points and the number of events within a tile (n_k) are recorded. The first order intensity of the process is discretised to be constant within tiles (λ_k). The tile centers are defined to be $\{c_k\}$. Based on

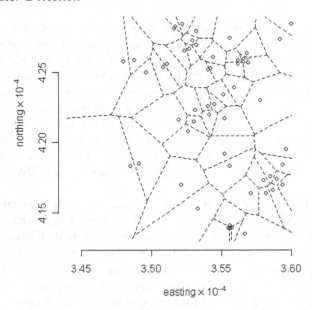

FIGURE 6.11: Lancashire larynx cancer data: Dirichlet tesselation produced with four external dummy points using the DELDIR package on R.

this definition a posterior distribution can be defined where

$$L(\mathbf{n}|\lambda) \propto \prod_{k=1}^{K} \lambda_k^{n_k} \exp\{-\lambda_k a_k\}$$

$$K \sim Poiss(\nu)$$

$$\{\lambda_k\}, k = 1, ..., K | K \ iid \ Ga(a, b)$$

$$\{c_k\}, k = 1, ..., K | K \ iid \ U(T).$$

In this definition, the centers and areas are not given any stochastic dependency, whereas the areas are really dependent on the center locations. In addition, the number of centers is not fixed in general. This lead to a posterior distribution, within the general case, which does not have fixed dimension, but assuming ν, a, b fixed, is proportional to :

$$\frac{\nu^K}{K!} \prod_{k=1}^{K} \lambda_k^{(n_k+a-1)} \exp\{-\lambda_k(a_k + b)\}.$$

In general, a reversible jump MCMC algorithm or metropolized Carlin-Chib algorithm must be used to sample from this posterior distribution unless K is fixed. The focus of this work was the estimation of λ_k. Of course, in general, λ_k will vary over iterations of a converged posterior sample and won't be allocated to the same areas. Hence, any summarization of the output would have to overlay the realizations of λ_k for a predefined grid mesh of sites (possibly the data points), at which the average intensity would be estimated. Hence a smoothly varying estimate of the intensity would result. In addition to simple intensity estimation, the authors also include a binary inclusion variable (d_k), which has a Bernoulli prior, and categorizes the tile as being in a high intensity area $(d_k = 1)$ or not. This allows a form of crude intensity segmentation (between areas of high and low intensity). In that sense the method provides a clustering algorithm, albeit where only two states of intensity delineate the "clusters." Mixing over the posterior allows for gradation of risk in the converged posterior sample (see for example Figure 6.3 b) of Byers and Raftery (2002)).

Note that in their application, Byers and Raftery (2002) have no background (population) effect which would be needed in an epidemiological example. In application to disease cases it may be possible to estimate a background effect using a control disease and to use this as a plug-in estimate (i.e. replace λ_k by $\widehat{\lambda}_{0k}\lambda_k$ in the likelihood, where $\widehat{\lambda}_{0k}$ is a background rate estimate in the kth tile). Costain (2009) provides an example of an application to case event data. Alternatively, if counts of the control disease are available within tiles then it would be possible to construct a joint model for both counts. Hegarty and Barry (2008) have also introduced a variant where product partitions are used to model risk.

6.4.1.3.2 Local Likelihood An alternative view considers the use of a grouping variable which relates to a sampling window. The sampling window is a subset of the study window. For example, a sampling window (lasso) is defined to be controlled by a parameter (δ). This parameter controls the size of the window. Usually, (but not necessarily) the window is circular so that δ is a radius. First consider cases of disease collected within a window of size δ, and denote these as n_δ. Second, denote cases of a control disease as e_δ within the lasso. Now assume that the case disease and control disease are observed at a set of locations and denote these as $\{x_i\}$, $i = 1, ..., n$ and $\{x_i\}$, $i = n+1, ..., n+m$. The joint set of $\{x_i\}$ can be described jointly by a Bernoulli distribution with case probability $p(x_i) = \lambda(x_i)/(\lambda_0(x_i) + \lambda(x_i))$, conditional on $\lambda_0(x_i), \lambda(x_i)$ and their parameters. Further, assume that within the lasso there is a risk parameter θ_{δ_i} and that $\lambda(x_i) = \rho\theta_{\delta_i} \; \forall x_i \in \delta_i$. Now assume that within the lasso the probability of a case or control is constant. In that case we can write down a local likelihood of the form

$$\prod_{i=1}^{n+m} \left[\frac{\rho\theta_{\delta_i}}{1 + \rho\theta_{\delta_i}}\right]^{n_{\delta_i}} \left[\frac{1}{1 + \rho\theta_{\delta_i}}\right]^{e_{\delta_i}}.$$

Note that the lasso depends on δ_i, defined at the i th location, and different assumptions about these can be made. Attention focuses on the estimation of θ_{δ_i}, rather than δ_i, which can yield information about clustering behavior. Based on this local likelihood (Kauermann and Opsomer, 2003), it is possible to consider a posterior distribution with suitable prior distribution for parameters. For example, the δ can have a correlated prior distribution (either a fully specified Gaussian covariance model or a CAR model). Alternatively it has been found that assuming an exchangeable gamma prior appears to work reasonably well. In addition the dependence of θ_{δ_i} on δ_i across a range of $\delta_i s$ should be weak a priori and so we assume a uniform distribution. Assuming a CAR specification, the prior distributions are then:

$$[\theta_{\delta_i}|\delta_i] \sim U(a,b) \; \forall i \qquad (6.5)$$
$$\rho \sim IGa(3, 0.01)$$
$$[\delta_i|\delta_{-i}; \tau] \sim N(\overline{\delta}_{\Delta_i}, \tau/n_{\Delta_i})$$
$$\tau \sim IGa(3, 0.01)$$

where Δ_i is a neighborhood of the i th point, $\overline{\delta}_{\Delta_i}$ is the mean of the neighborhood δs and n_{Δ_i} is the number of neighbors, $N(,)$ is an (improper) Gaussian distribution, τ is a variance parameter and ρ is a rate both with reasonably vague inverse gamma (IGa) distributions. An advantage of this approach is that a fixed dimension posterior distribution can be specified, albeit with a local likelihood. Figure 6.12 displays a smoothed version of the posterior average value of the exceedance value of θ_{δ_i}: the function shown is $1 - \widehat{\Pr}(\theta_{\delta_i} > 1)$ for the converged sampler with the prior specification shown (6.5). The case only map is shown with only the convex hull of the case distribution contoured. Further details of this model are given in Lawson (2006a). Note that it is clear that the southern area in the vicinity of (3.55 × 10e-4, 4.15 × 10e-4) demonstrates a very high exceedance probability (<0.01).

The main advantage of this approach is in the simplicity of programming (compared to birth-death MCMC methods) and the ease with which interpretation of output can be made. The cluster detection performance of these methods in simulations is impressive (see e.g. Hossain and Lawson, 2006).

6.4.2 Count Data

6.4.2.1 Hidden Process and Object Models

As in the situation where case events are analyzed, it is possible to specify object models for count data. Essentially these assume a hidden process exists and must be estimated. The hidden process could take a variety of forms. In the earliest examples, these forms followed those for case events where a hidden process of cluster centers was posited. The aggregation effects of accumulating case events into small areas (census tracts, ZIP codes, etc.) leads to integrals of the first order case intensity in the expectation of the count. Hence, denoting the count of disease in the i th area (within area a_i)

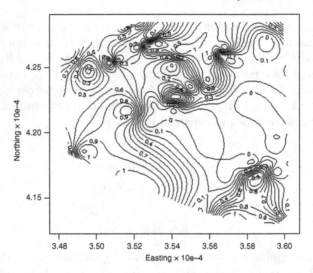

FIGURE 6.12: Lancashire larynx cancer data exceedance probability map showing smoothed posterior average value, from a converged sampler, of $1 - \widehat{\Pr}(\theta_{\delta_i} > 1)$ for the case data only. The smoothing was done using the MBA (R) package.

as n_i, then

$$E(n_i) = \int_{a_i} \lambda(u) du$$

and an "exact" hidden process model (HPM) would be defined as

$$n_i \sim Poiss(\int_{a_i} \lambda(u) du) \qquad (6.6)$$

$$\lambda(u) = \lambda_0(u|\psi_0)\lambda_1(u|\psi_1)$$

$$\lambda_1(u|\psi_1) = \sum_{j=1}^{K} \exp(\phi_j).h(u - c_j; \tau_h)$$

$$\phi_j \sim N(0, \tau_\psi).$$

Here, $h(u - c_j; \tau_h)$ is a cluster distribution function as before and $\{c_j\}$ is a set of hidden cluster centers and the background population function $\lambda_0(u|\psi_0)$ must be estimated as before (Lawson and Clark, 1999a). Often, a simplified (approximate) version of (6.6) is assumed where the expectation is simply a function of an expected rate (e_i) and relative risk parameter (θ_i):

$$n_i \sim Pois(e_i\theta_i),$$

$$\log \theta_i = \alpha_0 + \alpha_1 \sum_{j=1}^{K} \phi_j h(C_i - c_j; \tau_h),$$

where $\{C_i\}$ is a set of centroids of the small areas, α_0 is an intercept and α_1 is a linking parameter. The $\{\phi_j\}$ play the same role as before as random effects and the $\{c_j\}$ denotes hidden centers to be estimated. The expected rates are usually calculated based on an external standard population. An example of a variant of this model is given in Lawson (2006b), Chapter 6, where $\theta_i = \exp(\phi_j).[1 + \sum_{j=1}^{K} h(C_i - c_j; \tau_h)]$ with $\phi_j \sim N(0, \tau_\psi)$. Another variant of this model was specified by Gangnon (2006) where the log relative risk in a small area is assumed to be specified by $\log \theta_i = \alpha_0 + \Gamma_i + e_i$, where

$$\Gamma_i = \sum_{j=1}^{K} \phi_j h(C_i - c_j; \tau_h) \tag{6.7}$$

and ϕ_j is now a relative risk component associated with the j th cluster and $h(.)$ is a cluster membership function assumed to be uniform on a disc. The intercept term is α_0, while the e_i term is a random effect with zero mean Gaussian prior distribution. The function $h(.)$ associates small areas to centers. The cluster centers are assumed to be centroids of the small areas (to avoid empty clusters which could arise due to the membership function definition). The main difference between these two variants is the inclusion of relative risk components in the latter model (6.7).

6.4.2.2 Data-Dependent models

As in case event applications, it is possible for count data to be modelled without recourse to HP models. Two alternatives to such modeling are the use of partition models and splitting methods, and the local likelihood. A common feature of these methods is that spatial membership must be computed within an area. In both cases the methods require computation of distances within the spatial configuration of the data. Within MCMC algorithms these distances have to be recomputed as parameters are sampled.

6.4.2.2.1 Partition models and Regression Trees One of the first examples of applying partition modeling to small area count data was proposed by Knorr-Held and Rasser (2000). In their modeling approach, at the first level of the hierarchy, the small area count y_i is assumed to be conditionally independent Poisson with expectation $e_i h_j$ where e_i is the usual standardised expected rate. The relative risk for a given area is chosen from a discrete set of risk levels (which are called "clusters" by the authors). These "clusters" are a set of contiguous regions $\{C_j\}$ $j = 1, ..., n$, which have associated constant risk $\{h_j\}$. The relative risk assigned to the i th area is just h_j if $i \in C_j$. Hence, $y_i \sim Pois(e_i h_j)$. The number of clusters (k) is treated as unknown: $k \in \{1, ..., n\}$. Hence this method seeks a non-overlapping partition of the map into areas of like risk. In this definition of clustering, clusters are areas of *constant* risk and of course this is a different definition of clustering from the more usual definitions using "elevated risk." The essential difference is that the clusters are discrete partitions in this case. However, as in other cases, posterior averaging can lead to a more continuous relative risk estimate. Computational issues related to sampling partitions led the authors to consider reversible jump MCMC. In the examples cited the number of partitions was quite large (40 to 45) in the posterior realizations for male oral cavity cancer in 544 districts in Germany.

Partition models have since been extended (Denison and Holmes, 2001; Denison et al., 2002; Ferreira et al., 2002; Wakefield and Kim, 2013). In application to count data in small areas, the usual Poisson assumption is made at first level: $y_i \sim Pois(e_i \varepsilon(C_i))$. The partition is defined to be $\varepsilon(C_i) = \mu_{r(i)}$, the relative risk level associated with a region of a tesselation in which the centroid C_i of the small area lies. In this approach a discrete tesselation is used as a partition. The prior distribution of the levels is assumed to be independent gamma: $[\mu_j | \gamma_1, \gamma_2] \sim Ga(\gamma_1, \gamma_2)$, $j = 1, ..., k$. A prior model is also assumed for the locations of tile centers:

$$p(\mathbf{c}) = \frac{1}{K} \frac{1}{\{Area(T)\}^k},$$

where K is the maximum number of centers or tiles and k is the current number. The marginal likelihood of the data given the centers is

$$\text{constant} + \sum_{j=1}^{k} \{\log \Gamma(\gamma_1 + n_j \bar{y}_j) - (\gamma_1 + n_j \bar{y}_j) \log(\gamma_2 + n_j \bar{N}_j)\}$$

where n_j is the number of points in the j th region and $\bar{y}_j$ is the mean of the observations in that tile and $\bar{N}_j$ is the mean expected rate in the j th tile. An inclusion rule must be specified for the small areas included within any given tile. The usual rule is to include if the centroid falls within the tile. Unlike the Knorr-Held-Rasser model this allows different partition shapes. Computation for the tesselation model is based on a birth-death MCMC algorithm where the centers are added, deleted, or moved sequentially.

Of course, there are considerable edge effect issues with partition models and these apply to all partition models that use tesselations (rather than groupings): at the external study region boundary tesselations will inevitably be distorted due to censoring adjacent regions *outside* the study window. Grouping algorithms (such as proposed by Knorr-Held and Rasser, 2000,

Charras-Garrido et al., 2012; Forbes et al., 2013, or Wakefield and Kim, 2013) could also be affected as no information outside the boundaries has been reported and this could distort the edge region allocations. These issues seem to be largely ignored in the literature on partition modeling.

Extensions to tree models could be imagined (Denison et al., 2002). For example, all the "clustering" partition models proposed assume a flat level of partitioning. However we could conceive of a form of splitting where a tree-based hierarchical cluster formation is conceived. It could be possible to construct a multi-level partition by allocation of higher levels of partition based on residuals from lower levels. This could be conceived as a multivariate partition function with L levels and K_l partitions at the l th level. There could be ordering constraints on the different levels. This could provide an approach to multiscale feature identification in the data. A recent proposal, involves a two stage approach where an agglomerative clustering of prior data is used before modeling current data via a log linear mixed model (Anderson et al., 2014).

Connections between partition models and mixture models for relative risk are also evident. For example, the model of Green and Richardson (2002) assumes that there are a small number of risk components $\{\theta_j, \ j = 1, 2, ..., k\}$ and the small areas are allocated one of these levels via the allocation variables $\{z_i, \ i = 1, ..., n\}$. Hence $y_i \sim Pois(e_i \theta_{z_i})$ where θ_{z_i} is the allocated risk level for the i th area. Unlike other approaches, the allocation variables are given a spatially-structured Potts prior distribution. This allows grouping at a higher level in the hierarchy. In contrast, Fernandez and Green (2002) describe a mixture model where the count is assumed to be governed by a weighted sum of Poisson distributions and the weights have spatial structure.

One general question concerning these methods is whether they really can be considered "clustering" methods or not. While all these methods could yield a variety of posterior information concerning risk, and some part of that information could be employed to detect clusters, they do not directly address the detection of clusters; rather they seek to find underlying discrete risk levels that characterize the map. Often, the authors regard their methods as competitors with disease map smoothing models such as the convolution model of Besag et al. (1991) (see for example discussions in Knorr-Held and Rasser, 2000; Ferreira et al., 2002, or Fernandez and Green, 2002) and they have been compared in simulations as such (Best et al., 2005). Essentially, mixture models are closely identified with latent structure models and these are discussed more fully in Sections 5.7.5 and 12.3.3.

6.4.2.2.2 Local Likelihood As in the case event situation (Section 6.4.1.3.2), local likelihood models could also be applied. For small area counts within a lasso centered at the centroid of the area of dimension δ_i with a Poisson assumption, the probability of y_{δ_i} counts with expected rate e_{δ_i}, is just:

$$f(y_{\delta_i} | e_{\delta_i}, \theta_{\delta_i}; \delta_i) = [e_{\delta_i} \theta_{\delta_i}]^{y_{\delta_i}} \exp(-e_{\delta_i} \theta_{\delta_i}) / y_{\delta_i}!.$$

This can be employed within a local likelihood as:

$$L(\theta|\mathbf{y}) = \prod_{i=1}^{m} f(y_{\delta_i}|e_{\delta_i}, \theta_{\delta_i}, \delta_i).$$

Note that, the factorial term $(y_{\delta_i}!)$ must be included in the likelihood as it varies with δ_i. In this case the data and expected rate are accumulated within the lasso. A rule must be assumed for the inclusion of small areas. Centroids falling within the lasso are common assumptions. Then $y_{\delta_i} = \sum_{i \in \delta_i} y_i$,

and $e_{\delta_i} = \sum_{i \in \delta_i} e_i$. In this definition the counts and expected rates within

neighboring lassos can overlap and hence they can be correlated. Usually, the focus is on the estimation of θ_{δ_i} rather than δ_i. The linkage between the risk and the lasso is specified via:

$$\log \theta_{\delta_i} = \delta_i + \varepsilon_i$$

where $\varepsilon_i \sim N(0, \tau_\varepsilon)$ is an unstructured component and δ_i is assumed to have a spatially-structured prior distribution. In examples, a CAR prior distribution has been assumed for this purpose, i.e.

$$[\delta_i|\delta_{-i}; \beta_\delta] \sim N(\overline{\delta}_i, \beta_\delta/d_i)$$

where $d_i = \sum_{j}^{m} I(j \in \delta_i)$, and $\overline{\delta}_i = \sum_{j \in \delta_i} \delta_j/d_i$. This allows the correlation

inherent in the local likelihood to be modeled at a higher level of the hierarchy. Covariates can also be added to this formulation if required (see Hossain and Lawson (2005) for discussion). Figure 6.13 displays an application of the local likelihood model to a simulated realization where clusters of risk are introduced into the district geographies of the former East Germany. The counts are generated from a Poisson model with mean $e_i.\theta_i^{true}$ where the θ_i^{true} are the true risks. The left hand panel depicts the true risks and the right hand panel depicts the estimated risks under the local likelihood model with CAR prior distribution for the lasso parameters and inverse gamma distributions for the precision parameters (τ). A comparison of the estimation ability of a random effect convolution model compared to the local likelihood model is also made in Figure 6.14. It is noticeable that the local likelihood model produces a greater shrinkage than the convolution model. In full simulation comparisons based on operating characteristic curves the LL model performs well in recovering true risk as well as some features of clusters (see e.g. Hossain and Lawson, 2006).

6.4.3 Markov-Connected Component Field (MCCF) Models

A number of alternative approaches to count data clustering have been proposed and should be mentioned as they relate to methods previously discussed.

FIGURE 6.13: East German lip cancer mortality data (1980 through 1989). Simulated 'true' risk map (left panel) and local likelihood posterior expected relative risk estimates (right panel).

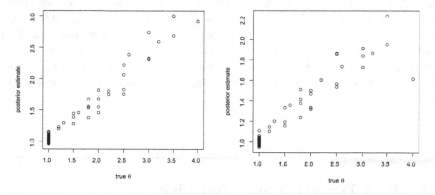

FIGURE 6.14: Comparison of true and posterior estimated relative risks for simulated data for East Germany district geographies. Left panel: BYM model; right panel: local likelihood model.

First a cluster modeling approach based on Markov-connected component fields has been proposed (Gangnon and Clayton, 2000) This approach also seeks to find a grouping of small areas into "clusters." The cluster set k clusters has associated a risk level λ_j , $j = 1, .., k$. The membership of the clusters is defined for each of the N small areas ($i = 1, ..., N$), via a membership (allocation) variable: $\mathbf{z} = (z_1, ..., z_N)$. Hence the counts in the N areas have distribution, conditional on the cluster assignment ($\mathbf{z}$):

$$y_i \sim Pois(e_i \lambda_{z_i})$$

where λ_{z_i} is the disease risk for the cluster associated with the i th area. Clusters are assumed to be aggregations of the basic small areas, and so the allocation assigns different disease risks to each area depending on prior distributional constraints. This setup was later used by Green and Richardson (2002) but they assumed a spatially-structured prior distribution for the **z**.

Here, the disease risks are assumed to have a $Ga(\alpha, \beta)$ prior distribution given the **z**, where $Ga(\alpha, \beta)$ has mean α/β. However the **z** are assumed to have a Markov-connected component field prior governing their form (Móller and Waagepetersen (1998)). This construction allows various cluster form specifications to be included in the prior structure. This construction allows a potential function to describe the clustering:

$$p(\mathbf{z}) \propto \exp\{-\sum_{j=1}^{k} S_j\}$$

where the S_j is a score function for the j th cluster. The authors use properties of the clusters such as circularity (shape) and size to yield a composite score: $S_j = \alpha + S_{j1}(size) + S_{j2}(shape)$. Once specified the authors use randomized model search criteria to evaluate different models. In a similar development, but different application, Móller and Skare (2001) further apply these priors to imaging problems. Note that this MCCF approach has many advantages over purely spatially-structured distributions for component assignment. It also allows the models to address specific features of the clusters that are not considered by other partitioning approaches. A recent partial review and evaluation of cluster modeling for count data can be found in Rotejanaprasert (2014).

6.5   Edge Detection and Wombling

A closely related area of concern to disease clustering is the idea that discontinuities in maps are the focus. These could be the edge of some uniform risk area or the edge of a variable risk cluster of some kind. In cluster modeling usually there is some criteria defining the cluster and its boundary. In edge detection the boundary is usually defined by a jump or discontinuity in risk, and so this in a sense takes the approach whereby the focus is the difference in risk rather than finding areas of elevated risk.

Wombling is closely related to the edge effect problem of Section 5.8, in that the focus is on boundaries of regions. There is a growing interest in the ability to locate boundaries within a spatial domain. This has been developing within geography for some time (see e.g. Oden et al., 1993 amongst others). These boundaries often have some natural context, such as catchments of health providers (Ma et al., 2010) or areas which vote predominantly for one party in elections (O'Loughlin, 2002). Basically these methods are edge detection methods, and many such methods are commonly found in the image processing literature (see e.g. the Canny operator, Mars-Hilldreth, and Gaussian derivative kernels in Bankman, 2007). Often they are intended for object recognition and use derivative-based methods for locating discontinuities.

The discovery of discontinuities in disease maps may be of interest in detection of natural boundaries for health provision (recent Bayesian examples are Ma et al. (2010), Lu and Carlin (2005), and Liang et al. (2009)). One simple approach to this problem can be examined within a Bayesian hierarchical model. Assume for small areas with observed tract counts y_i and $y_i \sim Poiss(e_i \theta_i)$.

Within a posterior sampling algorithm, θ_i^g is the estimate of θ_i at the gth iteration. It is possible to estimate the posterior expected value of the absolute difference between relative risks, $\Delta\theta_{ij}$, by simply computing:

$$\Delta\theta_{ij} = \sum_{g \in d} |\theta_i^g - \theta_j^g|/n(d),$$

where d denotes the converged sample set and $n(d)$ is the number in that set. Hence this estimator is available for all region boundaries. The parameters $\{\Delta\theta_{ij}\}$ really detect discontinuities between regions and there is no information in this approach that informs the edge detection beyond the average difference between *modeled* estimates of adjoining area risks. Hence cluster or object information that could "tie" the areas together is not used. Presmoothing the estimates of risk (e.g. via a convolution model) may lead to reduction in discontinuity, of course, and so the choice of model for the θ could be crucral. More sophisticated approaches have also been proposed (see e.g. Ma et al., 2010).

It is not clear how discontinuities are to be modeled for applications in disease incidence studies. Underlying most models of risk is the assumption of continuity of risk over space. The convolution smoothing models assume this as do most other risk models. The fixed mixture model of Lawson and Clark (2002) and also the mixture models of Green and Richardson (2002) do attempt to honor discontinuities within a general framework of smooth risk and allow the smoothness to be selected differentially over space. It is a matter of debate whether pure discontinuity models have advantages in incidence studies. Gelfand and Banerjee (2015) provide a recent review of Bayesian wombling.

7

Regression and Ecological Analysis

7.1 Basic Regression Modeling

The term *regression modeling* is used loosely here to denote a situation where a disease outcome variable is geo-referenced and is to be related to predictors and/or covariates. These predictors can be observed at different levels of spatial aggregation and can be spatially-referenced or not. Ecological analysis is special case of regression modeling where a relation is estimated at an aggregation level and inference is to be made at a lower aggregation level. In what follows, I do not discuss misaligned (MIDP) or modifiable areal unit problems (MAUP) directly as these are discussed in Chapter 9.

I define here an outcome variable, y_i, $i = 1, ..., m$. In general this outcome would be either at an individual or aggregate count level. Initially the discussion will focus on general issues related to covariate modeling with such an outcome. Later I will consider the special case of ecological analysis (Section 7.7).

7.1.1 Linear Predictor Choice

An initial example will serve to motivate the discussion of modeling issues. Some of these issues are purely general in that they relate to Bayesian modeling. Whereas others are specifically related to either spatial or spatial epidemiological contexts. The first example is from a study of county level birth outcomes in the US state of Georgia for the year 2007. A number of outcomes pertaining to births are available including low birth weight (LBW) (<2500g), very low birth weight (VLBW) (<1500g) and also total births count. Initially we will consider the single outcome LBW (y_i) as a proportion of total births (n_i) for $i = 1, ..., m$ counties. Also recorded are the county level predictors: population density (*popden*), proportion black population (*black*), median income (*inc*), proportion in poverty (*pov*) and unemployment rate (*uer*). These predictors were obtained from the Area Resource File (ARF) database (from the U.S. Department of Health and Human Services: http://arf.hrsa.gov). The ARF is a collection of county-level data sets from more than 50 sources such as the American Hospital Association, American Medical Association, National Center for Health Statistics, and US Census Bureau.

Hence an initial model for the spatial variation in LBW could be defined as

$$y_i \sim bin(n_i, p_i)$$

with a suitable link to the probability of low birth weight. Often a logistic model is assumed so that $logit(p_i)$ is a function of predictors and other effects. The focus here is to explain variation in LBW risk at the county level by inclusion of predictors thought to affect the incidence. We do not consider individual level ecological inference in this section (see Section 7.7).

Here we first consider a logistic regression where

$$logit(p_i) = \eta_i$$
$$\text{where } \eta_i = \boldsymbol{x}_i^T \boldsymbol{\beta}$$
$$\boldsymbol{x}_i^T \boldsymbol{\beta} = \beta_0 + \sum_{j=1}^{p} \beta_j x_{ij}.$$

In the example of LBW we consider socio-economic and ethnic indicators that could provide some explanation for county level LBW incidence. Low income and poverty both could lead to a higher LBW rate in counties due to poor prenatal care access or cultural factors. In addition, ethnicity may also confound due to behavioral and cultural modalities in different ethnic groups. There may also be masking of effects when racial factors are correlated with low income and/or poverty.

7.1.2   Covariate Centering

In the following models we consider covariate effects, via a centering specification. Covariates are often measured on different scales and these scale differences can cause computational problems within sampling algorithms. To improve convergence and allow for algorithmic speeding up, a centering or standardising transform is often recommended. The simplest form is the use of a mean to center the vector of covariate values i. e.

$$x_i^* = x_i - mean(\{x_i\})$$

where x_i^* is then used instead of x_i in the model. Note that the fitted parameter estimate will be unaffected by the centering, although the intercept will change. A further stage is to standardise the covariate using division by the standard deviation also. In that case we would have

$$x_i^* = (x_i - mean(\{x_i\}))/sd(\{x_i\}).$$

The fitted parameter on the original scale can be recovered by transformation. Note also that because functionals of parameters are estimable within posterior samplers we can simply compute these transformations within the sampling iterations. The above recommendations relate mainly to continuous

Model	DIC	pD
model 1	1139.9	5.9
model 2	1071.6	59.8

Table 7.1: DIC model comparisons for logistic linear models with and without a random intercept for county level LBW counts and a set of five county level covariates

covariates, and it is not usual for discrete covariates to be centered or standardized, although it is sometimes advocated (for example, its use in some WinBUGS help examples).

7.1.3 Initial Model Fitting

Initially, I have considered two basic models for these data:

1. A logistic linear regression with LBW as outcome and with all covariates (*popden (1)*, *black (2)*, *inc (3)*, *pov (4)* and *uer (5)*). This is a 'fixed' effect model, but with prior distributions for the regression parameters. I assume non-informative zero mean Gaussian distributions for these parameters, and standard deviation (SD) uniform hyperprior distributions:

$$\beta_* \sim N(0, \tau_{\beta_*}) \tag{7.1}$$
$$\sqrt{\tau_{\beta_*}} \sim U(0, c)$$

with $c = 100$.

2. A mixed effect model, as per model 1, but with an additive uncorrelated unit level random effect in the linear predictor: $\eta_i = x_i^T \beta + v_i$ with non-informative zero mean Gaussian prior distribution:

$$v_i \sim N(0, \tau_v) \tag{7.2}$$
$$\sqrt{\tau_v} \sim U(0, c).$$

This is effectively a random intercept model. Both the above models were fitted using mean centered covariates, and convergence was reached with burn-in of 10000 iterations. These models' DIC statistics are displayed in Table 7.1.

It is clear that the introduction of the random intercept has produced a considerable improvement in DIC ($\Delta DIC = 68.3$) even when the pD has increased. Hence, for these data, a random intercept model provided a better explanation of the data variation but is less parsimonious. Table 7.2 displays the regression parameter estimates for model 2.

It is notable that β_0, β_2, β_4 do not include zero within their range of credible values, unlike the other regression parameters. Hence, it might be concluded that ethnicity (*black*) and percentage below poverty (*pov*) are both important predictors in relation to LBW. Whereas, population density (*popden*), median

Parameter	Posterior Mean	2.5%	97.5%
β_0	-2.216	-2.25	-2.18
β_1	8.33E-06	-6.13E-05	8.06E-05
β_2	0.006571	0.003252	0.009978
β_3	0.001209	-0.005533	0.007787
β_4	0.01545	0.005792	0.02516
β_5	-0.006217	-0.05129	0.0338
τ_v	26.24	0.3686	101.1
τ_{b_0}	0.5107	0.2551	1.331

Table 7.2: Posterior mean estimates and 95% credible interval limits for five-covariate model for county level LBW with random intercept

income (inc) and unemployment rate (uer) are not, at least within a full model with five predictors.

It would be of interest at this point to examine subsets of predictors to establish whether there is redundancy in the predictor set. In a later section (Section 7.6) I examine a Bayesian variable selection method in relation to this data set.

I now examine two extensions to this analysis. First, I consider extending the random effect set to include spatial correlation. Second, I will examine very low birth weight (VLBW) to see if the same factors appear related to VLBW and LBW.

A straightforward extension of the log linear model with uncorrelated unit level random effect is to include an additive spatial component, so that

$$\eta_i = \boldsymbol{x}_i^T \boldsymbol{\beta} + v_i + u_i$$

where u_i has an improper CAR prior distribution (as first discussed in Section 5.4) This prior is defined by neighborhood relations and is an example of a Gaussian Markov random field (GMRF), where the conditional distribution of any u is defined by

$$u_i | \{u_l\}_{l \neq i} \sim N(\overline{u}_{\delta_i}, \tau_u / n_{\delta_i})$$

where δ_i is the neighborhood set of the i th unit, $\overline{u}_{\delta_i}$ is the mean of the neighboring u values, n_{δ_i} is the number in the neighborhood. The standard deviation is assumed to have prior distribution $\sqrt{\tau_u} \sim U(0, c)$. The DIC for this convolution model is 1286.35 with pD = 152.77. In this case we have a considerably larger DIC than found in either of the simpler models and so we could conclude that these LBW data are suitably modeled via an uncorrelated random effect model.

The second extension is to examine VLBW, which is a more extreme form of birth weight anomaly. In this case the three models fitted above with the same predictors yielded the following DICs and pDs: model 1: DIC = 769.7,

Parameter	Posterior Mean	2.5%	97.5%
β_0	-3.989	-4.043	-3.935
β_1	3.06E-05	-3.48E-05	9.99E-05
β_2	0.00987	0.00583	0.01382
β_3	7.87E-05	-0.00799	0.00785
β_4	0.00995	-0.000169	0.02138
β_5	0.001082	-0.06768	0.06927
τ_{b_0}	0.2773	0.0471	0.7221

Table 7.3: Parameter estimates for county level VLBW outcome for Georgia in 2007: posterior mean and 95 percent credible limits

$pD = 5.88$; model 2: $DIC = 768.1$, $pD = 18.10$; model 3: $DIC = 769.2$, $pD = 21.67$. Hence there is little difference between the models in this case and the simple model with no random effects appears to be sufficient. Table 7.3 displays the results of fitting the basic model to VLBW data.

In the VLBW case, unlike LBW, it appears that only one predictor (ethnicity: *black*) is significant, and the percentage living below poverty is no longer important. This is apparent for the full five-predictor model fitted and this does not mean that other predictors on their own or in other combinations could not have higher explanatory power.

7.1.4 Contextual Effects

Contextual effects arise due to grouping or classifying of measurement units. The context of the group membership is inherited by the unit. For example, an individual is a member of a family, and so a contextual family effect could be assumed for that individual, in common with others in that family. Another example in a spatial application would be an individual who lives within a census tract and also a county. In that case the individual could have a tract contextual effect and also a county contextual effect. In Chapter 11, we demonstrate the use of contextual modeling of survival experience of cancer registry patients in the context of their county of residence. Essentially, contextual effects are random factorial effects, and can occur at different levels within spatial analyses. They can form truly hierarchical sets of effects and hence truly hierarchical models.

An example of this, which will be discussed more fully in Chapter 9, is county level oral cancer mortality in Georgia for 2004. In this example, counts of deaths from oral cancer are available within counties. There are also 18 health districts within Georgia that subsume sets of counties. Health districts usually have for or five counties within them. These districts are used for administration of health resources. At the county level, a model can be posited whereby the relative risk is structured around both county(C) and health district (PH) effects. Hence assume that

$$\log(\theta_i) = \alpha_0 + \boldsymbol{x}_i^T \boldsymbol{\alpha} + v_i + u_i + v_{1j_{i \in j}} + u_{1j_{i \in j}} \tag{7.3}$$

where i denotes the county and j denotes the health district. The effects $v_i + u_i$ are county level effects whereas $v_{1j_{i \in j}} + u_{1j_{i \in j}}$ are a sum of PH level effects assigned to the county found within the given health district $(i \in j)$. The PH level effects can be specified as uncorrelated and correlated so that

$$v_{1j_{i \in j}} \sim N(0, \tau_{v1})$$

and

$$u_{1j_{i \in j}} | \{u_{1l}\}_{l \neq j} \sim N(\overline{u}_{\delta_j}, \tau_{u1}/n_{\delta_j}).$$

An example of the application of this model to the Georgia oral cancer mortality data for 2004 follows. In this example I have included certain important county-level socio-economic variables as covariates (ethnicity: *black*; percent under the poverty line: *Pov;* median income*: inc*). These are available from the area resource file (ARF: http://arf.hrsa.gov/). The results of fitting this model with relative risk given by (7.3), expressed as posterior mean standard deviation (sd), are α_0 :0.01117 (0.2286), α_1 :-0.02944 (0.0915), α_2 :0.07101 (0.1377), α_3 :-0.08883 (0.1068). None of these parameter estimates is well estimated and it could be concluded that the selected covariates do not contribute significantly to the oral cancer variation.

Figures 7.1 through 7.4 display the resulting effects when contextual assignment is included at the PH level in a county level model. The UH county level effect appears to be largely random while the county level CH effect tends to display elevated clustering in the northwest. On the other hand the PH district level UH and CH seem to display similar patterning and may suggest limited identification.

As a comparison, a model fitted to the same data but without the PH level effects yielded a DIC of 488.8 with pD of 24.4. This is slightly lower than the full model and suggests that in this case, the effect of PH districts is limited.

7.2 Missing Data

While most publicly available public health data is fairly complete, it is still possible that due to research design or recording error, outcomes or predictors or covariates could have missing values. I will discuss outcome and predictor error separately.

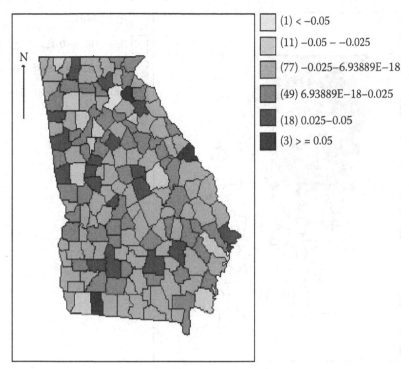

(1) < −0.05

(11) −0.05 – −0.025

(77) −0.025–6.93889E−18

(49) 6.93889E−18–0.025

(18) 0.025–0.05

(3) > = 0.05

N

FIGURE 7.1: Georgia oral cancer mortality, 2004: uncorrelated heterogeneity (UH) effect.

7.2.1 Missing Outcomes

Missing outcomes, while not common, could occur, and be potentially the most common form of missingness likely to be encountered in disease mapping. For example, for a particular year or period of years, a particular region may not have recorded an outcome of interest by a health department. Another example would be where a health survey is carried out in only a subset of areas, due to financial or administrative restrictions, with other areas having no information gathered.

Missing outcomes in Bayesian hierarchical models can be accommodated via the use of predictive inference. The predictive distribution of a new observation was introduced in Chapter 2 (Section 2.4). Essentially, if one treats the missing observation as a parameter to be estimated within a sampler then a predicted value of the missing observation can be sampled at each iteration and averaged over the converged sample. In the Georgia oral cancer 2004 example (Section 1.1), I removed a count from county 2 and reran the model in WinBUGS. WinBUGS automatically estimates missing outcomes as parameters using the predictive distribution. Figures 7.5 and 7.6 display the

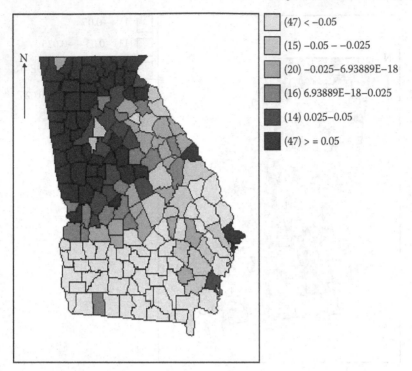

FIGURE 7.2: Georgia oral cancer mortality, 2004: correlated heterogeneity (CH) effect.

posterior sampling trace and summary for county 2 for the missing oral cancer count.

In a more sophisticated real example, where there is outcome missingness I examined the spatial variation in low birth weight (LBW: <2500 g) and very low birth weight (VLBW: <1500 g) for the 85 land-based census tracts in the Charleston County area of South Carolina for 2010.

In this example I used available census tract level predictors from the 2010 US census assumed to be related to adverse birth outcomes: median income (*medI*), unemployment rate (*Uemp*), population density (*pop*), percent below poverty (*pov*) and percent with less than high school education (*educ*) (Figure 7.7 and 7.8). In the analysis these covariates were mean centered and standardized. As LBW and VLBW are both subsets of total births we have assumed that they can be jointly modeled with the total births finite population denominator.

Denote y_{1i} and y_{2i} as the LBW and VLBW count in the i th census tract and n_i as total births for 2008. We assume a binomial model for each outcome with denominator n_i and so

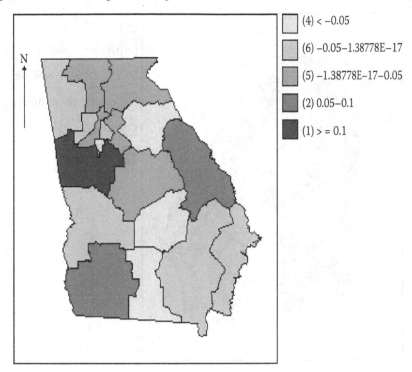

FIGURE 7.3: Georgia oral cancer mortality, 2004: health district (PH) uncorrelated heterogeneity (UH) effect.

$$y_{1i} \sim bin(p_{1i}, n_i)$$
$$y_{2i} \sim bin(p_{2i}, n_i)$$

with

$$\text{logit}(p_{1i}) = \alpha_{10} + \boldsymbol{x}_i^T \boldsymbol{\alpha}_1 + v_{1i} + u_{1i}.$$
$$\text{logit}(p_{2i}) = \alpha_{20} + \boldsymbol{x}_i^T \boldsymbol{\alpha}_2 + v_{2i}$$

Note that all regression parameters were assumed to have zero mean Gaussian prior distributions and the precisions with non-informative gamma prior distributions. The missing outcomes (census tract 22 for VLBW, LBW and total births) was assumed to have $Poiss\ (\mu)$ distribution with $\mu = 2$ for the total births. The same five covariate model was fitted to both LBW and VLBW but the random effect model for LBW has an added correlated spatial effect (an intrinsic CAR model). The very low birth weight was thought a priori likely to be more isolated in distribution and so we assumed only an uncorrelated effect. See Figures 7.9 through 7.11.

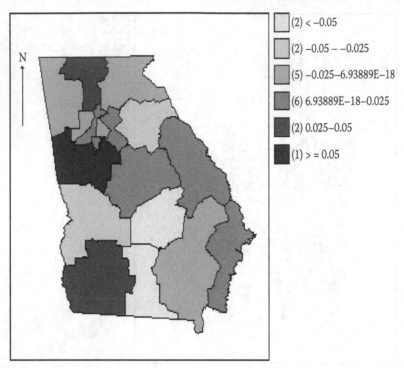

FIGURE 7.4: Georgia oral cancer mortality, 2004: health district (PH) correlated heterogeneity (CH) effect.

7.2.2 Missing Covariates

Missingness in covariates is also a common characteristic of medical studies and could arise in studies with spatial components. In this case, it is necessary to assign a prior distribution to the covariate concerned, even when other values of the covariate are fixed. Once a prior distribution is formed then the covariate can be imputed as a missing parameter within the model and averaged over the sampler. A similar approach would also be used for covariates with measurement error where a latent variable (the true covariate) is to be estimated.

7.3 Non-Linear Predictors

In much of this discussion above and elsewhere in this work I have assumed a linear predictor as the mode of linkage between covariates and outcomes.

FIGURE 7.5: Trace plot of estimated missing count for the second county in the Georgia oral cancer 2004 example.

Node statistics

Node	Mean	sd	MC error	2.5%	Median	97.5%	start	sample
y[2]	0.261	0.5576	0.01059	0.0	0.0	2.0	10001	2000

FIGURE 7.6: Node statistic for y[2] for county 2 when missing from the Georgia oral cancer 2004 data set.

However in many studies it may be more appropriate to assume a non-linear link to covariates. Spline-based models (Section 5.7.2) offer a possible solution in these cases. An alterative is to consider simpler smoothing functions such as random walks of order 1 or 2. This has been proposed by Fahrmeir and Lang (2001) and since implemented in BayesX, WinBUGS and INLA. Another option when spatial smoothing is required is to examine Gaussian process convolution models (Higdon, 2002) which can be easily implemented with a fixed spatial kernel function. In that case replace $\boldsymbol{x}_i^T \boldsymbol{\alpha}$ with $S(\boldsymbol{x}_i^T) = S_1(\boldsymbol{x}_{1i}) + S_2(\boldsymbol{x}_{2i})$.....where $S_*(\boldsymbol{x}_{*i}) = \Sigma_{j=1}^p \nu_j k(||s_i - w_j||)\boldsymbol{x}_{*j}$ where $\{w_j\}$ is a p set of sample site locations, $\{\boldsymbol{x}_j\}$ is the corresponding sampled value of covariate $\boldsymbol{x}$, ν_j is a zero mean Gaussian random effect, and $k(.)$ is a kernel function. Often a standard Gaussian form is assumed for $k(.)$. Further, the R package spikeSlabGAM allows the application of different smoothing terms for each predictor in a Bayesian posterior sampling setting (see also Section 7.6).

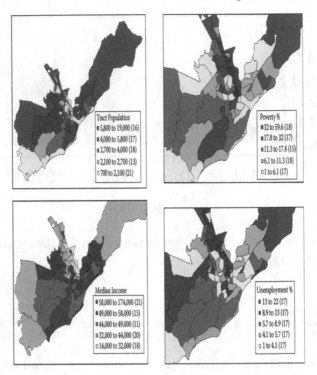

FIGURE 7.7: Four covariate maps for the Charleston census tract birth weight example. Top row: population, percent poverty. Bottom row: median income, unemployment rate.

7.4   Confounding and Multi-Colinearity

Confounding in observational studies is ubiquitous and can even arise in controlled trials as well. Confounding in outcome can occur when a covariate is related to the outcome but is not measured because a relation is unknown or the covariate could not be included in the study for whatever reason. A prime example would be a study of lung cancer where smoking status is not measured on individuals or in aggregate, and a distance covariate is used to assess exposure to an environmental air pollutant. Smoking could clearly affect lung cancer incidence in competition with an air pollutant source. Clearly smoking is a known confounder. Unknown confounders also exist of course, as our knowledge is not necessarily complete with respect to the etiology of the disease in question. Hence we usually need to make some allowance for confounding effects in our models. The classic approach to confounding is to assume that random effects will absorb the effects. For example, extra

% less than high school graduation
■ 28.2 to 67.8 (17)
■ 17 to 28.2 (17)
■ 9.2 to 17 (17)
▨ 2.6 to 9.2 (17)
□ 0 to 2.6 (17)

FIGURE 7.8: Education covariate for Charleston low birth weight example showing percent of population with less than high school education.

variation or noise in the data may be due to relative susceptibility of the observation unit. Individuals could display differential *frailty* in response to an insult, for example. Some may be more likely to contract a disease than others.

At the county level, particular county-specific health initiatives may induce differentials in risk. If these effects are reasonably random then a random effect which depicts uncorrelated extra noise or variation could be employed at the unit level. A classic example is to add an effect with a zero mean Gaussian prior distribution in the linear predictor, e.g. $v_i \sim N(0, \tau_v)$. Usually the addition of such an effect should not have any impact on the mean level parameter estimates within a model (for example, the regression parameters). It is essentially a partitioning of the residual variance. Hence, this will have an impact on the standard error estimates for the parameters.

It is possible that confounding effects could be structured and so the addition of $v_i \sim N(0, \tau_v)$ may not adequately capture these effects. Spatially structured confounding can be accommodated via the inclusion of a spatially structured (correlated) effect. Often an improper CAR prior distribution is assumed for this purpose: $u_i | ..... \sim N(\overline{u}_{\delta_i}, \tau_u / n_{\delta_i})$, though other choices are

FIGURE 7.9: Posterior mean fields for the probability of low birth weight (p_{1i}) and very low birth weight (p_{2i}) for the full five-covariate model for the Charleston study.

clearly available. In conjunction with the unstructured effect, this effect can accommodate a range of potential structures in data. Spatial structuring can include clustering of a variety of types.

With a structured effect there is a possibility, however, that the effect could display colinearity with regression predictors and so can impact the estimation of these effects. Recently a range of studies have shown that this effect is important. Essentially with the structured effect, if a covariate or predictor has a strong spatial structure, then it is possible that part of the explanation of the dependent variable will be absorbed by the structure random effect and not the predictor. This implies that a model without a structured effect could display a significant effect for a predictor, but when a structured effect is included the predictor could appear to be not significant.

Remedies for this problem have not been worked out to any great extent, and so this remains an open problem. Reich et al. (2006), Hodges and Reich (2010) and Ma et al. (2007) allude to this problem and the first authors propose an orthogonalization approach to remedying the correlation between random effect and parameter estimates. Alternative approaches have

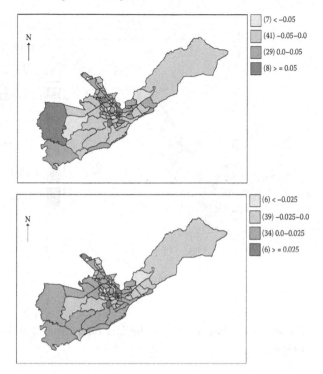

FIGURE 7.10: Posterior mean fields for the UH and CH effects for low birth weight for the full five-covariate model for the Charleston study.

been suggested, including comparison of DIC measures in staged model fitting where random effects are introduced sequentially. Recently, a two stage approach has been proposed (Lawson et al., 2012) whereby fixed effects are fitted in stage 1 and the residual is modelled with random terms, followed by refitting of fixed effects in stage 2.

7.5 Geographically Dependent Regression

It is also possible to extend models by allowing some extra variation in the regression relationship and dependence on spatial location. Originally, Fotheringham et al. (2002) defined the geographically weighted regression (GWR) extension, for a Gaussian setting, as

$$y_i(s) = \beta_0(s) + \beta_1(s)x_{1i} + \beta_2(s)x_{2i}........ + \varepsilon(s)$$

(12) < −0.25	
(42) −0.25–0.0	
(17) 0.0–0.25	
(10) 0.25–0.5	
(1) 0.5–0.75	
(2) 0.75–1.0	
(1) > = 1.0	

FIGURE 7.11: Posterior mean field for the UH effect very low birth weight the Charleston study.

where s is a spatial location. It is assumed that x_1, x_2 have spatial expression also, but this is not defined. Clearly to allow all parameters to be spatially varying within a Bayesian context, we would assume suitable prior distributions. Note that this is a special case of a random-intercept random-slope model. In the original development a distance weighting scheme was used within a weighted (generalized) least squares setting to estimate parameters. In a Bayesian setting, and assuming a single predictor (x_1), we could assume a generalized linear mixed model, such as

$$
\begin{aligned}
E(y_i) &= \mu_i \\
\mu_i &= e_i.g(\eta_i) \\
\eta_i &= \beta_0 + \beta_{1i}^* x_{1i} + R_i \\
R_i &= \beta_{01i} + u_i + v_i
\end{aligned}
\tag{7.4}
$$

where

$$
\begin{aligned}
\beta_0 &\sim N(0, \tau_0) \\
\beta_{01i} &\sim N(0, \tau_{01}) \\
\beta_{1i}^* &= (\beta_1 + \beta_{11i})
\end{aligned}
$$

and

$$\beta_1 \sim N(0, \tau_1)$$
$$\beta_{11i} \sim N(0, \tau_{11}) \text{ or } \beta_{11i}|..... \sim N(\mu_{\beta_{11}}, \tau_{11}/n_{\delta_i})$$
$$\mu_{\beta_{11}} = \overline{\beta}_{11\delta_i}.$$

In this formulation the intercept has an uncorrelated random component (β_{01i}) whereas the slope has an offset with a fixed parameter and a zero centered random parameter. The random slope could be unstructured or spatially structured. Often it is more sensible to give β_{11i} a spatially structured prior distribution (e.g. an improper CAR model) as it leads to greater parsimony and gradual spatial variation in the slope seems better justified than random jumps. This can be dubbed *geographically adaptive regression* (GAR). Two other terms are listed under the random component in Equation (7.4): $u_i + v_i$. These could be specified to have the usual random effect prior distributions which would allow the residual error to have unstructured and structured spatial effects. However, some caution should be advised when some multiple random effects are included as identifiability issues could arise. Hence it may be prudent to exclude $u_i + v_i$ when a GWR or GAR model is assumed. I have applied a GAR model in comparison to a standard mixed model with convolution, to the asthma mortality data for Georgia counties for the year 2000. The convolution and GAR models were specified for a Poisson data model as

$$y_i \sim Poiss(e_i \theta_i)$$

$$\log(\theta_i) = \alpha_0 + \alpha_1 x_{1i}^* + v_i + u_i \quad \text{convolution}$$

$$\log(\theta_i) = \alpha_0 + v_i + (\alpha_1 + \alpha_{11i}) * x_{1i}^* \quad \text{GAR}$$

where x_{1i}^* is the mean standardized population of the county, u_i and α_{11i} have improper CAR prior distributions, v_i and α_0, α_1 have zero mean Gaussian prior distributions whose standard deviations have uniform hyper-prior distributions. In this example the convolution model yielded a lower DIC (289.8) but larger pD (17.78) than the GAR model (DIC-292.1, pD = 11.54). Hence the GAR model is more parsimonious but has a slightly higher DIC. Given that the DIC difference is relatively small it may be concluded that the GAR model is preferred for reasons of parsimony. Geographically adaptive models have found wide use in environmental applications where, for example, *downscaling* of computer generated gridded pollution measures was used (see e.g. Berrocal et al., 2010; Chang, 2016; Lee et al., 2017).

7.6   Variable Selection

So far I have discussed a variety of aspects of model building when there are fixed sets of predictors and covariates. However in many applications it is important to be able to select a subset of predictors that best explains the outcome of interest. In this situation, it is important to be able to evaluate sub-models and to consider strategies for model or variable selection. A useful review and evaluation of variable selection approaches can be found in O'Hara and Sillanpää (2009).

First it is clear that different strategies can be employed varying from completely automatic selection to staged sequential selection methods. For small collections of predictors it is a simple matter to consider 'all possible regressions' and to make comparisons using, for example, DIC and pD. DIC can be used to compare non-nested models. An approximate posterior model probability can also be computed from a set of models as

$$Pr(M = i|data) = \exp(-\frac{1}{2}\Delta DIC_i)/ \sum_{l \in \{M\}} \exp(-\frac{1}{2}\Delta DIC_l)$$

where $\Delta DIC_i = DIC_i - DIC_{\min}$ and $DIC_{\min}$ are the lowest DICs amongst the models fitted (following the proposal for BIC by Burnham and Anderson (2010)).

More automated variable selection methods usually assume a full model and introduce special prior distributions to allow variables to be removed during MCMC sampling algorithms. An exception to this is reversible jump MCMC (Green, 1995), which is sampler based but allows the number of predictors to vary over iterations. It is not the intention here to review the range of approaches possible. Note that WinBUGS code for a range of approaches is available from the Bayesian analysis web site: https://doi.org/10.1214/09-BA403SUPP. See also Lesaffre and Lawson (2012), Chapter 11 for a more extensive review.

Instead, I will illustrate variable selection via an example. For the Georgia county level low birth weight (LBW) example discussed in Section 7.1.3, I have implemented the Kuo and Mallick method (Kuo and Mallick, 1998; Dellaportas et al., 2002) with 'entry parameters'. In this approach a full model is assumed but parameters are introduced in combination with the regression parameters, so that we assume the model:

$$y_i \sim bin(p_i, n_i)$$

$$\text{logit}(p_i) = \beta_0 + \sum_{j=1}^{p} \psi_j \beta_j x_{ij} + v_i,$$

where $\{\psi_1, ..., \psi_p\}$ are assigned independent Bernoulli prior distributions so that they can be assigned 0 or 1 values randomly. More formally,

node	mean	2.50%	97.50%
b[1]	0.005286	-2.528	2.533
b[2]	0.1126	0.07182	0.1521
b[3]	-0.00255	-2.453	2.378
b[4]	0.1154	0.07403	0.1637
b[5]	0.01564	-2.387	2.528
b0	-2.214	-2.247	-2.181
psi[1]	0.037	0	1
psi[2]	1	1	1
psi[3]	0.1325	0	1
psi[4]	1	1	1
psi[5]	0.05	0	1

FIGURE 7.12: WinBUGS posterior mean estimates and 95% credible limits for the regression parameters (β) and entry parameters (ψ) for the Georgia low birth weight model with an uncorrelated random effect.

$$\psi_j \sim Bern(p_j)$$
$$p_j \sim Beta(a, b)$$

with $a = b = 0.5$ for the Jeffrey's prior distribution and $v_i \sim N(0, \tau_v)$. Under this model some clear results arise. Figure 7.12 displays the converged parameter estimates for the regression parameters (b[]) and the entry parameters (psi[]).

It is clear that not only are b[2] and b[4] well estimated but they also have entry probability mean estimates (psi[]: ψ) of 1 which supports their inclusion on the model. For other parameters, the psi[] estimates are 0.037, 0.1325, 0.05. There is a clear separation of the values between 'close to 0' and 'close to 1' for $\Pr(\psi = 1|data)$. Usually a cut-off for acceptance of a predictor in a model is an entry probability around 0.5 (Barbieri and Berger, 2004) and so these variables do not meet a conventional cut-off for inclusion. These results are similar to the earlier analysis, which did not include entry parameters, and found that ethnicity (*black*) and percentage below poverty (*pov*) were significant. Note that Barbieri and Berger (2004) also stress that for normal linear models the model with only predictors with $\Pr(\psi = 1|data) > 0.5$ is often better for prediction than the model with highest posterior probability (or lowest DIC).

Recent developments in Bayesian variable selection including big data and sparse regression can be found in Balakrishnan and Madigan (2006), Petralias and Dellaportas (2013), Rockova and George (2014), and in such R packages as SPLS (sparse partial least squares) (Chung and Keles, 2010), sparsereg (Bayesian sparse regression) and lasso penalised methods (Park and Casella, 2008). Particular applications to environmental predictor mixtures can be found in Molitor et al. (2010), Park et al. (2014), and Bobb et al. (2015). Bayesian variable and model selection in special contexts, including spatial selection dependence in small area health applications, have also been explored (Cai and Dunson, 2008; Cai et al., 2013; Carroll et al., 2015; Choi and Lawson, 2016; Lawson et al., 2017).

7.7 Ecological Analysis: General Case of Regression

The setting to be discussed here, in the simplest case, assumes a dependent variable (outcome) for a small area as y_i and this is to be related to predictors. The predictors can be geo-referenced or not, depending on context. For example, we might measure a disease rate within a census tract and want to relate it to the level of poverty in the tract. Hence the dependent variable is an aggregate count of disease (y_i) and the covariate is an aggregate measure (percentage) of numbers of family units or people who fall below the defined poverty level in the small area (x_{1i}). This is an example of ecological regression: the disease count is thought to relate to poverty but we have no direct measure at a lower level of aggregation (e.g. the individual level) whether poverty is associated with the disease outcome. A typical mean model would be of the form:

$$y_i \sim f(\mu_i)$$
$$\text{where} \quad E(y_i) = \mu_i$$

with

$$g(\mu_i) = S(x_{1i})$$

where $g(\mu_i)$ is a link function and $S(x_{1i})$ is a linear or non-linear function of the covariate. In a simple case, with a Poisson data model and $\mu_i = e_i\theta_i$, with a log-linear setup, with expected rate as offset, we would have

$$y_i \sim Pois(\mu_i)$$
$$\mu_i = e_i\theta_i$$
$$= e_i \exp\{\alpha_0 + \alpha_1 x_{1i}\}.$$

In this formulation, $exp\{\alpha_0\}$ acts as an overall scaling parameter, and $\exp\{\alpha_1 x_{1i}\}$ adds the modulation of the covariate. Inference would focus on

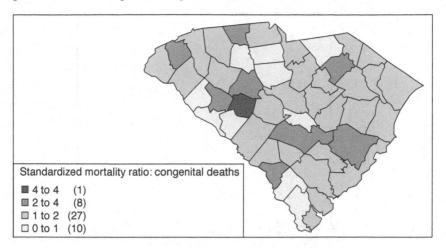

Standardized mortality ratio: congenital deaths
- ■ 4 to 4   (1)
- ■ 2 to 4   (8)
- □ 1 to 2   (27)
- □ 0 to 1   (10)

FIGURE 7.13: South Carolina congenital death standardized mortality ratio, 1990. Standardisation based on statewide rate for same year.

the parameters α_0 and α_1. This is just a log-linear Poisson regression model and in a Bayesian context we would assign prior distributions for α_0 and α_1 to complete the specification. As the focus is only on regression parameter inference rather than on estimation of the relative risk per se, it may be tempting to fit this log-linear model as defined, in the belief that, for example, unobserved confounding would have little effect on such parameter estimates. County level counts of congenital deaths for the state of South Carolina are available for a number of years. Here I focus on 1990. Figure 7.13 displays the standardized mortality ratio (*smr*) for the congenital anomaly counts for the counties of South Carolina in 1990.

Thematic map of the crude incidence ratio of low birth weight to total births for the counties of Georgia in 2007. The expected rate was computed using the age × sex standardized rate for South Carolina. Also available for this example, from the 1990 US census, is the percentage of family units living in poverty. This is determined by comparison to a threshold income level. It is reasonable to suppose that some relationship might hold between congenital deaths and poverty or deprivation in a population. Of course this is an ecological relation as individual poverty and disease outcome are not measured directly. Figure 7.14 displays the relationship found.

While there is some evidence for a positive relation at higher poverty levels, there is considerable noise in the relation and at lower levels there may appear to be an inverse relation. This form of relation may be apparent for a number of reasons. First, many areas in South Carolina are largely rural and these may display a positive relation between the standardised mortality ratio and poverty since they may be predominantly areas of low income. However some urban fringe areas have high concentrations of high income units mixed with

FIGURE 7.14: Scatterplot of the standardized mortality ratio for congenital abnormality outcome for the counties of South Carolina versus percent poverty from the 1990 US census. A LOESS line fit has been added.

rural poor populations. In these fringe areas the *smr* may be relatively high but the income will average out to be relatively high also. Second, in areas with a small percentages of poverty units it may be that health 'bootstrapping' can take place in the sense that in predominantly high income areas the health experience of low income units may be improved so that there could be a reduction in the positive relation. A negative relation might even arise. This is termed 'ecological inversion'. There is some evidence for ecological inversion in this display.

An appropriate model for these data might include a spline fit to mimic the relation. However as a first model, I will assume a basic log-linear Poisson regression model with diffuse Gaussian prior distributions for the regression parameters and uniform prior distributions for the associated standard deviations (Gelman, 2006). Model results are found in Table 7.4. The models fitted were as follows. Model 1 is a simple Poisson log-linear regression, model 2 is the same but with an added uncorrelated zero mean Gaussian random effect (with precision τ_v) and model 3 is the same as model 2 but with an added correlated random effect with a CAR prior distribution and precision τ_u. Note

	Model 1	**Model 2**	**Model 3**	**Model 4**
α_0	0.0859(0.248)	0.0236(0.191)	0.068(0.243)	0.192(0.209)
α_1	-0.0034(0.016)	0.0002(0.0119)	-0.0032(0.0155)	0.044(0.1198)
α_2	-	-	-	-
τ_0	43.87(393.2)	325.2(4159)	107.6(1338)	3425.(63910)
τ_1	35100.0(4.6E+4)	15210.0(3.5E+5)	48510.0(8.0E+5)	2611(49710)
τ_v	-	242.8(321.5)	68.43(143.9)	4.845(9.03)
τ_u	-	-	525.3(678.1)	-
DIC	171.17	171.01	173.16	167.95

Table 7.4: Model fitting results for a variety of models for South Carolina congenital mortality data.

that for the simple regression none of the parameters is well estimated and even when conventional random effects are added the model is not improved significantly. In fact, the addition of the correlated effect increases the DIC.

Overall the random effects do not appear to improve the model fit to any significant degree. Instead, noting that a non-linear model seems to be more appropriate as a description of the relation, I consider a low rank spline model fitted to the predictor:

$$\log \theta_i = \alpha_0 + \alpha_1 x_{1i} + \ldots\ldots + \alpha_p x_{1i}^p + S(x_{1i}) \tag{7.5}$$

$$S(x_{1i}) = \sum_{k=1}^{K} \psi_k (x_{1i} - c_k)_+^p$$

where $\{c_k\}$, $k = 1, ..., K$ is a set of knots and $\{\psi_k\} \sim N(0, \tau_\psi)$. Model 4 is a simplified version of (7.5) which is simply $\log \theta_i = \alpha_0 + \alpha_1 \sum_{k=1}^{K} \psi_k (x_{1i} - c_k)_+$.
Further model fits using spline only and a full second degree polynomial with added second degree spline ($p = 2$) gave DICs of 169.8 and 173.8 respectively. This supports the contention that model 4 provides the best fit in this case, for the models considered. It is also interesting to note that when a convolution model was fitted to the data with a linear predictor (model 3), a large outlier appeared in the estimated $\widehat{\theta}_i$. This was for Berkeley County on the urban-rural fringe of Charleston (see Figure 7.15), where the exceedance probability $(\widehat{\Pr}(\theta_i > 1))$ is 0.903. This is considerably higher than the value for any other county. For the best fitting spline model none of the counties showed excessive $\widehat{\theta}_i$s nor $\widehat{\Pr}(\theta_i > 1)$ exceeding 0.7.

To further highlight this effect, a simulation has been carried out, where, at the individual level, there is a strong positive relation between a binary disease outcome and socio-economic status (income below or above poverty threshold). It is assumed that there were 100 individuals within 100 areas. An income distribution was assumed for each region based on whether individuals were categorized below or above poverty level ($30,000). Conditional

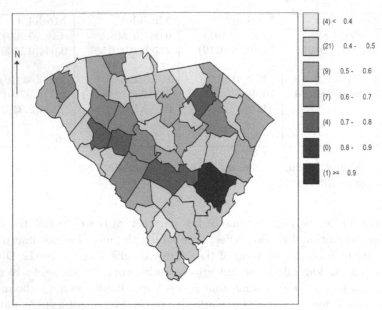

FIGURE 7.15: South Carolina congenital deaths and percent poverty by county for 1990 showing probability of exceedance of posterior expected relative risk.

on this given binary variable (x_{ij}: poor or not) the probability of disease was simulated via a logistic transform with added binomial noise. This transform allows the specification of the individual relation between poverty and outcome: $y_{ij}^s < -bin(1, p_{ij})$ where logit $p_{ij} = \alpha_0 + \alpha_1 x_{ij}$.

For the simulation shown the relation was logit $p_{ij} = 0.2 + 1.5x_{ij}$, a reasonably strong positive relation between outcome and poverty state. Counts of disease were then aggregated across the individuals within areas ($j = 1, ..., 100$), as was the number of poor (to yield percent poverty). To demonstrate the ability of such aggregation to yield a relatively complex relationship between outcome and poverty, Figure 7.16 displays the simulated aggregate relation between average income and disease count and poverty proportion and disease count.

It is noticeable that while a general increase in poverty seems to relate to an increase in disease count, this relation does not hold strongly. For some levels of poverty the relation is not strong and is even reversed. The noise in the relation is quite high of course. In further simulations, uncorrelated heterogeneity (zero mean Gaussian noise) was introduced to the linear predictor, along with further variation in income distribution. These changes led to greater noise in the relation and changes in the overall gradient or linear form of the relation.

Certainly from this output it would appear that there is no strong indication of a non-linear positive relation at the individual level. However, if the focus is the estimation of the aggregate level relation then it is clear that the overall relation at the aggregate level is weakly positive (or weakly negative with average income). It is also clear that a simple log linear model may not represent the variation well.

The addition of extra variation in a model in the form of random effects may help to reduce the noise but does not necessarily improve the estimation of the covariate relation. If the covariate relation is mis-specified then addition of modelled heterogeneity may not lead to a better model. This was demonstrated in the data example above where a spline model without random effects yielded a better empirical model fit, based on DIC than a convolution model with log linear predictor. Some authors (e.g. Clayton et al., 1993) have advocated the inclusion of spatially-structured(CH) random effects to make allowance for biases induced by the ecological nature of the analysis. In some cases this may be important, especially when making inferences at a different aggregation level. However, the above example demonstrates that the use of convolution models (which include CH and UH effects) can yield poor empirical fits when the aggregate relation is mis-specified.

In addition to this warning, there is now some evidence that convolution models, particularly those which have an (improper) CAR model specification for the CH effect, can lead to very poor estimation of certain covariate effects (Reich et al., 2006; Ma et al., 2007). In fact, the use of CAR random effects in linear combination with linear predictors with spatially-referenced covariates (trend surface components), can yield very poor estimates of the linear parameters even under a strong linear relation. Figure 7.17 displays two examples of the empirical power within a simulation of the estimation of a distance covariate parameter. Case event data were simulated under a variety of models and then a fine grid mesh was used to bin the events to form counts. The count in the i th bin (y_i) was assumed to have a Poisson distribution with expectation $e_i\theta_i$. Various models were assumed for θ_i. Model A is $\log\theta_i = \alpha + \log(1 + \exp(-\beta d_i))$ where d_i is the distance from a fixed location to the centroid of the i th bin; model B is $\log\theta_i = \alpha + \log(1 + \exp(-\beta d_i)) + v_i$ where $v_i \sim N(0, \tau_v)$ a UH component; and model C is $\log\theta_i = \alpha + \log(1 + \exp(-\beta d_i)) + v_i + u_i$ where $u_i|u_{-i} \sim N(\overline{u}_{\delta_i}, \tau/n_{\delta_i})$, a CAR prior distribution. The latter is a convolution model with added covariate term.

The simulations were carried out under a variety of scenarios. Two of these involved variant forms of background heterogeneity in risk. The left panel in Figure 7.17 has no additional heterogeneity while the right panel is under a binned log Gaussian Cox model with a generating process that is a spatial Gaussian process with exponential covariance: $\sigma \exp\{-d_{ij}/\omega\}$ with $\sigma = 0.1$ and $\omega = 0.5$ where d_{ij} is the distance between i th and j th points. It is noticeable that the convolution model appears to have poor performance under both scenarios in the estimation of the distance effect compared to either a simple log linear model or an UH component model.

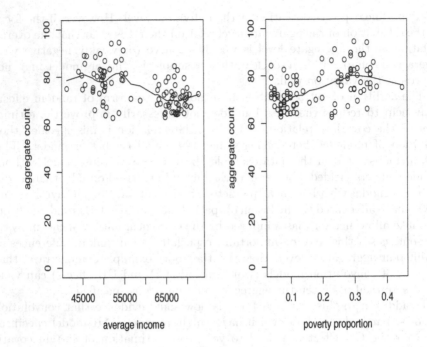

FIGURE 7.16: Simulation-based aggregate relation between total disease count and percent poverty for 100 areas with 100 individuals in each area. The LOESS fit is shown at the aggregate level. Left panel: count versus average income, right panel: count versus poverty proportion.

(This performance is also found when a Poisson simulation is made directly into the bins, and also when a multiplicative log link is defined.) Hence, it is also important to consider carefully the use of CAR-based convolution models when covariates are to be estimated. While convolution models are robust against mis-specification and are useful for general relative risk estimation (see e.g. Lawson et al., 2000; Best et al., 2005), there can be considerable aliasing of long-range spatial effects. Of course, covariates that are not directly spatial in form but are aliased with such spatial effects (which can be mimicked by CAR models) may also be affected.

7.8   Biases and Misclassification Errors

Besides the considerations discussed above (Section 7.7), it remains important to consider aggregate ecological analysis simply to provide descriptions of

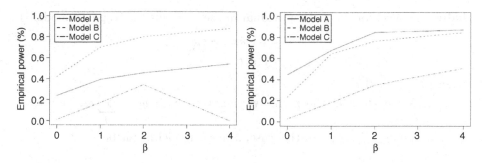

FIGURE 7.17: Empirical power curves estimated from credible intervals for the β parameter from a distance covariate model which includes additive random effect terms. Model A = covariate model only. Model B = Model A with UH term added. Model C = Model B with CH term added. Left panel: no background heterogeneity. Right panel: log Gaussian Cox process with Gaussian spatial Gaussian generating process.

aggregate level relations. Often the biases apparent when aggregate inference is to be pursued are much reduced (see e.g. Greenland, 1992; Greenland and Robins, 1994; Wakefield and Smith, 2016). When inference at different aggregation levels is to be considered, however, additional problems arise. The 'gold standard' for inference in medical studies is often the individual level, in that it is often the aim to be able to infer an outcome from individual level data. This is true for clinical or intervention trials where individual responses are used as the basis of group (aggregate or population) summarization or inference. A distinction should be drawn here between inference to be applied to an individual and inference made from individual data. The former can be attempted from various levels of data aggregation (with varying levels of success), whereas inference from individual level data can be used make inferences about individuals and also aggregated levels in the population. On the other hand, with aggregated data (such as county-level disease count data) is it possible to make individual inference? This is a much more difficult undertaking.

7.8.1 Ecological Biases

When inference is made from an aggregated study to a lower level of aggregation then bias can occur. This bias can have a number of component biases. A good example of the extent of such bias is given (for a non-medical example) in In the simplest example, assume a linear regression relation with $j = 1, ..., n_i$ individuals in $i = 1, ..., m$ units (areas). Assume first that at the individual level for the j th individual in the i th unit the response model is

$$y_{ij} = a + bx_{ij} + e_{ij}$$
$$= f(a, b, x_{ij}) + e_{ij}$$

where $E(e_{ij}) = 0$.

If we aggregate over the n_i individuals then with $y_i = \sum_j y_{ij}$ and $x_i = \sum_j x_{ij}$ and $e_i = \sum_j e_{ij}$. In this case, a linear model might be

$$y_i = a^* + b^* x_i + e_i^*.$$
$$= f^*(a^*, b^*, x_i) + e_i^*.$$

Now the question essentially is whether we can make inferences from $a^*, b^*, \{e_i^*\}$ at the disaggregated individual level. In one view, this can be interpreted as an example of the modifiable areal unit problem (see Section 9.1). However, inference at a lower level of aggregation is the sole focus here. In general, it is important to consider the model

$$E(y_i) = E(f^*(a^*, b^*, x_i)) \tag{7.6}$$

where expectation is with respect to y. Here it is often assumed naively that $E(y_i) = f(x_i) = a + bx_i$. However if the within-area distribution of x_{ij} is heterogeneous then this will not hold. This also assumes that the individual relationship has the same *form* (as well as parameter values) as the aggregate relationship. This is the naive ecological model of Salway and Wakefield (2005). We are interested in how (a^*, b^*) relates to (a, b), in particular how the slope parameter b^* relates to b. Under a logistic model when the response is binary then b would be the *odds ratio* for the exposure.

In general the ecological bias is the difference between the estimated b^*, say $\widehat{b^*}$, and the true individual parameter b. If the individuals did not vary with their exposures i.e. $x_{ij} = x_i$ then there is no bias. Biases can arise from a variety of sources: 1) bias due to confounding: variables missing on individuals or at group or area level; 2) bias due to effect modification: exposure effect varying between groups and areas; 3) contextual effects: areal and group level variables which are unmeasured (Greenland and Robins, 1994); 4) measurement error: there may be error in the classification of discrete exposures or measured confounders, as well as error in continuous covariates, Within-area variation in exposure or confounders is a major contributor to ecological biases. Other sources of bias are of course not unique to ecological studies (measurement error and unobserved confounding). These will be discussed later.

7.8.1.1   Within-Area Exposure Distribution

If it is possible to specific the within-area (group) exposure distribution then it is possible to try to assess the bias from this source. The aggregated model that corresponds with the basic individual level model integrates over the distribution of x (assuming the exposures are independent) i.e.

$$f^*(a^*, b^*, x_i) = E_x\{E_y(y_{ij})\} = \int f(a, b, x_{ij})p(x)dx \qquad (7.7)$$

where $p(x)$ is the distribution of the exposure. Salway and Wakefield (2005) cite a range of approximations to within-area distributions when there is independence or no spatial correlation in the data. Wakefield et al. (2011) provide a review of the effects of different within area sampling schemes. Of course, in spatial applications exposures could easily be correlated and we would instead be interested in aggregation of events from (say) a point process to a count process with small areas. It has been shown that for a log Gaussian Cox process with stationary covariance and log linear model for covariates (in this case assume that the intensity is $exp(x(s)\beta)$ where $x(s)$ is a spatially-referenced covariate value at s), then the spatial moments of the within-area distribution of can be computed if the within-area distribution of the covariate is known, i.e.

$$E^A = \frac{1}{|A|}\int_A exp(x(\mathbf{s})\beta)d\mathbf{s} = \int_{-\infty}^{\infty} \exp(x\beta)dF^A(x)$$

where $F^A(x) = \frac{1}{|A|}\int_A I(x(\mathbf{s}) \le x)d\mathbf{s} \qquad (-\infty < x < \infty)$,

the spatial cumulative distribution function of $x(\mathbf{s})$ (Cressie et al., 2004). This implies that partial knowledge of F^A (e.g. bounds, mean, variance) could be used to characterize the within-area distribution. For a single binary covariate (x_{ij}) and the area average $\bar{x}_i$ $(\bar{x}_i \approx x_i/n_i)$, the expected count for the i th area is

$$E^A = 1 + \bar{x}_i(e^\beta - 1).$$

More complex situations could arise (with, for example, continuous spatial fields). Assume the case of a bivariate continuous covariate with $\exp(x_1(s)\beta_1 + x_2(s)\beta_2)$ where separate spatial means and sample variances are available from surveys $(\bar{x}_1, \bar{x}_2, S_1^2, S_2^2)$ with the spatial covariance given by:

$$C_{12}(A) = \frac{1}{|A|}\int_A (x_1(\mathbf{s}) - \bar{x}_1)(x_2(\mathbf{s}) - \bar{x}_2)d\mathbf{s}.$$

The approximation to E^A is then

$$E_0^A = \exp(\bar{\mathbf{x}}\beta + \frac{1}{2}\beta^T \mathbf{\Gamma}\beta)$$

where $\bar{\mathbf{x}} = \{\bar{x}_1, \bar{x}_2\}$, $\beta = \{\beta_1 + \beta_2\}$, and

$$\mathbf{\Gamma} = \begin{bmatrix} S_1^2 & C_{12}(A) \\ C_{12}(A) & S_2^2 \end{bmatrix}.$$

Hence as long as the means, variances and covariances are known then it is possible to improve on $\exp(\overline{\mathbf{x}}\beta)$ by addition of covariation information. Can adjustments be made in these cases? If the within-area distribution is known or can be approximated to a reasonable level then the ecological model can be used directly with these ingredients. For highly skewed distributions, numerical integration may be required (see e.g. Salway and Wakefield, 2005). For spatial dependence the maximum entropy approach appears to work well. Another approach to dealing with a range of ecological problems is to try to include individual level data within the model so that the linkage between the aggregated and disaggregated data is modeled. An example of the use of the spatial approximation was applied to the percent poverty variable (which is an average of the binary covariate at the individual level) and the total count of abnormalities by county. For this situation, we assumed a Poisson data likelihood for the county anomaly count with expectation $e_i\theta_i$ and $\log\theta_i = \beta_0 + \log(1 + \overline{x}_i(e^{\beta_1} - 1))$ where $\overline{x}_i$ is the percent poverty and β_1 is the slope parameter. This model was fitted with prior distributions as follows:

$$\beta_0 \sim N(0, \tau_0)$$
$$\beta_1 \sim Ga(1, 1)$$
$$\tau_0 = 1/\sigma_0^2$$
$$\sigma_0 \sim U(0, 100)$$

with the restriction placed on the distribution of β_1 due to the possible sampling singularity when $\overline{x}_i(e^{\beta_1} - 1) < -1$. This model yielded a DIC of 165.68 which is lower than the result for spline model already cited. Of course, this does not necessarily imply that this is the best model for these data. The addition of a term for uncorrelated heterogeneity (UH), however, yields a higher DIC: 172.54, and in this case does not improve the model fit.

7.8.1.2 Measurement Error (ME)

Clearly another source of considerable error in regression models is the possibility that predictors, covariates or exposures are measured with errors. In the case of a discrete covariate this is called misclassification error (Gustafson (2004)). For example, if the data on disease outcome is related to individual income (as in the example above) with income dichotomised into below or above poverty level then we have a binary covariate. If someone was wrongly categorized as poor (1) when he or she should be 'not poor' (0) then this would be a misclassification. Of course if the outcome is also binary (disease or no disease) then misclassification could occur if diagnosis was prone to false positives or false negatives. A number of methods are available for such discrete error problems within a Bayesian paradigm and they are discussed in detail in Gustafson (2004).

For continuous variables it is usual to assume different types of errors depending on the form of error appropriate in context. In a simple binomial formulation assume that an individual has distribution given by $y_{ij} \sim bin(1, p_{ij})$ and $logit(p_{ij}) = \alpha_0 + \alpha_1 x_{ij}$. Assume that exposure variable x_{ij} is observed with error. In our example, assume that x_{ij} is the self-reported income for the individual. In a self-report, context error may creep in due to various psycho-social (contextual) effects. Under-reporting of income may happen when someone does not want to appear 'too well-off'; on the other hand someone else may want to brag and exaggerate income. We might assume this error is additive as a first assumption. Hence, a model for the observed income x_{ij} could be

$$x_{ij} = x_{ij}^T + e_j$$

where x_{ij}^T is the true income and also note that the error (e_j) has a person-specific component. This is regarded as the *classical ME* specification. Now the relationship with the outcome y_{ij} is via a logit link to the covariate. However we would usually assume that the outcome is related to the true covariate (and not the error corrupted version. Hence we would want a model such as

$$y_{ij} \sim bin(1, p_{ij}) \tag{7.8}$$
$$logit(p_{ij}) = \alpha_0 + \alpha_1 x_{ij}^T. \tag{7.9}$$

Now ME could be included in this model in a number of ways. First we could assume a reverse model for error where $x_{ij}^T = x_{ij} + e_j$, which assumes that by adding noise to the observed variable the true value will be obtained. Substituting this into (7.9) we have

$$logit(p_{ij}) = \alpha_0 + \alpha_1 (x_{ij} + e_j). \tag{7.10a}$$

This is known as *Berkson* error (see e.g. Carroll et al., 2006). A suitable distributional assumption for the random effect e_j would be $e_j \sim N(0, \tau_e)$. Two other alternatives can be considered for this error. One is a general random effect model which decouples the random effect from the covariate to yield a simple frailty model:

$$logit(p_{ij}) = \alpha_0 + \alpha_1 x_{ij} + e_j.$$

While this model has less justification than the Berkson model with respect to ME, it does appear to often demonstrate better goodness-of-fit, presumably because there is much noise within the model fit in general between y_{ij} and x_{ij}, compared to the noise in x_{ij} itself (Kipnis et al., 1999, 2001, 2003). The final option is to jointly model the covariate and the outcome in the sense that both the disease outcome and the observed covariate depend on the unobserved true value of the covariate. For example, if the disease outcome were binary and a binomial likelihood model was assumed with logit link and if the observed data are regarded as having classic ME, a reasonable model would be:

$$y_{ij} \sim bin(1, p_{ij})$$
$$logit(p_{ij}) = \alpha_0 + \alpha_1 x_{ij}^T,$$
$$x_{ij} \sim N(x_{ij}^T, \tau_x),$$

where τ_x is a variance term. In this case we now have a latent variable x_{ij}^T underlying both likelihoods and this can be regarded as an example of a latent variable or structural equation model (SEM). Hence, a Bayesian SE model (Stern and Jeon, 2004) could be defined once prior distributions for the parameters $\{\alpha_0, \alpha_1, \tau_x\}$ were defined. Further hyperprior distributions could be assumed for parameters defined in the prior distributions.

7.8.1.3 Unobserved Confounding and Contextual Effects in Ecological Analysis

Clearly, one major source of error in any regression study, let alone ecological study is the possibility of unobserved confounding. Confounding variables could create different responses in the outcome and so, if not accounted for, may influence the result. For example, environmental insults (such as air pollution) could affect asthma outcomes. In addition, smoking could affect this outcome. Hence a study which did not look at smoking or other respiratory-challenging lifestyle variables but simply looked at the relation between air pollution and asthma might draw erroneous conclusions. Such confounders could act to elevate the risk of the disease outcome, possibly in tandem with the exposure of interest (air pollution). There are two situations that should be considered.

First of all, direct correlation with the exposure variable may serve to alter the relation observed. For example, the combination of an observed enhanced exposure (e.g. air pollution) and (say) low socioeconomic status (via un-observed average income or an unobserved smoking indicator) could lead to spurious disease elevation due to the combination of effects (one of which is unmeasured). This often happens when, for example, industrial sites are studied and elevated disease risk is found in the vicinity of these sites. However the vicinity is often also a low socio-economic status area. This of course supports the use of deprivation indices (Diggle and Elliott, 1995) or other indices of risk to make allowance for such effects in environmental epidemiology studies.

Second, it is quite common for unobserved confounders to leave a degree of variation in risk unexplained in the resulting model fit. It may be that a confounder present in an area leads to higher disease risk, but the exposure is low in that area. For instance, areas with high numbers of smokers could yield high asthma mortality but could be far from air pollution sources. If smoking status was not measured these areas would appear as large residuals or outliers. To combat these unobserved confounder problems it has been suggested that random effects should be introduced into the analysis to 'soak up' this extra variation(see e.g. Lawson, 1996). In general, this supports the use of generalized linear mixed models (GLMMs) in these analyses, and these are quite commonly applied now.

Third, unobserved confounders could induce spatially-correlated effects in the risk variation (Clayton et al., 1993) and so the extension to spatially correlated (CH) random effects has been recommended. In general, the recommendation would be that both UH and CH effects should be added in any study, to allow for different possible forms of extra variation. It has also been emphasized that forms of ecological bias can be, to a degree, accommodated, by the inclusion of CH effects (Clayton et al., 1993). Hence for a Poisson data likelihood model for a small area disease count we would have

$$y_i \sim Pois(e_i \theta_i)$$
$$\log(\theta_i) = \beta_0 + x_i'\beta + u_i + v_i$$

where u_i, v_i are the CH and UH random effects respectively. While this is now a general panacea, the caution must be given that a) CH and UH terms may not improve overall model fit, b) can lead to highly biased estimates of covariate terms (depending on the prior model assumptions), especially if aliased with the long range spatial variation, and c) ecological within area distributional considerations can lead to better aggregate models (which can fit better than random effect models).

Finally, *contextual effects* (Goldstein and Leyland, 2001; Voss, 2004; Chaix et al., 2006) are considered variables that specify the socio-environmental context of an individual are special cases of confounding. Contextual effects are defined as 'aspects of the social and economic milieux of an area which engender an area outcome effect'. Often these are found at an aggregate level. For instance, an individuals's outcome on a clinical trial might be related to the area of residence. Hence, for example, the county of residence might be a contextual variable for that individual. Another important example would be the ecological inversion example. If a person of poor socio-economic (*se*) status lives in a high *se* area, that can lead to reduced health risk to that individual. Hence the person's outcome may be pulled towards the area level expected outcome. The *se* status of the area of residence could be an important variable in explaining health outcome at the individual level and may explain ecological inversion. A typical model for an individual binary outcome for the i th individual, y_i, might be modeled via a logit link to a probability such as

$$logit(p_i) = \beta_0 + \beta_1 x_{1i} + \beta_2 x_{cj} \atop {i \in j}$$

where x_{1i} is the individual *se* status and x_{cj} is the *se* status of the jth small
$${i \in j}$$
area to which the i th person belongs. Of course a range of such effects could be envisaged where hierarchies of regional or other clustering effects could be added to an individual level model.

7.9   Sample Surveys and Small Area Estimation

For most applications of Bayesian disease mapping the complete set of outcomes is usually available within spatial units. For example, counts of asthma mortality or morbidity provided by local government health agencies are usually complete for all areas, barring under-ascertainment or under-detection. However there are situations where only partial information is available. One such situation is when sample surveys are carried out and they do not provide complete information about all spatial units. Simple random sampling for surveys, where location is not used as part of the sampling algorithm, can lead to sparse representation of the characteristic of a study area. For example, a simple random digit telephone dialing sample of people within the age range of 20 to 30 of size 100 (n) out of a total population of 500,000 (N) within 150 (m) sub-regions is carried out. A binary outcome is elicited: have you ever attended a primary care physician for asthma?: The possible answers are *yes* or *no*. Denote this outcome by y_{ij} $i = 1, ..., n$, $j = 1, ..., m$ having sorted the sampled people into the m sub-regions. The population size in the j th sub-region is n_j and $N = \sum_j n_j$. A number of issues arise when using this type of data. Of course self-report is used and so there could be misclassification bias (measurement error). There is also the problem that with 150 spatial units and only 100 in the sample, some sub-regions will have no sampled persons. At the individual analysis level this does not necessarily pose a great problem if we are not interested in the sub-region effects then the sample can still be representative of the population as a whole. However if any sub-regional contextual effects were to be introduced in the individual analysis then there would be missing support for some sub-regions. A contextual model such as

$$y_{ij} \sim Bern(p_i) \tag{7.11}$$

$$\text{with } \operatorname{logit}(p_{ij}) = x_i^t \alpha + v_i^a + v_j^b \tag{7.12}$$

where $x_i^t \alpha$ is a linear predictor and v_i^a an individual frailty and v_j^b a contextual effect of sub-region could be considered. Note that a sampling weight adjustment would usually also be included in the model either as a covariate (Vandendijck et al., 2016) or as an adjustment to the outcome (Mercer et al., 2014; Chen et al., 2014; Vandendijck et al., 2016). The contextual effect could be unstructured as $v_j^b \sim N(0, \tau_{v^b}^{-1})$, in which case missing regions could contribute to the precision estimate of the effect. It might then be appropriate to consider augmenting the sampled outcomes within sub-regions with non-sampled outcomes (y_{kj}^*) in the n_j set. However with highly sparse sampling this could be a very ambitious idea as the whole vector of outcome and associated predictors are missing. Further, the contextual effect could be structured so that $v_j^b \sim N(\overline{v_{\delta_j}^b}, \tau_{v^b}^{-1}/n_{\delta_j})$ in which case the spatial structure would be missing sub-regions.

In an extreme case, the study region could consist essentially of *dis*connected graphs rather than a completely connected network (see e.g. Freni-Sterrantino et al., 2017). The question then must be to consider whether local estimators should be used only for areas where there is support, or to assume a global model for the sparse situation. If a global model is assumed, then v_j^b will be estimated from a subset of regions only. While a model such as Equation (7.11) is feasible, it does not allow for the fact that sampling has taken place. Usually, sampling weights (w_{ij}) are assumed where $w_{ij} = (n_j^*/n_j)^{-1}$ with the set of sampled individuals in the j th sub-region comprising s_j and $n_j^* = card(s_j)$. These weights are used to modify the outcome variable or as a predictors with a suitable link function. A possible model could be

$$\text{logit}(p_{ij}) = x_i^t \alpha + f(w_{ij}) + v_i^a + v_j^b \tag{7.13}$$

where $f(.)$ could take a variety of forms including splines or other smoothing devices (Chen et al., 2014; Vandendijck et al., 2016). If non-response occurs, and hence missingness is present, additional weights can be introduced (Watjou et al., 2017).

Other forms of sampling, with stratification based on sub-region for example, could be utilised. However, in some cases sub-regions can be excluded from such surveys due to cost limitations. Hence the sub-region missingness would be built into the problem. In the case of sampling based on sub-regions there will be a set of sampled regions (sa) and non-sampled regions (ns). The sampling could be akin to cluster sampling with random selection within the regions. Observed outcomes would be in the set $\{y_{ij}\}_{i \in s_j, j \in sa}$, and the unobserved $\{y_{ij}^*\}$ would be in ns. In this case, if the ns set is reasonably small and prevalance is to be estimated only (i.e. no covariates are present), it may be possible to impute the $\{y_{ij}^*\}$ from the predictive distribution based on data augmentation of the missing $p_{ij}s$. If so then the regional effects could be estimated iteratively from the 'complete' data realisations: $\{y_{ij}\}$ and $\{y_{ij}^*\}$. However this would require sampling weights to be estimated or generated for the non-sampled areas.

7.9.1 Estimation of Aggregate Quantities

Often based on survey data some aggregate quantity is to be estimated. For example, the binary outcome y_{ij} above could be used to estimate a prevalence within the j th sub-region. This problem has been studied by many authors, albeit without much reference to the spatial nature of the geographical problem. Rao (2003) and more recently Rao and Molina (2015) provide extensive overviews of a wide range of approaches to estimation within small areas from surveys. In fact this area of focus is called *small area estimation*. In the case of a binary outcome a crude estimate of the prevalence in the j th sub-region is

$$p_j = \frac{1}{n_j} \sum_{i=1}^{n_j} y_{ij}.$$

However, this assumes that the whole population is available. To handle the fact that a sample has been used, sampling weights proportional to the inverse probability of being sampled are set up and used in the estimation. For example, the Horowitz-Thomson estimator,

$$p_j^{HT} = \frac{1}{n_j^*} \sum_{i \in s_j} w_{ij} y_{ij}, \tag{7.14}$$

where w_{ij} is the sample weight and s_j is the set of sampled individuals and $n_j^* = card(s_j)$ is commonly used. Often models are used instead to estimate what is essentially the probability of a positive response: a crude naive model that ignores the sampling could be $y_j = \sum_{i \in s_j} y_{ij}$ and $y_j \sim bin(n_j^*, \widetilde{p}_j)$ and logit$(\widetilde{p}_j) = \beta_0 + v_j + u_j$ which is the classic convolution model for a binomial data likelihood. Various modifications of this model have been proposed whereby the design weights modify the outcome (see e.g. Mercer et al., 2014; Raghunathan et al., 2007; Chen et al., 2014). For example, $y_j^{LN} = \text{logit}(p_j^{HT})$ or $y_j^{AS} = \sin^{-1}(\sqrt{p_j^{HT}})$, amongst others, have been proposed. This leads to a binomial convolution model at the area level:

$$y_j^* \sim bin(n_j^*, \widetilde{p}_j)$$
$$\log \text{it}(\widetilde{p}_j) = \beta_0 + v_j + u_j$$

where $*$ can denote LN or AS. The comments above concerning the missing sub-areas also apply here and it may be useful to consider imputation of outcomes and parameters for missing areas.

8

Putative Hazard Modeling

In this chapter I focus on the analysis of a specific application area: the modeling of disease risk around a known location or locations. This focus is a particular example of a regression application which can have ecological elements. Some of the discussion will focus on case event level modeling, which is not at an aggregated level. However, many of the issues discussed in Chapter 7 are relevant to aspects of this modeling.

In putative source analysis, the location(s) of potential (putative) source(s) of health hazard (pollution or other insult) are known, and it is the task of the analysis to determine whether the source or sources affect health risk in their vicinity. Hence the term *putative* is used to mean "suspected" in this case. Many examples come from environmental epidemiology where a location is the focus of the risk assessment (Lawson and Cressie, 2000; Lawson, 2002) or pollutants serve as the focus (e.g. incinerators, chimneys, road networks). If the exposure pathway were water ingestion then the focus might be water sources or supply networks (e.g. groundwater wells, rivers). Usually the risk is assumed to be related to location of residence of the population. This is termed residential exposure or risk, and measures of the relation between the source location and residence are used in the analysis.

Analyses will be formulated depending on whether a disease is of interest or whether a source is of potential interest. For instance, if we are interested in acute asthma risk, then we might monitor emergency room admissions for asthma in the vicinity of an air pollution putative source (Anto and Sunyer, 1990). On the other hand, if a public report of a general (non-specific) fear of an elevation of disease risk in the vicinity of a putative source is made, then focus may be on the source characteristics and diseases that may be affected. Hence multiple diseases may be analyzed in this case.

For example, the Sellarfield nuclear reprocessing plant in Northwest England was the focus of studies in late 1980s. This led to a variety of radiation-related disease studies (mainly for radiation-related outcomes, e.g. childhood leukemia) (see e.g. Gardner, 1989). However, risk from such a site may be from a variety of sources (water pollution, air pollution, occupational radiation risk, etc.) besides simply residential air pollution exposure. Hence it is not always clear what the main effects are that should be modeled. In a study of larynx cancer in Northwest England around a putative source (incinerator), evidence for residential exposure was assessed via a distance covariate

FIGURE 8.1: Larynx cancer and lung cancer incident cases in the vicinity of an incinerator in NW England for the period 1974-1983. The incinerator is marked with '+'.

measured from residential addresses of death certificates and putative sources (Diggle, 1990).

In Figure 8.1, the left panel displays the residential locations of incident cases of larynx cancer for 1974 through 1983. The right panel displays the distribution of lung cancer cases for the same period. At location 35450, 41400 is an incinerator which is the putative focus in this case. It could be considered that an incinerator could elevate disease risk around it and respiratory disease could be a target.

The evidence for the effect of the incinerator could be manifold. The primary effect might be elevated incidence near the site of the putative source. Hence one might be tempted to consider a distance decline effect around the source. This would be a primary form of evidence for a linkage. Of course, confounding due to correlation between deprivation and distance would need to be considered if such information were available. For a variety of source type and exposure pathways, distance decline is a fundamental piece of evidence. A secondary form of evidence is directional in nature. With air pollution as the primary putative exposure the effect of wind direction and strength should be considered. There are many examples where directional effects can be important. Often within putative health studies a retrospective analysis of incident cases or mortality events is carried out. This is often needed as the existence of a putative source is often noted only after some exposure period.

8.1 Case Event Data

Diggle (1990) used larynx cancer case residential addresses as the outcome of interest in a post hoc study of that disease around a putative source (incinerator). The study is post hoc as elevated incidence of larynx cancer was registered as a concern by the local residents in the vicinity of the incinerator. This concern motivated the study. The impact of the post hoc nature of the study is largely a design issue and is discussed more fully in Lawson (2006b), Chapter 7.

The original data used in that study is shown in Figure 8.1. The cases of larynx cancer (58) within a rectangular study window for the 1974 through 1983 are shown in the left panel. As part of the study, case residential addresses of respiratory cancer for the same study period were collected. These were to be used as a type of *control* disease that could allow for the spatial distribution of the background "at risk" population. This essentially acted as a geographical control at a fine resolution level. Any areas where there are lots of *at risk* people are more likely to yield cases and so we must adjust for this effect.

The right hand panel of Figure 8.1 displays the map of these 978 control cases. Some discussion has focused on whether respiratory cancer is a valid control disease for larynx cancer in a putative air pollution study. Here we assume that the control is valid, but in general the choice of control disease is important in any particular application.

Assume we observe within a study region (W), a set of m cases, with residential addresses given as $\{s_i\}$, $i = 1, ..., m$. Here the random variable is the *spatial location*, and so we must employ models that can describe the distribution of locations. Often the natural likelihood model for such data is a heterogeneous Poisson process (PP). In this model, the distribution of the cases (points) is governed by a first order intensity function. This function, $\lambda(s)$ say, describes the variation across space of the intensity (density) of cases. This function is the basis for modeling the spatial distribution of cases. Denote this model as

$$\mathbf{s} \sim \mathbf{PP}(\lambda(\mathbf{s})).$$

The unconditional likelihood associated with this model is given, barring a constant, by:

$$L = \prod_{i=1}^{m} \lambda(s_i) \exp\{-\int_W \lambda(\mathbf{u})d\mathbf{u}\}$$

where $\lambda(s_i)$ is the first order intensity evaluated at the sample locations $\{s_i\}$. This likelihood involves an integral of $\lambda(\mathbf{u})$ over the study region.

The definition of the intensity of cases must make allowance for the effect of the background at-risk population. Often the intensity is specified with a multiplicative link between these components:

$$\lambda(s) = \lambda_0(s)\lambda_1(s|\theta).$$

Here the *at-risk* background is represented by $\lambda_0(s)$ while the modeled excess risk of the disease is defined to be $\lambda_1(s|\theta)$, where θ is a vector of parameters. In putative source modeling we usually specify a parametric form for $\lambda_1(s|\theta)$ and treat $\lambda_0(s)$ as a nuisance effect that must be included. Usually some external data is used to estimate $\lambda_0(s)$ non-parametrically (leading to profile likelihood). In the larynx cancer example, the respiratory cancer distribution would be used to estimate $\lambda_0(s)$.

It is possible to reformulate this problem by viewing the joint realization of cases and controls and, conditional on that realization, examining the probability that the binary label on a point is either a case (1) or a control (0). If this approach is taken the background nuisance function disappears from the problem (Diggle and Rowlingson, 1994). This depends implicitly on a control disease being available and relevant (i.e. matched well) to the problem.

Assume the problem can be reformulated as a binary logistic regression where $\lambda_0(s)$ drops out of the likelihood. Denote the control disease locations as $\{s_j\}$, $j = m+1,...,m+n$, and with $N = n+m$, a binary indicator function can be defined:

$$y_i = \begin{cases} 1 \text{ if } i \in 1,..,m \\ \\ 0 \text{ otherwise} \end{cases}$$

$$\forall i, \ i = 1,...,N$$

and the resulting likelihood is given by:

$$L(\mathbf{s}|\theta) = \prod_{i=1}^{N} \frac{[\lambda_1(s_i)]^{y_i}}{1 + \lambda_1(s_i)}.$$

By conditioning on the joint set of cases and controls, the resulting logistic likelihood does not require the evaluation of a spatial integral nor the estimation of a background population function. The definition of the form of $\lambda_1(s_i)$ will be important in inference concerning putative sources of hazard.

Parametric Forms

Often we can define a suitable model for excess risk within $\lambda_1(s)$. In the case where we want to relate the excess risk to a known location (e.g. a putative source of pollution) then a distance-based definition might be considered first. For example,

$$\lambda_1(s) = \rho \exp\{\mathbf{F}(s)\alpha + \gamma d_s\} \tag{8.1}$$

where ρ is an overall rate parameter, d_s is a distance measured from s to a fixed location (source), γ is a regression parameter, $\mathbf{F}(s)$ is a design matrix with columns representing spatially-varying covariates, and α is a parameter vector. The variables in $\mathbf{F}(s)$ could be site-specific or could be measures on the individual (age, gender, etc.). In addition this definition could be extended to include other effects. For example, we could have

$$\lambda_1(s) = \rho \exp\{\mathbf{F}(s)\alpha + \eta v(s) + \gamma d_s\} \tag{8.2}$$

where $v(s)$ is a spatial process, and η is a parameter. This process can be regarded as a random component and can include within its specification spatial correlation between sites. One common assumption concerning $v(s)$ is that it is a random field defined to be a spatial Gaussian process.

An example of the kind of specification typical in a putative source example would involve a range of variables or functions of variables thought to be indicative of risk association with the source. The variables included depend on the context. In retrospective studies where no information or direct measures of emission patterns are available then resorts must be made to exposure surrogates (i.e. variables that may show a retrospective linkage with the source). Distance from source is a prime example of a variable that might yield such information. Direction from source to residence may also be indicative of wind-related effects (particularly in air pollution studies). For prospective studies, direct measures of pollutant outfall (such as soil-sampled or air-sampled chemical or particulate concentrations) could be monitored over time. Without these direct measurements, surrogates would be required and often these would have to represent historical time-averaged effects in retrospective studies.

What form would a relevant exposure model take? The definition for $\lambda_1(s|\theta)$ often assumed is as follows (see Diggle, 1990; Diggle and Rowlingson, 1994; Lawson, 1995; Diggle et al., 2000; Wakefield and Morris, 2001; Lawson, 2006b for variants):

$$\lambda_1(s_i|\theta) = \exp\{A_{1i}\}.\exp\{A_{2i}\}.A_{3i} \tag{8.3}$$
$$A_{1i} = x_i'\beta + z_i'\gamma$$
$$A_{2i} = \rho_1 \cos(\phi_i) + \rho_2 \sin(\phi_i).$$
$$A_{3i} = [1 + \alpha_0 e^{-\alpha_1 d_i}]$$

$\theta = \{\beta, \gamma, \alpha, \rho\}, \alpha = \{\alpha_0, \alpha_1\}, \rho = \{\rho_1, \rho_2\}$. Here the distance variable is defined as $d_i = ||s_i, c||$ where c is the putative source location and the angle to the source is defined as ϕ_i. A generalization allows multiple sources and we can include these in one model by adding further distance or direction variables with parameters. This is not pursued here. The rationale for each of the terms (A_1, A_2, A_3) is as follows. All terms are exponentiated to ensure positivity, although term A_{3i} has a link parameter (α_0) which requires a constraint so that $\alpha_0 e^{-\alpha_1 d_i} < 1$. The term A_{1i} consists of covariates and random effects. The row vector of covariates (x_i') can consist of personal covariates although within a logistic likelihood model these would have to be available for the control as well as case disease. The corresponding regression parameters are the vectors β. The covariates can include functions of Cartesian coordinates for trend estimation and these are available for all locations. The individual level random effects can be included via the row vector z_i' with the corresponding unit vector γ. These effects could include individual frailty terms (with, for example, zero mean Gaussian prior distributions) or correlated effects where the prior distribution includes some form of spatial correlation.

The general specifications above in Equations (8.1) and (8.2) demonstrate a variant of this specification. The term A_{2i} specifies the directional dependence in the outcome. By including functions of the trigonometric functions (cos, sin) it is possible to recover the mean angle of the exposure. In this case only linear functions are assumed.

More complex variants are possible (see e.g. Lawson, 1993b) that allow for angular distance correlation or peaked distance effects. An alternative specification for the angular effect could be $A_{2i} = \rho_1 \cos(\phi_i - \mu_0)$. Here the ρ_1 plays the role of an angular concentration parameter and the angle is measured relative to an overall mean (μ_0). If a predominant time-averaged wind direction is found to affect a source then the estimation of μ_0 might be important in determining a link to a source. Finally, the term A_{3i} defines the distance effect.

The rationale for the hybrid-additive form, $[1 + \alpha_0 e^{-\alpha_1 d_i}]$, is the idea that risk at distance from the source should not affect the background disease risk. If a multiplicative model were assumed (such as $A_{3i} = e^{-\alpha_1 d_i}$) this would lead to a reduction in risk at great distances which is not appropriate. It should be mentioned however that often it is much more difficult to estimate α_0, α_1 under the hybrid-additive model as the parameters are not well identified and constraints must be placed on $\alpha_0 e^{-\alpha_1 d_i}$ (see also Ma et al., 2007).

More details of possible model variants are given in Lawson (2006b) chapter 7. Step function forms have been proposed by Diggle et al. (1997), but the underlying rationale for these, that there could be a zone of constant risk around a source, is not borne out by dispersal models or empirical studies of source dispersion. On the other hand, peak-decline models are supported by time-averaged dispersal models (see e.g. Arya, 1998). A simple example of this general approach is given in Wakefield and Morris (2001), albeit for an aggregated small area application. In that work $A_{2i} = 0$ with no assumed directional effects, and $A_{1i} = \beta_0 + \beta_1 x_{1i} + u_i + v_i$ is a single deprivation index covariate (x_{1i}) with two random effects (one correlated u_i and one uncorrelated v_i). The third term is defined as $A_{3i} = [1 + \alpha_0 e^{-(d_i/\alpha_1)^2}]$ which gives a Gaussian distance rather than exponential effect. However the comments above also apply to this model form.

A general specification for the logistic example applied to the larynx cancer data has been specified in Section 6.2.1.2. In that section the focus was on cluster detection. However the underlying model used there is also relevant here. The model assumed for the case probability was:

$$p_i = \frac{\lambda(\mathbf{s}_i|\theta)}{1 + \lambda(\mathbf{s}_i|\theta)}$$
$$\lambda(\mathbf{s}_i|\theta) = \exp\{\beta_0 + v_i\}.\{1 + \exp(-\alpha_1 d_i)\}$$

where d_i is distance from the incinerator, β_0 is an intercept term, and v_i $\sim N(0, \tau_v)$ an uncorrelated random effect, and zero mean Gaussian prior distributions for the β_0 and α_1 parameters. There is no directional term. In this case this is justified given the choice of study area: a rectangle with the putative source close to one region boundary. This means that much of the directional data is censored (outside the boundary of the region). Hence there is limited use for including a directional model here. A correlated random effect could also be included within this model, though this is not reported here.

For case-control data this is possible either by assuming a full multivariate Gaussian process prior distribution for the correlation (with covariance specified as a function of inter-point distances). It is also possible to specify neighborhoods via the construction of a Dirichlet tesselation of the complete realization and the derivation of tile neighbors. Care must be taken in this latter case to avoid edge effects, although these should not be great for the definition of neighborhoods (rather than distances). In our example the posterior expected estimates of β_0 and α_1 (with *sds* in brackets) were -6.35 (0.831) and 0.695 (3.054). Hence, in this example the overall rate was well estimated whereas the distance effect was not. The posterior expected estimate of the precision of the uncorrelated random effect was 0.1229 (0.07413) in the model where $v_i \sim N(0, \tau_v)$ and $\tau_v = a^2$ where $a \sim U(0, 100)$ following the suggestion of Gelman (2006).

Some comments concerning analyses of putative source data should be made in light of the general discussion above concerning ecological bias, confounding, contextual effects, and measurement error. While some of these comments are most appropriate to the aggregated data situation, we discuss many issues here which are common to both.

First of all, it is important to critique the model components included above. Should correlated random effects be included? Would they absorb the effects of unobserved confounders? In general it may be important to include both uncorrelated and correlated effects, from the standpoint that confounders could induce noise effects of both kinds. However it should be borne in mind that confounders correlated with the distance or directional effects are not likely to be removed by random effect inclusion.

The inclusion of variables in the analysis that inform about context could also be important. For example deprivation indices available at a level aggregated above residence (such as at census tract or ZIP code level) could help to inform about regional excess risk. Of course deprivation might be correlated with distance or direction. In Wakefield and Morris (2001) this was certainly true. In the same work, LOESS smoothers of the distance effect suggest that an irregular decline occurs and it may be more appropriate to consider spline models for the distance and or direction. In fact a 2-D spline model for the distance and directional effect could be a useful inferential tool. Of course splines may not yield unequivocal evidence for a risk gradient. The possibility that ecological bias exists in aggregate data will be discussed in the next section.

The possibility that measurement error (ME) exists in outcomes or covariates can also be important. For example misdiagnosis could occur where a control could in fact be a case or vice versa. This would be more likely if the two diseases were linked by progression. For example, early stage breast cancer could be used as a control for late stage breast cancer. Clearly the staging could be subject to misclassification. ME could exist in any covariates whether the location of an address, the socio-economic status of an individual or the deprivation status of a region. One solution for covariates is to assume Berkson error and a model such as

$$\beta_1(x_{1i} + \varepsilon_i)$$
$$\varepsilon_i \sim N(0, \tau_\varepsilon)$$

where x_{1i} is a covariate or utilize the classical ME approach and specify a joint model for the covariate and the outcome.

8.2 Aggregated Count Data

It is often relevant or feasible to consider the analysis of count data within aggregated spatial units (small areas). These units will usually be arbitrary political administrative units (e.g. census tracts, ZIP codes, counties, municipalities, postal zones, etc.). The definition of these units should have little or no impact on the health outcome observed.

Assume we observe counts $\{y_i\}$, $i = 1, ..., m$ in m small areas and we also observe expected rates $\{e_i\}$, $i = 1, ..., m$. While we usually assume the expected rates to be fixed for our purposes, it could be useful to consider them to be random quantities also (see e.g. Best and Wakefield, 1999). Here we mainly focus on fixed expected rates.

A typical model at the data level is often

$$y_i \sim Pois(\mu_i)$$
$$\mu_i = e_i \theta_i$$

and the focus is on the modeling of the relative risks $\{\theta_i\}$. Usually the log relative risk is the focus and we often formulate a model akin to that in Equation (8.3) where the i th small area is "located" at its centroid. Of course this assumes an average effect over the small area rather than direct modeling of the risk aggregated from the point process model. Direct aggregation from a Poisson process would give $y_i \sim Pois(\int_{a_i} \lambda(\mathbf{u}|\theta)d\mathbf{u})$, where a_i is the physical extent of the i th small area. Now if both $\lambda_0(s)$ and $\lambda_1(s|\theta)$ were constant over the area (a strong assumption) the result would be $\lambda_{0i}.\lambda_{1i}.|a_i|$ (where $|.|$ denotes "area of") which is almost the same as $e_i\theta_i$ barring the area effect. The expected rate is usually standardized over the population rather than area. However if you make the (strong) assumption that the population is uniform, of course if the e_i is specified for the local population, the assumption is that $\lambda_{0i}|a_i| \approx e_i^*n_i = e_i$ where e_i^* is the externally standardized unit population rate. This *decoupling approximation* as it is called is often made as the starting point of an analysis. It is not usually unreasonable when non-spatial region-specific covariates are included but it can be important when spatially-dependent covariates (such as interpolated pollution measures) are involved.

Making the simple assumption of $\mu_i = e_i\theta_i$, we can specify the general model as

$$\theta_i = \exp\{A_{1i} + A_{2i} + \log(A_{3i})\}$$
$$A_{1i} = x_i'\beta + z_i'\gamma$$
$$A_{2i} = \rho_1 \cos(\phi_i) + \rho_2 \sin(\phi_i)$$
$$A_{3i} = [1 + \alpha_0 e^{-\alpha_1 d_i}]$$

where the i th small area is located at the centroid or other suitable associated point, d_i is the distance from the centroid to the source location, and ϕ_i is the angle from the centroid to the source location. Often it is assumed that $A_{1i} = \beta_0 + u_i + v_i$ where the typical convolution model with a CAR prior distribution of Chapter 5 is assumed:

$$u_i|u_{-i} \sim N(\overline{u}_{\delta_i}, \tau_u/n_{\delta_i}).$$
$$v_i \sim N(0, \tau_v)$$

With small area data and neighborhoods defined, a CAR is a convenient and reasonable assumption. An alternative specification could be of the form of a full multivariate Gaussian with a covariance matrix, thus

$$\mathbf{u} \sim \mathbf{N}(\mathbf{0}, \sigma^2\mathbf{\Gamma})$$

where the i, j th element of the covariance matrix is $\gamma_{i,j} = \exp(-d_{ij}/\phi)$. This has the advantage of directly modeling distance effects, has a distance-dependent covariance and also has a zero mean vector, and so models a stationary process.

The CAR specification is not stationary and can suffer from aliasing with long range spatial effects (see e.g. Ma et al., 2007). One option is to use a proper CAR with trend specification. Unfortunately, the full MVN specification requires inversion of an $m \times m$ covariance matrix whenever new parameters are evaluated, e.g. within a posterior sampling algorithm. This could be a major computational disadvantage.

In the example below, respiratory cancer incidence for 1988 in the counties of Ohio was examined. The US Department of Energy Fernald Materials Processing Center is located in southwest Ohio (Hamilton County). The Fernald facility recycles depleted uranium fuel from the US Department of Energy and Department of Defense nuclear facilities. The facility is located 25 miles north west of Cincinnati. The recycling process can create a large amount of uranium dust which is radioactive. The period of greatest emission activity was between 1951 and the early 1960s when some dust may have been accidentally released into air.

Respiratory cancer is of interest in relation to a potential environmental health hazard. Exposure to radioactive contaminated air in the vicinity of a facility could, over a period of years, lead to increased risk for a variety of diseases. Exposure risk can be considered to be increased if residence were proximal to the facility during the highest activity years or in subsequent decades.

One disease of concern to evaluate would be respiratory cancer as it is the most prevalent form of cancer potentially associated with this exposure. An exposure pathway via inhalation would be considered. Data is available for Ohio counties for 1988 counts of respiratory cancer. This period is sufficiently lagged from the peak emission time and sufficient cancer lag time (20 to 25 years) should have passed. The expected rates used for standardisation are the Ohio state data for age a gender breakdowns of each county. Two covariates at the county level are also available.

The first covariate is the percent poverty for each county from the 1990 census. The 1990 census is used as it is the nearest to the year in question and the level should remain reasonably stable over two years. This covariate would be useful in allowing for deprivation effects that could confound the respiratory cancer outcome. They may include general health outcomes but also behavioral effects such as smoking or use of alcohol in lifestyle. The second covariate is the simplest exposure surrogate variable: distance from the site. This distance was computed to the centroids of the counties. A sequence of models was fitted to these data with different assumptions. First a basic model with a convolution prior distribution for spatial effects and measurement error for both covariates was considered with the form:

$$y_i \sim Poiss(e_i.\theta_i)$$
$$\theta_i = \exp\{\alpha_0 + \alpha_1(x_{1i} + \epsilon_{1i}) + \alpha_2 \log(f_i) + u_i + v_i\}$$
$$f_i = 1 + \exp\{-\alpha_3(d_i + \epsilon_{2i})\}.$$

The random effects $\epsilon_{1i}, \epsilon_{2i}$ have zero mean Gaussian prior distributions with standard deviations with uniform distributions on the range 0, 10 (Gelman, 2006). These represent Berkson error in the covariates. Different prior distributional assumptions for the random effects in the convolution component were also considered. The first option (model 1 in Table 8.1) was with fixed but

Model	DIC	α_0	α_1	α_2	α_3
1 Fixed precisions	656.23	-0.506 (0.275)	0.111 (0.002)	-2.043 (0.065)	2.296 (0.063)
2 Variance hyperpriors	520.49	-0.362 (0.108)	0.030 (0.009)	-0.242 (0.279)	4.993 (5.252)
3 No ME	616.75	-0.935 (0.350)	0.034 (0.006)	0.526 (0.400)	-0.156 (0.180)
4 No distance	619.35	-0.490 (0.074)	0.034 (0.006)	-	-
5 Proper CAR	557.03	-0.817 (0.059)	0.061 (0.006)	0.024 (0.004)	-6.204 (1.072)

Table 8.1: Results for a variety of models fitted to 1988 respiratory cancer incident counts for counties of Ohio

very small precisions (0.0001) for the uncorrelated random effects $(v_i, \epsilon_{1i}, \epsilon_{2i})$ and second with variance hyperprior distributions $(\tau_* = \sigma_*^2; \sigma \sim U(0, 10))$.

Also considered was a variant of the CH effect: a proper CAR model (Section 5) with $c_{ij} = \frac{1}{n_{\delta_i}}$ if $i \sim j$ and $c_{ij} = 0$ if $i \not\sim j$. This model with no measurement error yielded $\widehat{\gamma} = 0.729(0.142)$, $\widehat{\tau}_{pc} = 8902.0(26010.0)$ with a DIC of 557.03 for the model with fixed precision on the UH effect (0.0001). Overall, different precision specifications seem to affect the model fits considerably in that model 2 is much superior to model 1. The inclusion of ME appears also to be important as is the distance effect, even when it is not well estimated (model 4). Interestingly, and as a caution, model 1 yields a significant distance effect $(\widehat{\alpha_1}, \widehat{\alpha_2})$ and supports a possible source effect. However amongst the models fitted, the lowest DIC is for the model with a CAR component and precision hyper-prior distributions, with measurement error, where the distance effect is not significant.

Other analyses of these data, especially in the more general space-time context, are found in Xia et al. (1997), Waller et al. (1997), Carlin and Louis (2000), Knorr-Held and Besag (1998), and Knorr-Held (2000). Measurement error was considered by Xia et al. (1997) in a space-time context.

8.3 Spatio-Temporal Effects

When data are observed with a time label then it is possible to extend modeling by considering spatio-temporal effects. Perhaps the most convenient way to do this is to consider a breakdown between *main* effects of space and time separately and the interaction between space and time.

In Section 12 a more general review of disease mapping models is made. Here we briefly consider how space-time data can be modelled with putative sources of hazard as the main focus. The extension of methods for spatial applications to where we have data observed in space and time is immediate.

8.3.1   Case Event Data

Assume we observe within a study region (W) and a time period (T), a set of m cases, with residential addresses given as $\{s_i\}, i = 1, ..., m$, and also time labels $\{t_i\}, i = 1, ..., m$. Here the random variables are the *spatial location and the time of occurrence*, and so we must employ models that can describe the distribution of locations and times. Recent reviews of a wide range of approaches to space-time point process data appear in Diggle (2007) and Diggle and Gabriel (2010). Time here could be a diagnosis date, date of death, or cure. The heterogeneous Poisson process (hPP) model assumed for spatial data can be extended to space-time readily. In this model, the distribution of the cases (points and times) is governed by a first order intensity function. This function, $\lambda(s,t)$, describes the variation across space and time of the intensity of cases. This function is the basis for modeling the spatio-temporal distribution of cases. Denote this model as

$$\mathbf{PP}(\lambda(s,t)).$$

As in spatial applications, the unconditional likelihood associated with this model is given, barring a constant, by:

$$L = \prod_{i=1}^{m} \lambda(s_i, t_i) \exp\{- \int_W \int_T \lambda(\mathbf{u}, \mathbf{v}) d\mathbf{u} d\mathbf{v}\}$$

where $\lambda(s_i, t_i)$ is the first order intensity evaluated at the sample locations $\{s_i, t_i\}$. This likelihood involves an integral of $\lambda(\mathbf{u}, \mathbf{v})$ over the study region and time period.

The definition of the intensity of cases must make allowance for the effect of the background at-risk population, which in this case will be time-varying. Often the intensity is specified with a multiplicative link between these components:

$$\lambda(s,t) = \lambda_0(s,t)\lambda_1(s,t|\theta).$$

Here the *at-risk* background is represented by $\lambda_0(s,t)$ while the modeled excess risk of the disease is defined to be $\lambda_1(s,t|\theta)$, where θ is a vector of parameters. As before a likelihood model can be derived from this likelihood and Bayesian methods could be based on this form. An example of using this form in cluster detection was given by Clark and Lawson (2002). The disadvantage of using this form is the need to integrate the intensity over space and time. This can be avoided if a control disease were available within the study region over the same time period. Once again the conditional logistic model could be derived. Assume that a control disease is governed by intensity $\lambda_0(s,t)$. The joint realization of case and control diseases for a Poisson process with intensity is $\lambda_0(s,t)[1+\lambda_1(s,t|\theta)]$. Then by conditioning on the joint realization, the binary labeling of the points will be governed by the case probability

$$p_i = \frac{\lambda_1(s_i,t_i|\theta)}{[1+\lambda_1(s_i,t_i|\theta)]}.$$

Given the set of locations, the point labels (y_i) can be considered at the data level to be independently distributed with a binomial distribution:

$$y_i \sim bin(1,p_i).$$

This is, again, just a logistic model for the binary outcome, where the probability is a function of space and time. Interest will focus on the definition of the excess or relative risk function $\lambda_1(s_i,t_i|\theta)$. The specification of $\lambda_1(s_i,t_i|\theta)$ will depend on the context and it is important to include covariates, ideally time-varying, as well as random effects. Variates that pertain to evidence for a link to a putative source of hazard may vary. First a general formulation could be as follows:

$$\lambda_1(s_i,t_i|\theta) = \exp\{A_{1i} + A_{2i} + A_{3i}\}$$
$$A_{1i} = \mathbf{x}_i'\beta + \mathbf{z}_i'(t)\gamma$$
$$A_{2i} = f(d_i, \phi_i, t_i)$$
$$A_{3i} = \Sigma_i + \xi_i + \psi_i.$$

Within A_{1i} are terms depending on fixed constant covariates ($\mathbf{x}_i'$) and their parameters β, and also terms depending on time-varying covariates $\mathbf{z}_i'(t)$ and their parameters γ. Time-varying covariates could be very important in these studies. For example in a prospective study, if pollutant concentration were available at different times the data would be time-varying. For term A_{2i} functions of distance to source (d_i) and angle to source (ϕ_i) could be important (as in the spatial case). However because a source may vary its output over time, the resulting spatial risk field would vary over time, and so time-varying effects should be included in A_{2i}. The final term includes random effects that can allow for spatial (χ_i), temporal (ξ_i), and spatio-temporal interaction (ψ_i). Note that in all analyses, covariates would have to be available for all case and control locations. Special methods would have to be developed when this is not the case.

An example of a possible model for a time-varying emission source (air pollutant) could be, for data given with polar coordinates (ϕ_i, d_i):

$$\lambda_1(s_i,t_i|\theta) = \exp\{\beta_0 + \beta_1 x_{1i} + \beta_2 x_{2i} + f_i(t) + \chi_i\}$$

where x_{1i} is the age of the person, x_{2i} is the socio-economic status of the person, and

$$f_i(t_i) = \log(1 + \alpha_0(t_i)\exp\{-\alpha_1(t_i)d_i\}) + \kappa(t_i)\cos(\phi_i - \mu(t_i))$$

$$\alpha_0(t_i) \sim Gamma(c_0\mu(t_i), c_0)$$
$$\alpha_1(t_i) \sim N(\alpha_1(t_{i-1})/\Delta(t_i, t_{i-1}), \tau_{\alpha_1})$$

where $\mu(t_i)$ could be defined as a time-varying risk function for example, $\alpha_1(t_i)$ is a form of Gaussian process and $\Delta(t_i, t_{i-1})$ is the time difference between the i th and the previous case/control. Note also that the directional component with precision $\kappa(t_i)$, and mean angle $\mu(t_i)$ will also in general vary with time. Essentially the time averaging that is assumed for a static spatial model must be dropped here in favor of a parsimonious dynamic model. While of course a convolution of Gaussian distributions could be employed for a directional component around a source (see e.g. Esman and Marsh, 1996, Arya, 1998), this is not parsimonious compared to a Von Mises-type formulation such as $exp\{\kappa(t_i)\cos(\phi_i - \mu(t_i))\}$.

The final component of the risk function could consist of spatial and temporal random effects and interaction effects. Care should be taken in the choice of such effects as the ability to detect exposure effects may depend on the specification of the random components. First we could consider a separate spatial component, such as χ_i where spatial dependence (fixed in time) could be specified via a Gaussian process specification with a distance-based spatial covariance (i.e. $\chi \sim \mathbf{MVN(0, \Gamma)}$ where $\Gamma_{ij} = \sigma^2 \exp\{-\alpha d_{ij}\}$. Further CAR alternatives could be considered if a suitable neighborhood structure were assumed. Temporal effects could be assumed whereby a conditional autoregressive Gaussian dependence is defined on the time lag between events:

$$\xi_i \sim N(f(\xi_{i-1}), \tau_\xi)$$
$$f(\xi_{i-1}) = \alpha\xi_{i-1}/\Delta(t_i, t_{i-1}).$$

Finally, a space-time interaction could be assumed. Various specifications could be imagined for this ranging from non-separable dependence structures (see e.g. Knorr-Held, 2000, Gneiting et al., 2007) to independent effects. The simplest and most parsimonious form might be

$$\psi_i \sim N(0, \tau_\psi).$$

This would at least ensure that aliasing between covariate effects varying over time would be minimized.

Finally it should be noted that often the binary outcome in space-time is an "end-point" event, say, for example, in an infectious disease situation where infection spreads within a finite population. In that case special survival-based methods can be used to examine the progression of the disease (Lawson and Leimich, 2000, Lawson and Zhou, 2005). Examples of space-time analysis around sources of pollution are few and this area is one that could be much further developed.

8.3.2 Count Data

In the situation where small area counts are recorded within fixed time periods in a sequence, then the modeling approach is a relatively straightforward extension of the spatial case. Define the counts of disease within $i = 1, ..., m$ spatial small areas and $j = 1, ..., J$ disjoint and adjacent time periods as $\{y_{ij}\}$. The corresponding expected rates with these space-time units are $\{e_{ij}\}$. Also assume relative risk parameters for each unit: $\{\theta_{ij}\}$. The basic data model is often again Poisson with

$$y_{ij} \sim Pois(e_{ij}.\theta_{ij}).$$

Inference focuses on terms within the specification of θ_{ij}. Also assume that the distance and direction (angle) from a source is known and can be computed as (d_i, ϕ_i). Assume a log linear form:

$$\theta_{ij} = \exp\{A_{1i} + A_{2j} + A_{3ij}\}.$$

Here the terms have explicit spatial (i), temporal (j), and interaction (ij) labels. An example of a typical specification could be:

$$A_{1i} = f(\mathbf{x}_i'\beta) + u_i + v_i \tag{8.4}$$
$$A_{2j} = \xi_j + g(\alpha_j d_i) + \kappa_j \cos(\phi_i - \mu_{\phi j}) \tag{8.5}$$
$$A_{3ij} = \psi_{ij}.$$

The first term includes fixed covariates within small areas (including distance and direction), and so $f(\mathbf{x}_i'\beta)$ could include functions of fixed areal covariates (poverty, socio-economic data, distance, direction) whereas u_i, v_i could be the usual CH and UH random effects (see e.g. Heisterkamp et al., 2000 for an early example). The second term has the temporally-dependent components. The random effect (ξ_j) often has an autoregressive dependence. The term $g(\alpha_j d_i)$ would be a function of the time-dependent parameter α_j which relates to distance. Again an autoregressive dependence could be assumed for this. The directional parameters κ_j and $\mu_{\phi j}$ also can have dependence on previous times. Time variation of output from sources may be modeled in this way. Finally the interaction a prior independence or can have prior non-separable structure (Knorr-Held, 2000).

In the example that follows I applied a general model to the variation over 10 years (1979 through 1988) of respiratory cancer in Ohio. I have assumed a general model of the form:

$$y_{ij} \sim Pois(e_{ij}.\theta_{ij})$$
$$\theta_{ij} = \exp\{A_{1i} + A_{2j} + A_{3ij}\}.$$

Here, we assume no directional effect as the spatial scale of the county level data is quite large and it is unlikely that a directional effect could be manifest

Model	DIC	α_0	α_1	α_2	α_3
1	5762.6	-0.393(0.082)	129.8(22.8)	0.003(3.69E-4)	0.047(0.072)
2	5759.8	27.16(0.086)	-362.4(3.14)	0.101(4.36E-4)	0.089(9.35E-4)
3	5739.9	-0.625(0.063)	1	1	-

Table 8.2: Ohio respiratory cancer (1979 through 1988): putative source model fits

at this scale. We also assume that a distance effect could still remain even via occupational exposure and so we model this here:

$$A_{1i} = \alpha_0 + \alpha_1 \log[1 + \alpha_2 \exp\{-\alpha_3 d_i\}] + u_i + v_i$$
$$A_{2j} = \xi_j$$
$$A_{3ij} = 0$$

with

$$u_i | u_{-i} \sim N(\overline{u}_{\delta_i}, \tau_u)$$
$$v_i \sim N(0, \tau_v)$$
$$\xi_j \sim N(\xi_{j-1}, \tau_\xi).$$

The variance parameter (τ_*) distributions are assumed to be defined with $\sqrt{\tau_*} \sim U(0, 10)$. The regression parameters are all assumed to have zero mean Gaussian distributions with large variances $(1/0.00001)$.

Alternative models have been considered. First the addition of $A_{3ij} = \psi_{ij}$ with $\psi_{ij} \sim N(0, \tau_\psi)$ was examined (model 2). Finally, temporal dependence in the regression parameters was considered. Specifically, an autoregressive prior distribution was assumed for α_3 of the form $\alpha_{3j} \sim N(\alpha_{3j}, \tau_3)$ which leads to

$$A_{1ij} = \alpha_0 + \alpha_1 \log[1 + \alpha_2 \exp\{-\alpha_{3j} d_i\}] + u_i + v_i.$$

Table 8.2 displays the results of the fitting process. The best fit overall is model 3 with no space-time interaction with fixed α_1, α_2 and temporally dependent α_{3j}.

Figure 8.2 displays the estimated temporal random effect (ξ_j) and 95% credible interval for model 1. Figure 8.3 displays the results for the same effect for model 2 with zero-mean Gaussian space-time interaction.

Figure 8.4 displays the posterior averaged temporally-dependent distance regression effect for model 3. Figure 8.5 displays the corresponding posterior averaged time-dependent random effect for all years for model 3. Note that in model 3 it was necessary to fix $\alpha_1 = \alpha_2 = 1$, due to the identifiability issues when time dependence is allowed for these parameters. It is clear from the limited number of models fitted here, that a time-varying regression on a simple model of distance is considerably better (in terms of DIC) than constant parameters. The time-dependent distance effect, α_{3j}, remains well

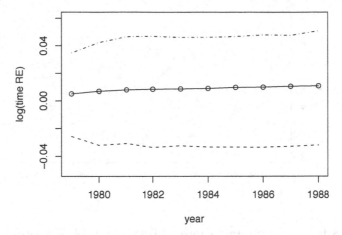

FIGURE 8.2: Ohio respiratory cancer, 1979 through 1988. Estimated temporal random effect with 95% credible interval for model 1.

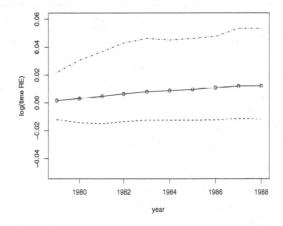

FIGURE 8.3: Ohio respiratory cancer, 1979 through 1988. Posterior average temporal random effect with 95% credible interval for model 2 with zero-mean Gaussian interaction.

estimated under this model whereas the temporal random effect is negligible. Model 3 did not include an interaction term and it would also be interesting to examine the effect of inclusion of such a term, though this is not pursued here. Note also that for many applications it would also be important to include a directional effect (possible time-varying) in the model (such as in (8.5)). Finally we have not presented the mapped output for the posterior averaged spatially-expressed random effects in this model (CH and UH). These may be

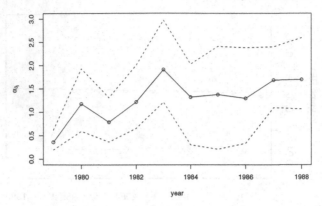

FIGURE 8.4: Ohio respiratory cancer, 1979 through 1988. Space-time model with time-dependent distance effects. Plot of posterior average distance effect over years with 95% credible interval.

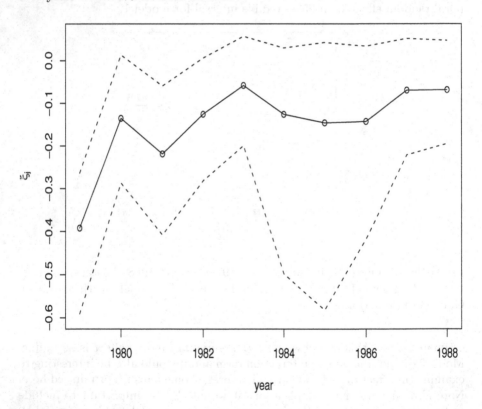

FIGURE 8.5: Ohio respiratory cancer, 1979 through 1988. Posterior average time random effect (ξ_j) with 95% credible interval.

of interest for the examination of unusual aggregations of risk as they appear or disappear over time. Of course, Bayesian residuals, predictive residuals, or even exceedance probabilities can be computed for space-time models in the form $q_{ij} = \widehat{\Pr}(\theta_{ij} > 1) = \sum_{k=1}^{K} I(\theta_{ij}^k > 1)/K$ where $\{\theta_{ij}^k\}$, $k = 1, ..., K$ denotes the posterior sampled values of the relative risk for each region and time period. Residual maps or maps of q_{ij} could also be very informative. Of course, as noted earlier, the reliability of q_{ij} heavily depends on the correctness of the model.

The main emphasis in this section has been in demonstrating the modeling of spatio-temporal effects when time is included. More recent developments have seen the estimation of unobserved or latent operating periods in space-time (Al-Hadhrami and Lawson, 2011) and other space-time effects (Blangiardo et al., 2011). Other issues that are not addressed here, but could be important are measurement error in covariates or outcomes, ecological bias when making inferences at lower aggregation levels from space-time data, and contextual or confounder effects.

9

Multiple Scale Analysis

The spatial analysis of single diseases is often sufficient. However, in some applications there is a need to consider different scales of aggregation within an analysis.

One such situation arises when it is of interest to consider a relationship at different aggregation levels. For example, if the relation of an outcome at county level to a covariate is examined, will the relationship hold true at lower aggregation levels (e.g. census tract) or at higher levels (e.g. state or country)? In general, it is unlikely that this would be the case as, if it were, there would be little need to consider different levels of analysis. In fact ecological bias would not occur. Scale change issues are often known as the modifiable areal unit problem (MAUP), whereby modification of the areal units could lead to different inferences. In geostatistics, this is known as the change of support problem (Cressie, 1996, Gelfand, 2010, Banerjee et al., 2014).

9.1 Modifiable Areal Unit Problem (MAUP)

The MAUP can be considered to include a variety of special cases. One of these is ecological bias (seen in Chapter 7.7) where the issue is whether inference can be made at a lower level of aggregation (individual level usually) from aggregate data. For example, can we make inferences from county or region level analysis to the individual level? We saw in that case the various aspects of this problem that include measurement error, knowledge of the within area distribution of exposures, contextual effects, and unobserved confounding.

9.1.1 Scaling Up

By scaling up, I mean trying to make inferences at a higher aggregation level than that used in the analysis. In general, aggregation leads to smoothing or averaging of data. For example, a spatial process is present at location s, $z(s)$ say, and when observed over a larger area A, the process will be a smoothed version. i.e. $z(A) = \int\limits_A z(u)du$. The integration is with respect to

the extent of A. Note that the mean of the process in A can be defined as

$\mu(A) = \int\limits_A z(u)du/|A|$, where $|A| = \int\limits_A du$. This can be regarded as an average

over the area. Note that this integration leads to a reduction in variability, and so at the aggregate level we would expect less variability. Cressie (1993), Section 5.2, noted this aspect in a geostatistical context. In an analogy with GIS operations, this is equivalent to zooming out in a map operation. One problem that this leads to is that processes operating at different aggregation levels may appear, or become important, at different scales. The possibility that the process observed at different scales will behave differently is clear. This suggests that "scale labelling" is useful when dealing with changes in support or aggregation. By scale labelling, I mean the allocation of a scale of operation of a process. Methods for incorporation of scale effects within a Bayesian analysis of small area health data are various.

First, it is clear that it is possible to consider aggregated variables as confounders within an analysis. Multilevel modeling (Leyland and Goldstein, 2001) often addresses the issue of multiple levels within an analysis and these can include spatially-aggregated covariates. This of course includes contextual effects as a primary example (see Section 7.8.1.3; also Goldstein and Leyland, 2001).

In general, the scaling up of health outcomes has been described for individual (point process) to small area (count) levels in previous chapters. Denote a scale level (integer) variable: $l_k, k = 1, ..., K$ where K is the number of levels. It is assumed here that aggregation levels can be discretized into such levels. Hence, an aggregation involves an outcome model indexed by the level: $f_k(y_{ik}; \mu_{ik}, l_k)$. Here, y_{ik} $i = 1, ..., m$ is the outcome variable for m units. (Assume here that this could be binary or a count or less often continuous.) Note that at different levels there could be different models and so subscripted f_k is appropriate. We need to establish the relation between levels of the scaling. Given a set of levels it is tempting to consider a general model formulation which links levels in the analysis. Assume that all units are aligned and that k is ranked from lowest to highest aggregation level. Define the set alignment as follows: there are m_k regions at the k th level and for $k = 2, .., K$ there are m_{k-1} regions at the lower aggregation. The allocation of the regions at m_{k-1} to the m_k regions is defined by $S_{i,k}$ which is the set of regions at the $k - 1$ level uniquely within the i th region at the k th level. For count data this would mean that $y_{ik} = \sum\limits_{l \in S_{i,k}} y_l$, $y_{ik-1} = \sum\limits_{l \in S_{i,k-1}} y_l$, ...for $k = 2, ..., K$. For example, we could specify a vector model of the form

$$
\mathbf{y} = \left\{ \begin{array}{l} \{y_{i1}\}, i = 1, ..., m_1 \\ \{y_{i2}\}, i = 1, ..., m_2 \\ \{y_{i3}\}, i = 1, ..., m_3 \\ . \\ . \\ \{y_{iK}\}, i = 1, ..., m_K \end{array} \right\} \sim \left\{ \begin{array}{l} f_1(\mu_1, l_1) \\ f_2(\mu_2, l_2) \\ . \\ . \\ . \\ f_K(\mu_K, l_K) \end{array} \right. \quad .
$$

Often the distribution at each level is the same and so $\mathbf{y}_k \sim f(\mu_k, l_k)$. An example of this approach is given in Section 9.1.3. Often the nesting of the data leads not only to a single distribution but also to a single likelihood. For example with nested Poisson counts then

$$
L(\mu_k | \mathbf{y}_k) = \prod_1 f(\mathbf{y}_1; \mu_1)....\prod_K f(\mathbf{y}_K; \mu_K).
$$

Linkage between the scales can be achieved by dependence between $\mu_1....\mu_K$. Of course for functions of counts with aggregation then the rates should sum across units within each cell and so $\mu_K = \sum_{l \in S_{i,k-1}} \mu_l$. Louie and Kolaczyk (2006) give an example of multiple scale analysis for disease data.

9.1.2 Scaling Down

By scaling down, I mean trying to make inferences at a lower aggregation level than that used in the analysis. The classic situation where ecological bias arises is an example of this scaling down: trying to make inference at the individual level from aggregate level analysis. Disaggregation is the reverse operation from aggregation and the parallel with the mathematical operation of integration carries over, so disaggregation is equivalent to differentiation. Hence the opposite of smoothing would be to add noise to an existing field. Plummer and Clayton (1996) essentially assume the knowledge of a distribution of noise at the lower level in an ecological application. In general, knowledge of the variation between the lower aggregation level units is important if attempting to make inference at a lower aggregation level.

9.1.3 Multiscale Analysis

By multiscale analysis , I mean where data is available at multiple resolution levels. For example, the focus of the analysis might be to include all data at levels of aggregation in an analysis of all the levels. This could be a joint analysis or could be separately carried out in the null case. Louie and Kolaczyk (2006) give an example of multiple scale analysis for disease data. More concretely, assume that we observe public health district data (in the US) and county level data. Figure 9.1 displays the public health districts (18) and counties (159) of the state of Georgia.

FIGURE 9.1: State of Georgia public health district boundary (thick lines) and county boundary (thin line) map.

The county set is a unique subdivision of the district set, i.e. each county falls uniquely within one public health (PH) district. Public health districts are administrative units within which certain health services are provided. It is therefore possible that some grouping effect based on health district could be found for counties that lie within a given district. For example, Dalton PH district includes the six counties of Whitfield, Murray, Fannin, Gilmer, Pickens, and Cherokee. These counties lie completely within Dalton and no other district. Hence in this case there is multiscale information which is completely aligned in the sense that the lower level county units fall completely and uniquely within the higher aggregation level units (districts).

In this particular case we could imagine the data defined with $K = 2$, with y_{i1} $i = 1, ..., 159$ where l_1 is the county level and y_{j2} $j = 1, ..., 18$ and l_2 is the district level. We could further assume a model of the form

$$y_{i1} \sim f_1(\mu_{i1}; l_1)$$
$$y_{j2} \sim f_2(\mu_{j2}; l_2).$$

Hence, under a null (separate) model we might have $\mu_{i1} = \exp\{\alpha_{10} + u_{i1}\}$, and $\mu_{j2} = \exp\{\alpha_{20} + u_{j2}\}$ where α_{10} and α_{20} intercepts and u_{i1} and u_{j2} are effects at levels 1 and 2. These can be random effects or functions of measured predictors. Clearly to ensure linkage between levels and so to model the joint behavior μ_{i1} and μ_{j2} may be linked. One natural way to do this is to consider the contextual effect of district on county and so we could have:

$$\mu_{i1} = \exp\{\alpha_{10} + u_{i1} + \underset{i \in j}{u_{j2}}\}$$

$$\mu_{j2} = \exp\{\alpha_{20} + u_{j2}\}.$$

In effect, because there is dependence between the mean levels then joint estimation of the latent factors (u_{i1}, u_{j2}) must be considered. Additionally, it might be considered that the district level should have a contribution from effects at the county level and so a further possibility could be to consider $\mu_{j2} = \exp\{\alpha_{20} + u_{j2} + \sum_{i \in j} \mu_{i1}\}$. In the latter formulation it would be useful to keep the separate effect of level (u_{j2}). Further extensions or variants of these linkages are possible.

9.1.3.1 Georgia Oral Cancer 2004 Example

As an example of analysis at multiple scales, I examined the Georgia PH district and county example. In this case, mortality counts from oral cancer were considered for 2004 in the state of Georgia. The state-wide expected rate for oral cancer was obtained and applied to the local county populations. The count of oral cancer mortality within public health (PH) districts is also available for the same period (as a sum of constituent county counts). Expected rates can also be summed from counties or directly calculated from the district population. We have applied the different two level models to the Georgia PH-county data. A separate model was fitted to each level along with a joint model with contextual effect. Table 9.1 displays the results in terms of DIC and pD for the different fitted models. Figure 9.2 displays the posterior average maps for the joint model with Poisson data model. The relative risks are defined as:

$$\theta_{i1} = \exp\{\alpha_{10} + v_{i1} + u_{i1} + \underset{i \in j}{v_{j2}} + \underset{i \in j}{u_{j2}}\}$$

$$\theta_{j2} = \exp\{\alpha_{20} + v_{j2} + u_{j2}\}$$

where each $v_{*1} + u_{*1}$ is a convolution of a UH and CH random effect.

Figures 9.3 and 9.4 display the posterior averaged maps for the $\widehat{\theta}$, and u_i, v_i effects obtained when separate models are fitted to county level and PH level. Overall it appears that the joint model mainly benefits the PH analysis as the DIC is marginally lower under the joint model. Some recent examples of multiscale modeling and their uses in discrete data situations can be found in Aregay et al. (2014), Aregay et al. (2015), Aregay et al. (2016c), Aregay et al. (2016b), Aregay et al. (2017), Aregay et al. (2016a) and Fonseca and Ferreira (2017).

Model	pD	DIC
County	97.68	507.01
PH district	18.72	124.96
Joint model: County	100.5	513.38
Joint model: PH district	18.07	123.77

Table 9.1: Goodness-of-fit results for separate and joint models for Georgia oral cancer county level public health (PH) data

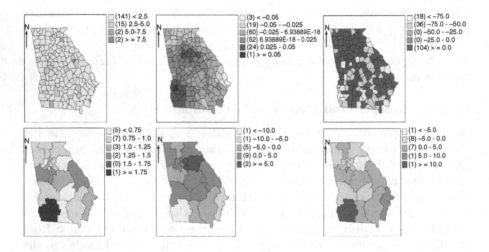

FIGURE 9.2: Multiscale model for the Georgia oral cancer data: posterior average effects. Top row (left to right) county level $\widehat{\theta}, u_i, v_i$; bottom row (left to right) PH level $\widehat{\theta}, u_i, v_i$.

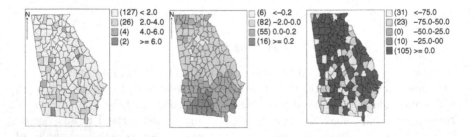

FIGURE 9.3: Georgia oral cancer model after a simple convolution model at county level is fitted. Posterior average maps of (left to right) $\widehat{\theta}, u_i, v_i$.

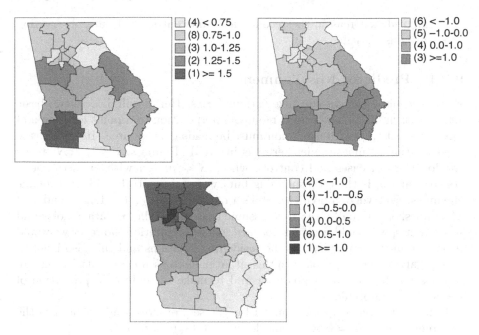

FIGURE 9.4: Georgia oral cancer PH district model. Posterior average maps of (left to right) $\widehat{\theta}, u_i, v_i$.

9.2 Misaligned Data Problem (MIDP)

While multiscale analysis can concern spatial units that are completely matched when aggregated there is also a situation where units are not matched and are termed *misaligned*. This often occurs when samplings at different spatial scales are not linked. A classic example of this scenario is where residential addresses of disease cases are to be related to measurements of environmental pollution obtained from a network of sites. The locations of the cases do not match the pollution measurement sites. Another example of misalignment is where data on disease is available in different administrative units that are not matched spatially. For example, census tracts, are not matched to postal codes in the UK or ZIP codes in the US. Thus in both cases some mechanism must be used to provide data on the same spatial scale within or at the same spatial region or location. In the first example, it is usually the case that *interpolation* of pollution measurements to residential locations would be required. In the second case, it may be that disease outcome data (counts) need to be available in the same spatial units. In the first situation a predictor is misaligned, whereas in the second case, different spatial data observation levels are misaligned.

9.2.1   Predictor Misalignment

Predictor misalignment can take various forms. Here I will discuss two basic situations: misalignment which requires interpolation to a point location and misalignment where interpolation must be made over an area. In both cases, interpolation or measurement error is involved. Define s_i $i = 1, ..., N$ to be the locations of cases and controls, where $N = m + n$ with m cases and n controls and y_i is the corresponding binary case or control label. Also define the measured level of a predictor at a set of sites as $z(s_l)$, $l = 1, ..., L$ and $z_l \equiv z(s_l)$ for short. Usually we would assume that the predictor data is noisy and so even at s_l we would want a smoothed value. In addition, however we would usually want to have $z(s_i)$, at the residential locations and this also involves interpolation. One approach to this situation is to assume a spatial Gaussian process for the measured predictor and to make a conditional prediction of the level of the predictor.

Define a Gaussian process model for the sets of sites, and define also the parameter vector $\theta = (\tau, \psi)^T$, and let $\mathbf{z}_s^T = (z(s_1), ..., z(s_L))$:

$$\mathbf{z}_s | \alpha, \theta \sim N(\mu_s, \Gamma)$$

where $\mu_{s_l} = \mu(s_l, \alpha)$, a predictor at the l th site, and Γ is a spatial covariance matrix. Often μ_s will consist of trend surface components and $\Gamma_{ll'} = \tau \rho(s_l - s_{l'}; \psi)$ where τ is a variance and $\rho(.)$ is a correlation function measuring the relation between values of z at separation distance $s_l - s_{l'}$. Choices of $\rho(s_l - s_{l'}; \psi)$ are many, (see Cressie, 1993; Diggle and Ribeiro Jr., 2007) and a simple choice could be an exponential form such as $\rho(s_l - s_{l'}; \psi) = \exp\{-\psi ||s_l - s_{l'}||\}$. More generally the powered exponential family defined by $\rho(s_l - s_{l'}; \psi) = \exp\{-(\psi ||s_l - s_{l'}||)^k\}$ can be assumed with $0 < k < 2$, the extra parameter allowing for a slower distance decline at short separation for $k > 1$, when $\psi << 1$.

For predicting a new set of locations $(z(s_i), i = 1, ..., N)$ the predictive distribution is

$$f(\mathbf{z}(s_i) | \mathbf{z}(s_l)) = \int f(\mathbf{z}(s_i) | \mathbf{z}(s_l), \alpha, \theta) f(\alpha, \theta | \mathbf{z}(s_l)) d\alpha d\theta.$$

Under a Gaussian process assumption, and denoting the set of original sites as s and the interpolant sites as $s\prime$ then this distribution is just

$$N(\mu_{s'} + \Gamma_{s,s'}^T \Gamma_s^{-1}(\mathbf{z}_s - \mu_s), \Gamma^*) \tag{9.1}$$
$$\Gamma^* = \tau[\Gamma_{s'} - \Gamma_{s,s'}^T \Gamma_s^{-1}\Gamma_{s,s'}].$$

This distribution can be sampled from directly assuming that the inverse Γ_s^{-1} can be calculated. It is therefore possible to assume a joint model for the binary outcome y_i and the interpolated (latent) predictor $z(s_i)$:

$$y_i \sim Bern(p_i) \ i = 1, ..., N \tag{9.2}$$

$$\log itp_i = \beta_0 + \beta_1 z(s_i)$$

$$\begin{pmatrix} \mathbf{z}_s \\ \mathbf{z}_{s'} \end{pmatrix} \sim N \left(\begin{pmatrix} \mu_s \\ \mu_{s'} \end{pmatrix}, \tau \begin{pmatrix} \Gamma_s & \Gamma_{s,s'} \\ \Gamma_{s,s'}^T & \Gamma_{s'} \end{pmatrix} \right).$$

Essentially, $z(s_i)$ is a latent variable and must be estimated via the joint model. One issue that remains is whether $z(s_i)$ should be estimated in the same model as p_i or should be estimated from $\mathbf{z}_s$ separately. Note that if a joint model is assumed then both β_1 and $z(s_i)$ are unknown and so there is likely to be some difference in the estimated $z(s_i)$ compared to an estimate simply obtained by Bayesian kriging. There is some support for the idea that Bayesian kriging via (9.1) should be used separately, as there seems a priori little reason to assume that the disease outcome should influence the estimate of the latent predictor.

Predictor misalignment can also occur where a value of the predictor is needed over an area rather than at a point location. For example, count outcome data are commonly available for small areas and we might want to interpolate measured covariates over small areas. This process is equivalent to block kriging in geostatistics and can be specified as follows. Define the counts in m small areas as y_i, $i = 1, ..., m$. We assume the usual Poisson relative risk model with $y_i \sim Poiss(e_i\theta_i) \ \forall_i$. Also assume a log linear model relating the relative risk to a block level covariate: $\log(\theta_i) = \beta_0 + \beta_1 z_i$. Here z_i denotes the value of z averaged over the i th small area. Assume a set of measurement site values as before: $\mathbf{z}_s^T = (z(s_l), ..., z(s_L))$ and also a set of small area average values: $\mathbf{z}_A^T = (z_1, ..., z_m)$ for areas $A_1, ... A_m$. We need the predictive distribution, which is now

$$f(\mathbf{z}_A|\mathbf{z}_s) = \int f(\mathbf{z}_A|\mathbf{z}_s; \alpha, \theta) f(\alpha, \theta|\mathbf{z}_s) d\alpha d\theta.$$

As in the point location case, there is a conditional distribution for these averages of the form:

$$N(\mu_A + \Gamma_{s,A}^T \Gamma_s^{-1}(\mathbf{z}_s - \mu_s), \Gamma_A^*) \tag{9.3}$$

$$\Gamma_A^* = \tau[\Gamma_A - \Gamma_{s,A}^T \Gamma_s^{-1} \Gamma_{s,A}]$$

where Γ_s^{-1} is the inverse of the site covariance, and

$$(\Gamma_{s,A})_{li} = |A_i|^{-1} \int_{A_i} \rho(s_l - u; \psi) du \qquad (9.4)$$

$$(\Gamma_A)_{ii'} = |A_i|^{-1} |A_{i'}|^{-1} \int_{A_i} \int_{A_{i'}} \rho(v - u; \psi) du dv$$

$$(\mu_A)_i = |A_i|^{-1} \int_{A_i} \mu(s) ds.$$

The integrals in (9.4) must be evaluated. Banerjee et al. (2004) suggest a Monte Carlo integration where the random locations are generated and the integrals replaced by sums. Finally the joint model could be specified as

$$\mathbf{y} \sim Pois(\mu(z_A)) \qquad (9.5)$$

$$\begin{pmatrix} \mathbf{z}_s \\ \mathbf{z}_A \end{pmatrix} \sim N \left(\begin{pmatrix} \mu_s \\ \mu_A \end{pmatrix}, \tau \begin{pmatrix} \Gamma_s & \Gamma_{s,A} \\ \Gamma_{s,A}^T & \Gamma_A \end{pmatrix} \right). \qquad (9.6)$$

Here, it could be assumed that $\mu(z_A)$ is simply a linear function of the latent interpolated predictor, assuming no confounders are present. Hence it is often assumed that $\mu(z_A)_i = e_i \exp(\beta_0 + \beta_1 z_i)$.

Note that, in general, integrating the first order intensity we get $\mu(z_A)_i = \int_{A_i} \lambda_0(u) \lambda_1(u) du$ where, if we assume $\lambda_0(u)$ to be constant over A_i, so $\lambda_0(u) \equiv e_i$. Denote the exposure of the j th individual within the i th area as $z_{j(i)}$, and an individual has an exposure $\lambda_1(u)_i = \exp(\beta_0 + \beta_1 z_{j(i)})$. Hence, $\mu(z_A)_i =$

$$e_i \int_{A_i} \exp(\beta_0 + \beta_1 z_{j(i)}) ds = e_i \exp(\beta_0) \int_{A_i} \exp(\beta_1 z_{j(i)}) ds \approx e_i e^{\beta_0} \sum \exp(\beta_1 z_{j(i)}),$$

where the sum is over the whole at-risk population within the small area.

Now usually individual level exposure is unavailable and indeed without knowledge of locations, this is not possible to directly estimate. However it is possible to use the region average (z_i) with some further assumptions to allow a closer approximation. For example we could assume that $z_{j(i)} \sim N(z_i, f(z_i))$ where the variance is assumed to be a function of the mean level. Wakefield and Shaddick (2006) suggest a function such as $f(z_i) = a + bz_i$. Hence it may be possible to consider a model whereby (9.3) could be used to predict block effects but within the Poisson model the individual effects could be used via

$$\mu(z_A)_i = e_i e^{\beta_0} \sum \exp(\beta_1 z_{j(i)})$$

$$z_{j(i)} \sim N(z_i, a + bz_i).$$

Other alternatives could be considered where point interpolation could be made to random points within each small area, instead of block kriging.

It can be important to consider how individuals within areas are related to exposures, at least from the view of making inference at the individual level. If individual inference is to be made from aggregate data then ecological bias should be considered. Of course, it still remains valid to make inference at aggregate levels from aggregate level models.

Some simple alternatives are often suggested to the above when using interpolated predictors. One simple approach is to use plug-in estimates z_i directly within the first level aggregated data model. For example,

$$y_i \sim Pois(e_i \theta_i) \ \forall_i,$$
$$\log \theta_i = \beta_0 + \beta_1 z_i,$$

can be assumed where z_i is estimated separately via a Bayesian kriging model. However in such a model, the error in the interpolated value is ignored. As discussed in Section 7.7, it is possible to make allowance for some biases by inclusion of random effects in models for example: $\log \theta_i = \beta_0 + \beta_1 z_i + \epsilon_i$ where ϵ_i is assume to be unstructured with distribution $\epsilon_i \sim N(0, \tau_\epsilon)$. In effect we are assuming that error in the interpolation is simply random noise. While this will indeed make some allowance for the interpolation error, it is unlikely to properly correct for any correlated error and will also be associated with the error in the relation between y_i and z_i. Alternatives involving forms of Berkson measurement error, such as $\log \theta_i = \beta_0 + \beta_1 [z_i + \epsilon_i]$, could be envisaged where either unstructured error with distribution $\epsilon_i \sim N(0, \tau_\epsilon)$, or spatially-structured error could be assumed. Given that z_i has already been kriged it is less justified to assume spatially-correlated error here.

Another alternative would be to employ a simplified model for the Gaussian field that is to be interpolated. For example, it would be possible to assume a proper CAR model for the variation in the $\mathbf{z}_s^T$ so that at a location j, say, where z_j is observed then assume

$$z_j \sim N(z_j^*, \tau)$$
$$z_j^* | z_{-j}^* \sim N(\overline{z_j}^*, \tau^* / n_{\delta_j})$$
$$\overline{z_j}^* = t_j + \phi \sum_{l \in \delta_j} (z_l^* - t_l) / n_{\delta_j},$$

where t_j is a trend component. The neighborhood (δ_j) defined for this example could be obtained as Dirichlet tesselation neighbors of the set of sample sites and outcome locations (if point-to-point interpolation is needed). The interpolated values are treated simply as missing. However this ignores some parts of the variation in the field due to the Markov neighborhood dependence that is assumed by the model.

Another simplifying assumption, in the case of point-to-block interpolation, is to consider the centroids of small areas as sampling points which augment the measurement at the sampling sites. In this case, the small area centroids could be considered to have missing data. More specifically, assume an augmented set $\mathbf{z}_{s'} : (\mathbf{z}_A, \mathbf{z}_s)$ where the z_A are missing:

$$y_i \sim Pois(e_i \theta_i)$$
$$\log \theta_i = \beta_0 + \beta_1 z_i$$
$$\mathbf{z}_{s'} \sim N(\mu_{s'}, \tau \Lambda)$$

where $(\mu_{s'})_i = \sum_{i \in \delta_i} z_i / n_{\delta_i}$ and $(\Lambda)_{ii} = n_{\delta_i}^{-1}$. This CAR-Voronoi model was proposed by Greco et al. (2005) in an application to interpolation of groundwater uranium to the centroids of ZIP codes in South Carolina.

9.2.1.1   Binary Logistic Spatial Example

In a study of a birth-defect disease outcome the residential addresses of births and births with defects were recorded within a US state. For an anonymized study area of 2×2 km the locations of this binary outcome were recorded (see Figure 9.5). Soil chemical measurements were also made on a range of chemicals. In this example, I examined arsenic (AS). Arsenic in soil is commonly found in agricultural areas where heavy pesticide use has occurred over many years. It is important to consider whether exposure to As (via a possible groundwater exposure pathway) during pregnancy could lead to adverse birth defect outcomes. Many residents of certain rural areas in the US have a groundwater domestic supply. In this example we know the residential location of outcomes (births with or without defects) during a particular month of pregnancy, but have a mismatched set of soil chemical sampling sites.

Figure 9.5 displays the distribution of the sites (left panel) and the distribution of the births with their binary marks (right panel). The example displays a point-to-point misalignment problem. It is appropriate here to interpolate the soil chemical values from the measurement sites to the residential locations so as to provide a local exposure estimate. Of course, issues relating to appropriateness of the residential location as exposure "location" do arise, but here we assume that this location is the most relevant for the given month of exposure. The first model assumed is closely related to that given in (9.2) except that the model is not jointly estimated. Instead, the interpolation is carried out separately via Bayesian kriging and a plug-in estimator is used ($\widehat{z}_{KB}$). The model assumed for this estimator is a Gaussian spatial process with an overall mean parameter (β_z) and a power exponential correlation function so that the covariance matrix is given by:

$$\sigma^2 \Sigma_{ij} = \sigma^2 \exp\{-(d_{ij}/\phi)^\kappa\} \tag{9.7}$$

with smoothing parameter $\kappa = 1$ which is equivalent to a Matérn covariance with exponential form. The prior distribution for β_z is assumed to be zero-mean Gaussian with variance 3.0, whereas the inverse variance prior distribution is assumed for σ^2. A discrete prior distribution on a fixed range

$(0, 2 * max(d_{ij}))$ is used for ϕ. The parameters within the kriging model were estimated using krige.bayes in the R package geoR. The posterior expected estimates under this model were $\overline{\beta}_z = 1.9903$, $\overline{\sigma}^2 = 5.00629$, and $\overline{\phi} = 0.0001$. This suggests a very small covariance distance effect. The overall logistic model used was

$$y_i \sim Bern(p_i)$$
$$\log p_i = \beta_0 + \beta_1 \widehat{z}_{KB}$$

with $\beta_* \sim N(0, \tau_*)$, $\tau_*^{-1/2} \sim U(0, 100)$. This is a simple Bayesian logistic regression model with zero-mean Gaussian prior distributions for the regression parameters. This is referred to as model 1 in Table 9.2. Models 2 and 3 are variants intended to account for the measurement error in using the Bayesian kriged values in the model (rather than a joint model). Model 2 assumes an additive unit level uncorrelated error (UH) which has zero-mean Gaussian prior distribution. Model 3 assumes that instead of a UH term the classical measurement error model is reversed and the kriged estimate is the true value observed with error. This is termed Berkson error. The final models utilize the original data and model the spatial covariation of that data directly.

Model 4 assumes that the constant mean estimated from the As data has an additive uncorrelated error term $(v_i \sim N(0, \tau_v), \tau_*^{-1/2} \sim U(0, 100))$ and an additive spatially-correlated error including the term $\mathbf{W} \sim \mathbf{MVN}(\mathbf{0}, \sigma^2 \mathbf{\Sigma})$ with covariance term $\mathbf{\Sigma}_{ij} = \exp\{-(d_{ij}/\phi)\}$, with prior distributions $\sigma^2 \sim Exp(0.1)$, $\phi \sim Exp(0.1)$ which are reasonably overdispersed while favoring small values.

Finally, model 6 consists of a joint model for the birth outcome and arsenic levels where the true arsenic level is a latent variable $(\widehat{z}_i)$ interpolated via the point to point Bayesian model in (9.2). Again a power exponential covariance was assumed with $\kappa = 1$ as in (9.7). The assumed prior distributions were $\sigma^2 \sim Exp(0.1)$, $\phi \sim Exp(0.1)$ as in the previous model, and a random level β_z was assumed to have a zero-mean Gaussian prior distribution with variance 3.0.

The resulting goodness-of-fit DIC measures for these models were computed and they are reported on Table 9.2. While these are only a subset of overall potential models and prior distribution choices which could be made, they represent a range of approaches often advocated for this type of modeling exercise. It is notable that the general random effect model (model 2) provides the lowest DIC amongst the plug-in ME models and by this measure performs better than the potentially more plausible Berkson model.

Amongst models making direct use of the data, the joint model (model 6) yields the lowest DIC (140.90) and, with the exception of model 2, this yields the lowest DIC.

However for this particular example the overall best model judging by DIC alone is the model 2. This might suggest that there is overall more noise in the relation between the outcome and the covariate than within the covariate

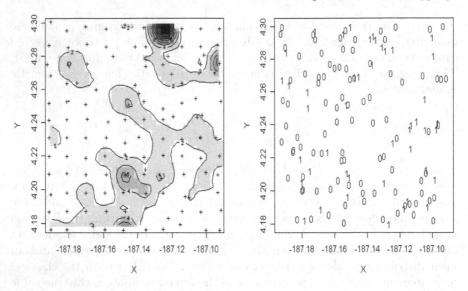

FIGURE 9.5: Spatial distribution of sampling sites (+) and multivariate B-spline smooth of arsenic (image and contourplot from R package MBA) for the synthetic data example (left panel); spatial distribution of binary mark (1: MRDD, 0: otherwise) in same window.

Model	Components	DIC
1	$\beta_0 + \beta_1\widehat{z}_{KB_i}$, $\beta_* \sim N(0,\tau_*), \tau_*^{-1/2} \sim U(0,100)$	178.33
2	$\beta_0 + \beta_1\widehat{z}_{KB_i} + v_i$, $v_i \sim N(0,\tau_v), \tau_*^{-1/2} \sim U(0,100)$	116.22
3	Berkson error: $\beta_0 + \beta_1(\widehat{z}_{KB_i} + v_i)$, priors as in 2	155.47
4	$\beta_0 + \beta_1(\overline{As} + W_i) + v_i$, $\mathbf{W} \sim MVN(\mathbf{0}, \sigma^2\mathbf{\Sigma})$	177.50
5	As in 4 but $\beta_0 + \beta_1(\overline{As} + W_i)$ only	176.10
6	$\beta_0 + \beta_1\widehat{z}_i$	140.90

Table 9.2: Model fit results for point misalignment example

itself. There are of course various criticisms that can be levelled at such models. First, approximate models, such as those using plug-in estimators do not have a priori justification based on Gaussian model or process theory. However from a pragmatic viewpoint, if a parsimonious description is achieved by such a model in a regression context then it is supported. The model is less justified when treating out-of-data prediction on a priori grounds. However joint modeling of the latent spatial effect could also be challenged in the sense that the true value of the spatial covariate is forced to depend on the outcome model parameters. This may also seem to be a criticism of that model. Of course prior distributional assumptions can be challenged for all models fitted and this also can impact model selection.

For the lowest DIC model the posterior mean estimates of the parameters $\beta_0, \beta_1, \tau_0, \tau_1, \tau_v$ are available after a burn-in of 60000 and a sample of 10000 taken. This yielded estimates of mean (sd) of $\widehat{\beta}_0 = -1.715(3.832)$, $\widehat{\beta}_1 = -0.6846(1.951)$, $\widehat{\tau}_0 = 106.1(4233.0)$, $\widehat{\tau}_1 = 76.16(3419.0)$, $\widehat{\tau}_v = 0.05634(0.1135)$. The overall conclusion of this analysis is that the arsenic has does not have a significant (linear) relationship with birth defects under a logit model. Further model variants could be conceived where changes to both the linear model and also to the interpolation model could be considered. Often environmental predictors have non-linear relations with health outcomes and so polynomial or spline models could provide better fits. In our experience, with other soil chemicals it is indeed the case that a non-linear model yields a significant fit when a linear model does not. Further alternative interpolants such as spline models could be considered (see e.g. French and Wand, 2004).

Finally, Zhu et al. (2003) describe the analysis of misaligned data when spatio-temporal analysis is to be considered . They also consider (log) linear relations between outcome and predictor within a small area setting (ZIP code units).

9.2.2 Outcome Misalignment

Outcome misalignment occurs when spatial outcome units are not matched. One basic and commonly occurring form is where counts are observed within small areas but expected rates are only available in other misaligned areas. For example, in the US, ZIP codes don't align with counties but are smaller-sized spatial (postal) units; in the UK postcodes don't align with census enumeration districts. When comparing diseases within study areas it can be the case that different aggregation units are found and matching must take place. When multiple diseases are concerned, multivariate misalignment could occur. This latter is not considered in this volume, but could be of considerable practical importance when multiple disease analysis is considered. Multiple disease analysis is discussed in Chapter 10.

Outcome misalignment leads to the consideration of the overlay of different spatial aggregations to yield mismatched outcomes. In cases where data must be matched on their spatial units then some form of allocation model must be considered. Banerjee et al. (2014), chapter 6, discuss in detail the issues related to interpolation of Gaussian processes to point locations and to areas. Note that in general, outcome misalignment can be treated as a problem in block-to-block interpolation in that areas of different sizes are to be matched, at least in terms of parameter estimation.

A common assignment rule is simply to use proportional allocation. Define a simple misalignment with a single large region (A) and a set of misaligned smaller regions which overlap $A : (B : \{b_1, ..., b_p\})$. Parts of the smaller regions overlap areas of A. It is possible to consider the problem first of the aggregation of count data from smaller to larger regions. Define M larger scale regions with counts $\{y_{A_i}\}$, $i = 1, ..., M$ and P smaller region counts $\{y_{b_l}\}$, $l = 1, ..., P$; also the corresponding expected rates: $\{e_{A_i}\}, \{e_{b_l}\}$, and area proportions $\{p_{b_{l_i}}\}$, where $p_{b_{l_i}}$ is the proportion of region b_l within the i th region.

First for aggregation to larger region counts, it is simple to assume proportional allocation and to derive the larger region counts from smaller regions:

$$y_i = \sum_l p_{b_{l_i}} y_{b_l}. \tag{9.8}$$

One consequence of proportionate allocation is that the resulting larger region y_i could be non-integer. Special rules could be devised to make sure the allocation resulted in integer allocation, but these can lead to edge effect problems. Note that many packages allow non-integer Poisson data (e.g. Win/OpenBUGS). Given two levels of count data a consideration might be given to employing a joint model for the different levels where:

$$y_i \sim Pois(\mu_i); \quad \mu_i = \sum_l p_{b_{l_i}} e_{b_l} \theta_{b_l}. \tag{9.9}$$

$$y_{b_l} \sim Pois(\mu_{b_l}); \quad \mu_{b_l} = e_{b_{l_i}} \theta_{b_l}.$$

Occasionally aggregation is required only for expected rates. The situation envisaged is where expected rates are available in smaller regions only and must be aggregated to match the counts in larger regions. For example, if a Poisson model for the larger region counts were assumed then

$$y_i \sim Pois(e_i.\theta_i) \tag{9.10}$$

$$e_i = \sum_l p_{b_{l_i}} e_{b_l}.$$

In consideration of disaggregation then smaller regions can be allocated counts by proportionate allocation downwards. For example, assuming the same complete misalignment as above, the derived count in a smaller region y_{b_l} could be derived via $y_{b_l} = \sum_i p_{il} y_i$ where the proportion of the larger region overlapping the smaller region is denoted by p_{il}.

9.2.3   Misalignment and Edge Effects

In Section 5.8, various issues were discussed relating to the importance of edge effects in disease mapping studies. These can be even more important in consideration of misaligned data at different aggregation levels. For example assume a simple proportionate aggregation as in (9.8) above. Close to the study area boundary, the neighborhoods of regions are censored. Assume that smaller regions misalign to a study area where the hull of the larger regions forms the outer boundary of the study area. This implies that the smaller regions can overlap the study boundary. To estimate the proportionate allocation, the area of these smaller regions must be known as well as the counts within the regions. Hence, knowledge of an area larger than the study area is important. Further when modeling small area counts where aggregation to larger regions has been assumed, then the effects of local neighborhoods on the spatially structured risk could be considerable. In model (9.9) above, any spatially-structured risk involving adjacencies will depend on knowledge of both the neighbors of the larger regions and the neighbors of the smaller regions. In model (9.10) the relative risk in larger regions is not a direct aggregation of the lower level risk as in (9.9). In (9.9) the smaller regions could have neighborhood effects and these aggregate to the larger level. In (9.10) only the expected rate is aggregated. The effect of neighborhood censoring within CAR random effect models has been highlighted by Vidal-Rodiero and Lawson (2005).

If the study area does not correspond to the hull of the larger regions then some regions will overlap the boundary. This could also cause edge effect problems due to the need to assign data from the larger region to the area within the study area. Clearly misalignment can have significant implications for inference about areas at or close to the edge of study boundaries.

10

Multivariate Disease Analysis

Often it is appropriate to consider the analysis of the geo-referenced distribution of more than one disease. For example, the focus may be on a group of diseases with similar etiology in an epidemiological study. Another example, in the context of public health, could be the examination of the general health status of a region (possibly following a cluster alarm signal). In the latter case a range of disease types might be considered to find out whether any show signs of unusual risk variation. In some cases, the focus is on the spatial distribution of the vectors of diseases (for example relative risk estimation for a vector). In other cases the diseases are to be contrasted or correlations between their spatial distributions are to be considered. In this chapter the focus will be on relative risk estimation and modeling of risk both in terms of correlation between diseases and in terms of comparison. The simplest situation where such analysis can be considered is where two diseases are to be examined.

10.1 Notation for Multivariate Analysis

Both case event and count data will be considered in this chapter.

10.1.1 Case Event Data

Assume there are L diseases. Assume a fixed study area T common to all diseases. Misaligned study areas are not considered here. Assume that the residential addresses of each disease type are known and given by $\{s_{l_i}\}$, $l = 1, ..., L$ and $i = 1, ..., m_l$. It is assumed that there are m_l cases of the l th type within the study region. In addition, there may be control realizations for each of the diseases. The number of these are denoted by m_{c_l} and the total number of cases and controls for a l th disease is $N_l = m_l + m_{c_l}$. Hence the controls will be denoted $\{s_{l_i}\}$, $l = 1, ..., L$ and $i = m_l + 1, ..., N_l$. Often a common control for all diseases is used but this can be discussed later.

10.1.2   Count Data

Here, the simplest situation is considered where fixed spatial units (small areas) have a range of L disease counts observed in a fixed time period. Define these counts as $\{y_{l_i}\}$, $l = 1, ..., L$, $i = 1, ..., m$. Hence, within each spatial unit there are L counts of disease. The corresponding expected rates and relative risks are defined as $\{y_{l_i}\}$, $\{\theta_{l_i}\}$.

10.2   Two Diseases

Often the simplest situation arises where two diseases are to be modeled and $L = 2$.

10.2.1   Case Event Data

For case events the observed data is now $\{s_{l_i}\}$, $l = 1, 2$ and $i = 1, ..., N_l$. Analysis will depend on the object of the study. If the object is to locally (in space) compare the probability of getting disease 1 or disease 2, then a *competing risk* formulation might be useful. This might be conditioned on the observed total of cases, $N_T = m_1 + m_2$. On the other hand it might be important to jointly model each disease (perhaps with linking parameters) without conditioning.

First define the first order intensity for the l th disease as $\lambda_l(s|\psi_l) = \lambda_{l0}(s|\psi_{l0}).\lambda_{l1}(s|\psi_{l1})$, $l = 1, 2$. If the competing risk scenario is important, it is possible to simply consider the joint realization of the cases conditional on N_T. Assume that the diseases can be considered to be independent heterogeneous Poisson processes (PPs) governed by $\lambda_1(s|\psi_1)$ and $\lambda_2(s|\psi_2)$. This independence assumption would be at least conditional given parameters (ψ_{l0}, ψ_{l1}, $l = 1, 2$). Hence we can assume that the joint realization is a PP and so by conditioning on the joint realization it is possible to consider the probability that given a case at s is of type 1 or 2. This is given by

$$\Pr(l(s) = 1|case(s)) = \frac{\lambda_1(s|\psi_1)}{\lambda_1(s|\psi_1) + \lambda_2(s|\psi_2)} \tag{10.1}$$

where $l(s)$ is the type label of the location and $case(s)$ means case at s. Note that the locations are now conditioned and it is possible to make inference about the labelling of the points. Substituting the full intensities into (10.1), gives

$$\frac{\lambda_{10}(s|\psi_{10}).\lambda_{11}(s|\psi_{11})}{\lambda_{10}(s|\psi_{10}).\lambda_{11}(s|\psi_{11}) + \lambda_{20}(s|\psi_{10}).\lambda_{21}(s|\psi_{11})}.$$

Note that in this case it is not possible to remove the background intensities (λ_{l0}) as these are different for each disease and the conditioning is on the case distribution. If it can be assumed that $\lambda_{10}(s|\psi_{10}) = \lambda_{20}(s|\psi_{10})$ then these cancel in (10.1) and so the type probability is just $\lambda_{11}(s|\psi_{11})/(\lambda_{11}(s|\psi_{11}) + \lambda_{21}(s|\psi_{11}))$. Hence there is some advantage in seeking a common control realization. Note that other conditional probabilities can be derived (such as $\Pr(case(s))$ given the joint case-control realization). Suitable likelihoods can be constructed from these probabilities and inference can proceed as usual. Lawson and Williams (2000) provide an early likelihood-based analysis of a competing risk scenario. As far as this author is aware, there are no published examples of the application of Bayesian approaches to these competing risk scenarios.

If an unconditional analysis is to be considered, albeit conditional in the case-control realizations, then a joint likelihood could be derived for the joint realization. Define the binary case-control label as

$$y_{l_i} = \begin{cases} 1 & \text{if the event is a case} \\ 0 & \text{otherwise} \end{cases}$$

and the joint likelihood as

$$L(\mathbf{s}|\psi_0, \psi_1) = \prod_{l=1}^{2} \prod_{i=1}^{N_l} p_{l_i}^{y_{l_i}} (1 - p_{l_i})^{(1-y_{l_i})}$$

$$p_{l_i} = \frac{\lambda_l(s_{l_i}|\psi_l)}{\lambda_{l0}(s_{l_i}|\psi_{l0}) + \lambda_l(s_{l_i}|\psi_l)}$$

$$= \frac{\lambda_{l1}(s_{l_i}|\psi_{l1})}{1 + \lambda_{l1}(s_{l_i}|\psi_{l1})}.$$

10.2.1.1 Specification of λ_{l1}

The specification of λ_{l1} $l = 1, 2$ can now be designed to model the relation between the two diseases. For example, simple functional dependence between the two could be modeled by a relation such as $\lambda_{11}(s|\psi_{11}) = \rho\lambda_{21}(s|\psi_{21})$. Often a log-linear specification is assumed for the intensity so that $\lambda_{l1}(s_{l_i}|\psi_{l1}) = \exp\{f(s_{l_i}, \psi_{l1})\}$. For descriptive models, often the linear component will consist of covariate terms and/or random effect terms. For example

$$f(s_{1_i}, \psi_{11}) = \beta_0 + \mathbf{x}_i'\beta_1 + v_{1_i}$$
$$f(s_{2_i}, \psi_{21}) = \beta_0 + \mathbf{x}_i'\beta_2 + v_{2_i},$$

could be assumed, where $\mathbf{x}_i'\beta_*$ is a covariate predictor with common covariates and separate parameters, and v_{1_i}, v_{2_i} are random effects for each disease. Often these random effects would be uncorrelated, but more complex structuring could be assumed. A common random effect could be added to this model so that

$$f(s_{1_i}, \psi_{11}) = \beta_0 + \mathbf{x}'_i\beta_1 + v_{1_i} + \delta w_i \qquad (10.2)$$
$$f(s_{2_i}, \psi_{21}) = \beta_0 + \mathbf{x}'_i\beta_2 + v_{2_i} + w_i/\delta$$

and then a common random component is included. This could be spatially-structured (as a zero-mean Gaussian process) or could be uncorrelated. A linked component model where a common component has been included for two diseases (for count data) has been proposed by Knorr-Held and Best (2001) and further, Held et al. (2005) suggested a more general model. The components suggested were included to mimic an ecological regression on an unobserved component. While this was proposed for count data there is no reason why this idea could not be applied with case event data. Of course other forms of common parameterisation can be considered and what is described here in (10.2) is a very basic proposal.

An alternative possibility is to consider including a bivariate effect in the models so that a defined correlation parameter is included. For example, the joint model

$$f(s_{1_i}, \psi_{11}) = \beta_{10} + \mathbf{x}'_i\beta_1 + W_{1i}$$
$$f(s_{2_i}, \psi_{21}) = \beta_{20} + \mathbf{x}'_i\beta_2 + W_{2i}$$
$$\begin{pmatrix} W_{1i} \\ W_{2i} \end{pmatrix} \sim \mathbf{N}(\mathbf{0}, \mathbf{\Gamma})$$
$$\text{with } \mathbf{\Gamma} = \begin{pmatrix} \kappa_1 & \sqrt{\kappa_1}\sqrt{\kappa_2}\rho_W \\ \sqrt{\kappa_1}\sqrt{\kappa_2}\rho_W & \kappa_2 \end{pmatrix}$$

with no spatial correlation could be considered, where ρ_W is a correlation parameter and κ_1, and κ_2 are variances. Further extension to cross-correlated spatially-structured prior distributions could also be considered.

10.2.2   Count Data

For counts the observed data is now $\{y_{l_i}\}$, $l = 1, 2$ and $i = 1, ..., m$. Analysis will depend on the object of the study. If the object is to locally (in space) compare the risk of getting disease 1 or disease 2, then a *competing risk* formulation might be useful. This might condition on the observed total of cases within spatial units $N_i = y_{1_i} + y_{2_i}$. On the other hand it might be important to jointly model each disease (perhaps with linking parameters) without conditioning.

In the simple case where we condition on a small area total then conditioning on the sum of two (conditionally independent) Poisson variates leads straightforwardly to a binomial model for either type count. For example we could model the disease 1 count as

$$y_{1_i} \sim bin(p_{1_i}, N_i).$$

This has sometimes been called the proportional mortality model (see e.g. Dabney and Wakefield, 2005). In this formulation the probability can be formulated to define the relation between the two diseases. For example the logit of the probability can be modeled with differences in risk terms:

$$\text{logit}(p_{1_i}) = \log(\frac{e_{1_i}}{e_{2_i}}) + \beta_0 + W_i^*$$
$$W_i^* = (W_{1_i} - W_{2_i})$$

where $W_i^* = (W_{1_i} - W_{2_i})$ is the difference between the random effects for the two diseases. These could have UH and CH components also. In this formulation the relative expectation of the two diseases is also included and the random effect W_i^* (which is a difference) can be directly modeled with prior distributions for its components.

As mentioned above it is possible to consider:

1. A joint model for the two diseases with common components:

$$y_{1_i} \sim Pois(\mu_{1_i}) \tag{10.3}$$
$$y_{2_i} \sim Pois(\mu_{2_i})$$
$$\mu_{1_i} = e_{1_i}.\exp\{\beta_{10} + \mathbf{x}_i'\beta_1 + W_{1i}\}$$
$$\mu_{2_i} = e_{2_i}.\exp\{\beta_{20} + \mathbf{x}_i'\beta_2 + W_{1i}\}$$

where $\mathbf{x}_i'\beta_1$, $\mathbf{x}_i'\beta_2$ is the covariate predictor for diseases 1 and 2 respectively, and W_{1i} is a common component. As before this component is a composite random effect term and could consist of UH and CH components. Instead, as in the case event scenario, separate random effects could be assumed which have a prior correlation so that

$$\mu_{1_i} = e_{1_i}.\exp\{\beta_{10} + \mathbf{x}_i'\beta_1 + W_{1i}\}$$
$$\mu_{2_i} = e_{2_i}.\exp\{\beta_{20} + \mathbf{x}_i'\beta_2 + W_{2i}\}$$
$$\begin{pmatrix} W_{1i} \\ W_{2i} \end{pmatrix} \sim \mathbf{N}(\mathbf{0}, \mathbf{\Gamma})$$
$$\text{with } \mathbf{\Gamma} = \begin{pmatrix} \kappa_1 & \sqrt{\kappa_1}\sqrt{\kappa_2}\rho_W \\ \sqrt{\kappa_1}\sqrt{\kappa_2}\rho_W & \kappa_2 \end{pmatrix}$$

where κ_1, and κ_2 are the prior variances of the two effects and as before the ρ_W is the prior correlation.

2. A shared component model which mimics an ecological regression on the unobserved shared component. In the example of Knorr-Held and Best (2001) this takes the form of

$$\mu_{1_i} = e_{1_i}.\exp\{\beta_{10} + u_{2_i} + \delta u_{1i}\} \tag{10.4}$$
$$\mu_{2_i} = e_{2_i}.\exp\{\beta_{20} + u_{1i}/\delta\},$$

where u_{2_i} is a separate random component for the first disease, u_{1i} is the shared component and δ is a scaling component that is necessary because of identifiability. Usually has the prior distribution $\log(\delta) \sim N(0, \kappa_\delta)$ with a small fixed variance of $\kappa_\delta = 0.17$ in Knorr-Held and Best (2001). In Knorr-Held and Best (2001), both the shared terms u_{1i} and u_{2_i} were assumed to have a CAR prior distribution with

$$p(\mathbf{u}|\kappa) \propto \kappa^{(m-1)/2} \exp\left\{ -\frac{\kappa}{2} \sum_{i \sim j} (u_i - u_j)^2 \right\}.$$

Dabney and Wakefield (2005) give examples of comparison of the proportional mortality and incidence models with shared component models in a particular application.

10.2.3 Georgia County Level Example Involving Three diseases

As an example of the application of the above models to we have examined a group of diseases which could a priori be considered to display the same or similar distributional features. These are ambulatory care sensitive conditions and counts of these conditions were recorded. They are chronic diseases which are to some measure exacerbated by adverse environmental air pollution: asthma, chronic obstructive pulmonary disease (COPD), and angina. While both asthma and COPD are respiratory in nature, whereas angina is a coronary outcome, all the diseases could be affected by poor environmental air quality. Therefore, in some part, there may be some evidence of this feature in the spatial distribution. A common component could emerge from a joint analysis or at least some evidence of correlation between the spatial distributions could be apparent. In the following, a range of models have been fitted to parts of these data.

A full multivariate analysis is examined in a later section (Section 10.3). Demonstrated here are the individual convolution models for each disease (model 1), the joint model with common CAR component with separate UH components (model 2), and the shared component model (model 3). The diseases compared in models 2 and 3 are asthma and COPD. Figure 10.1 displays the results for separate analyses of the three diseases (model 1). Table 10.1 displays the results for these models in terms of DIC. It is clear that a common component as in model 2 leads to improved goodness-of-fit as judged by lower DIC. The improvement seems to be mostly to the asthma (DIC from 1144.7 to 1137.3).

The form of the common component can be seen in Figure 10.3 and it appears that there is a considerable positive concentration of common risk in the south eastern portion of the state. The separate posterior mean uncorrelated components appear to be quite different in distribution (Figure 10.2).

FIGURE 10.1: Georgia county level data, 2005 for asthma, angina, and COPD (all ages standardised by the state rate).

Of course it should be borne in mind that there is some lack of identifiability in these components and so more reliance should be placed on the shared component model (model 3 as in (10.4)). The results of the shared compo-

Model	DIC	pD
1 (COPD, angina, asthma)	1239.8, 840.9,1144.7	145.9, 121.1, 127.8
1 Overall	3225.46	394.8
2 Common (COPD, asthma)	1238.3, 1137.3	144.6, 121.5
3 Shared (COPD, asthma)	1244.9, 1131.97	141.7, 119.0

Table 10.1: Model comparisons for three-disease example: joint, common, and shared

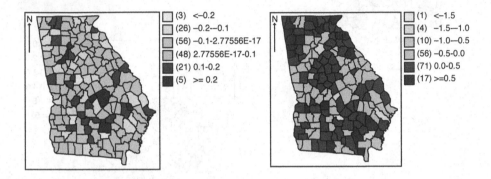

FIGURE 10.2: Georgia county level posterior expected UH components for asthma (left) and COPD (right) when fitted with a common CAR component.

nent model (model 3 in Table 10.1) with prior distribution $\log(\delta) \sim N(0, \kappa_\delta)$ with $\kappa_\delta = 0.17$, seem to suggest that the shared component does reduce the DIC close to that of the common component model (overall DIC: 2375.6 and 2376.9 for common and shared components respectively). However the shared component model yields a lower pD, as there is only one separate component, instead of two. The posterior expected estimate (sd) of the δ parameter is 0.6855 (0.03712). The model fitted was

$$\mu_{AS_i} = e_{AS_i} \cdot \exp\{\beta_{10} + u_{2_i} + \delta u_{1i}\}$$
$$\mu_{COPD_i} = e_{COPD_i} \cdot \exp\{\beta_{20} + u_{1i}/\delta\}$$

with both u_{2_i} and u_{1i} having CAR prior distributions with $\tau_{u*}^{-1/2} \sim U(0, 100)$. It is debatable whether there is an advantage in fitting the shared component model here as there is no separate component for COPD.

In Figure 10.4 the left panel displays the separate CAR component for asthma while the common shared component is displayed in the right panel. These components are remarkably similar in spatial distribution although the shared component had a greater concentration of common risk in the far south of the state. This is slightly different in distribution from the common component in model 2.

FIGURE 10.3: Georgia county level analysis with common component: posterior expectation of the common CAR component (model 2).

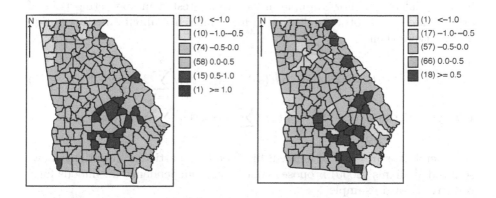

FIGURE 10.4: Georgia county level asthma and COPD: posterior expected estimates of the asthma CH component (left) and the shared component (right).

10.3 Multiple Diseases

Many of the considerations that apply to two diseases extend straightforwardly to more than two diseases. It is often the case that a range of diseases might be of interest in both public health applications and in epidemiological studies. However there are also more difficulties, as the possible comparisons increase with L. Multivariate disease modeling is now the focus.

10.3.1 Case Event Data

Assume that there are L diseases with $\{s_{l_i}\}$, $l = 1, ..., L$ and $i = 1, ..., N_l$ where $L > 2$. Here $N_l = m_l + m_{c_l}$ so that there are m_l cases and m_{c_l} controls for the l th disease. I will only consider models that are conditional on the realization of the case and control events. Hence the assumption is made that m_l, m_{c_l} and the locations of these cases and controls $\{s_{l_i}\}$, $l = 1, ..., L$ are fixed at the likelihood level. Often groups of disease are the focus. For example, it might be that a range of respiratory (asthma and COPD) and other chronic diseases (angina) are to be examined in relation to an air pollution source or to general pollution levels measured at sites. We are now interested in the case event intensities $\lambda_l(s|\psi_l) = \lambda_{l0}(s|\psi_{l0}).\lambda_{l1}(s|\psi_{l1})$, $l = 1, .., L$. Considerations similar to those cited in Section 10.2.1 lead to different forms of conditioning. First it is clear that conditional on an event at s, the probability it is of disease type k is a normalization: $\lambda_k(s|\psi_l)/\sum_{l=1}^{L}\lambda_l(s|\psi_l)$.

Other normalizations can be derived. It is important to focus on particular forms of inference. For example, if we are interested in competing risks of one disease over another and want to make inference about the distribution of case types then

$$\Pr(l(s) = k \text{ and } case(s)) = [\lambda_k(s|\psi_l)/\sum_{l=1}^{L}\lambda_l(s|\psi_l)].\Pr(case(s))$$

$$where \quad \Pr(case(s)) = \sum_l \lambda_l(s|\psi_l)/[\sum_l \lambda_l(s|\psi_l) + \sum_l \lambda_{l0}(s|\psi_{l0})].$$

In general, it is possible to define a likelihood for particular situations. Lawson and Williams (2000) proposed conditional independence likelihoods for a putative hazard example.

As far as this author is aware, there are few published examples of Bayesian analysis of multi-type spatial disease realizations. A multivariate analysis of the residential locations of death certificates for respiratory disease (bronchitis) and air-way cancers (respiratory, gastric, and oesophageal) was proposed by Lawson and Williams (2000). Data were obtained for the years 1966

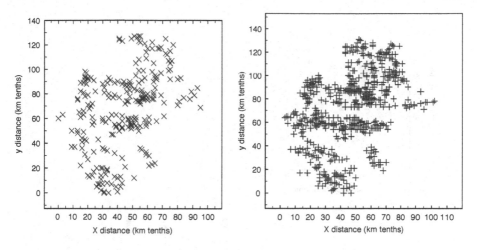

FIGURE 10.5: Arbroath mortality study control disease realisations: composite cancer control (left panel); CHD control (right panel).

through 1976 for a small industrial town in the UK. These diseases were chosen as potentially related to adverse air pollution. Control diseases examined were coronary heart disease mortality (which is age-related but not usually affected directly by air pollution), and a composite control of lower body cancers (prostate, penis, testes, breast, cervix, uterus, colon, and rectum). These latter cancers were useful as they are less affected by respiratory inhalation insult and so can be regarded as reasonable controls matched on age to the risk profiles of the case diseases.

Figures 10.5 and 10.6 display the location maps of the residences. It is notable how the composite control follows closely the CHD spatial distribution. In the example of Lawson and Williams (2000), the first order intensity was related to a fixed putative pollution source via a distance measure, so that $\lambda_{l1}(s|\psi_{l1}) = 1 + f_l(s|\psi_{l1})$ and for each disease the link was defined as $f_l(s|\psi_{l1}) = \alpha_{1l} \exp[-\alpha_{2l}d(s) + \alpha_{3l} \log d(s)]$ where $d(s)$ is the distance from the location s to a fixed point (putative source). A joint likelihood was derived and estimation of parameters proceeded via MCMC applied to the posterior distribution with uniform prior distributions for all parameters. The likelihood used conditional probabilities for different case types and a common control disease. Further, exploration of the possibility of weighting of different diseases was considered via the total intensity specification: $\lambda(s|\psi) = \sum_l w_l \lambda_l(s|\psi_l)$.

Subsequent development of a weighted likelihood was considered in the context of prior expert opinion about what weight each disease should have in defining evidence for an effect.

FIGURE 10.6: Arbroath mortality study data for gastric and oesophageal cancer (top); respiratory cancer (middle); bronchitis (bottom).

Further non-Bayesian analysis of multiple diseases has been developed within a non-parametric smoothing approach by Diggle et al. (2005), where estimation of the conditional disease probability:

$$\widehat{p}(s) = \lambda_k(s|\psi_l)/\sum_{l=1}^{L} \lambda_l(s|\psi_l)$$

is carried out non-parametrically to produce surfaces of these probabilities.

The basic likelihood derived for the multitype situation can be seen as a special case of a ordinal logistic formulation where a probability of a disease type is to be modeled. Different formulations of ordinal logistic regression can be considered, but the most common for nominal categories is the multinomial logit model. If a single common control is assumed it would be very convenient to consider that as a baseline category in the comparison. This is denoted as $L + 1$ below. Hence, a possible multinomial logit model would be

$$\log\left(\frac{p_k(s_{k_i})}{p_{L+1}(s_{k_i})}\right) = \alpha_k + f_l(s_{k_i}|\psi_{l1})$$

$$\text{where } p_k(s_{k_i}) = \lambda_k(s_{k_i}|\psi_k)/\sum_{l=1}^{L+1}\lambda_l(s_{k_i}|\psi_l).$$

Again it would be straightforward to define a Bayesian hierarchical model around this formulation. For example, $\log\left(\frac{p_k(s_{k_i})}{p_{L+1}(s_{k_i})}\right) = \alpha_k + f_l(s_{k_i}|\psi_{l1})$ where $f_l(s_{k_i}|\psi_{l1}) = w_{l_i} + v_{l_i}$ where w_{l_i}, v_{l_i} are spatially-correlated and uncorrelated random effects. This would allow separate random effects for each disease. Zhou et al. (2008a) gives an example of Bayesian formulation where spatial correlated effects are modeled with categorical ordinal outcomes, and these models could be modified for the simpler nominal case.

As an example of the application of this multinomial logit model, the Arbroath case event data has been examined. Only one control is assumed and it is regarded as the comparison group here. In the following, the composite lower body cancer is the control disease (label 1), followed by gastric and oesophageal cancer (label 2), respiratory cancer (label 3), and bronchitis (label 4). The multinomial logit model was fitted assuming a simple random effect model with y_{il} denoting a sparse indicator variable of dimension $N_T \times L$, where $N_T = \sum_l N_l$, which takes values as follows

$$y_{il} = \begin{cases} 1 \text{ if } l = 1 \\ 0 \text{ otherwise} \end{cases} i = 1, 250, l = 1, .., L \ .$$

$$y_{il} = \begin{cases} L \text{ if } l = L \\ 0 \text{ otherwise} \end{cases} i = 437, 630, l = 1, .., L \ .$$

In the Arbroath example, $L = 4$, and $N_T = 630$ with 250 control cases, 90 gastric and oesophageal cancer, 97 respiratory cancer, and 193 bronchitis case events. Assume that

$$y_i \sim Mult(p_i, 1)$$

where

$$\Pr(y_{il} = 1) = p_{il} = \lambda_l(s_{l_i}|\psi_l)/\{1 + \sum_{k=2}^{L}\lambda_k(s_{l_i}|\psi_k)\} \ l > 1$$

$$\Pr(y_{i1} = 1) = p_{i1} = 1/\{1 + \sum_{k=2}^{L}\lambda_k(s_{l_i}|\psi_k)\}$$

FIGURE 10.7: Arbroath posterior expected probability surface ($\overline{p}_{il}$) for composite control (left), and gastric and oesophageal cancer (right).

with $\lambda_l(s_{l_i}|\psi_l) = \exp(\alpha_l + w_{il})$ where

$$\alpha_l \sim N(0, \tau_l)$$
$$w_{il} \sim N(0, \tau_{w_l})$$

with a separate intercept and a simple uncorrelated effect for each disease. Suitably dispersed prior distributions were assumed for the variance parameters. More complex random effect structures could be envisaged of course. However this formulation serves to demonstrate the modeling approach. Following convergence the DIC for this model was 3054.97 with pD = 579.14. The model provides relative estimates of disease probabilities as well as scalar parameters. The posterior expected estimates of α_l, $l = 2, 3, 4$ with sd in brackets: $-2.455(0.2331)$, $-2.078(0.2208)$, $-0.234(0.1921)$. It appears from this that gastric and oesophageal cancer and respiratory cancer have significantly different overall levels compared to the combined control whereas the bronchitis level is not significant.

Figures 10.7 and 10.8 display the posterior expected estimates for the p_{il} for the control ($l = 1$), gastric and oesophageal cancer ($l = 2$), respiratory cancer ($l = 3$), and bronchitis ($l = 4$). It is noticeable that the control has a highly variable distribution, while, in relation to control, the gastric and oesophageal cancer appears with peaks in markedly different locations (the respiratory cancer has a similar patterning). The bronchitis distribution also differs from control but seems to have larger areas of elevated risk.

It is of course possible to extend this approach to models with more sophisticated random components (such as CH components based on full MVN prior specification or approximate MRF models based on Voronoi neighborhoods). Indeed models including correlation between diseases, with share components or cross-correlation could be specified. These are largely unexplored in this application area.

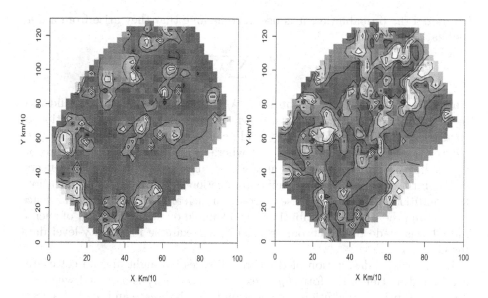

FIGURE 10.8: Arbroath posterior expected probability surface ($\bar{p}_{il}$) for respiratory cancer (left), and bronchitis (right).

10.3.2 Count Data

In the case of count data various possibilities exist. Assume that there are $\{y_{il}\}$, $l = 1, .., L$ and $i = 1, ..., m$, with $L > 2$. Hence in each area there is a vector of counts representing the L different diseases. Various approaches can be adopted depending on the focus. First, by conditioning on the total count within the small area: $y_{T_i} = \sum_{l=1}^{L} y_{il}$, it is possible to consider the multinomial distribution for the count probability vector $\mathbf{y}_i$. On the other hand, it is also possible to examine the unconditional distribution of the counts assuming conditional independence and a Poisson count distribution. In the first case, assume that

$$\mathbf{y}_i \sim Mult(\mathbf{p}_i, y_{T_i})$$

and, because of the constraint, the probability vector is defined to be $0 < p_{il} < 1$, and $\sum_{l=1}^{L} p_{il} = 1 \; \forall i$. The log-likelihood is then considered to be

$$l(\mathbf{y}|\mathbf{p}) = \sum_{i=1}^{m} \sum_{l=1}^{L} y_{il} \log p_{il}.$$

To model the probabilities, it is useful to assume that they arise from a normalization such as:

$$p_{il} = \frac{\lambda_{il}}{\sum_k \lambda_{ik}}$$

where the constant term in the rate (λ_{il}) cancels out. Here it would also be convenient to assume that rate terms consist of a log linear function of covariates or random effects. A typical general example could be $\lambda_{il} = e_{il}\theta_{il}$ where $\theta_{il} = \exp\{\alpha_l + \mathbf{x}_i'\beta + u_{il} + w_{il}\}$ where α_l is a disease-specific intercept, $\mathbf{x}_i'\beta$ a linear predictor, and $\mathbf{u}_l$, and $\mathbf{w}_l$ are disease specific random effects. Other forms are possible in specific applications. Clearly by normalization, the conditioning on the total disease count in each area yields relative inference concerning the disease distribution. An example of the application of such a model was made to the chronic three-disease example for county-level data for Georgia.

In this case a description of the three diseases is sought and no covariates are included. Hence the form $\theta_{il} = \exp\{\alpha_l + u_{il} + w_{il}\}$ is assumed where u_{il} has a CAR prior distribution specification for each disease and w_{il} has a zero-mean Gaussian specification. In that analysis the converged sampler yielded a DIC of 1879.9 with pD = 240.112. The resulting spatially-correlated random effects for the three diseases are shown in Figures 10.9, 10.10, and 10.11. It is clear that under this multinomial model the spatially-structured risk is quite different for each of the three cases. In fact the distribution of high risk areas of COPD seems to be inversely related spatially to those of angina.

Alternative formulations of multivariate risk can be envisaged. In fact the shared component models discussed in Section 10.2.2 have been extended to multiple diseases by Held et al. (2005). In their formulation a Poisson likelihood is assumed:

$$y_{il} \sim Poisson(e_{il} \exp[\eta_{il}])$$

and

$$\eta_{il} \sim N(\alpha_l + \sum_k \delta_{k,l} u_{ki}, \tau_l)$$

and

$$\sum_{l=1}^{n_k} \delta_{k,l} = 0$$

and the terms $\log \delta_{k,1}, \ldots, \log \delta_{k,n_k}$ have multivariate normal distribution with mean zero and given marginal variance. In an application to spatial health surveillance data, Corberán-Vallet (2012) proposed the use of L shared components for K diseases as:

$$\eta_{il} = \rho_k + \sum_{l=1}^{L} \phi_{l,k} \delta_{l,k} w_{l,i} + \psi_{ik}$$

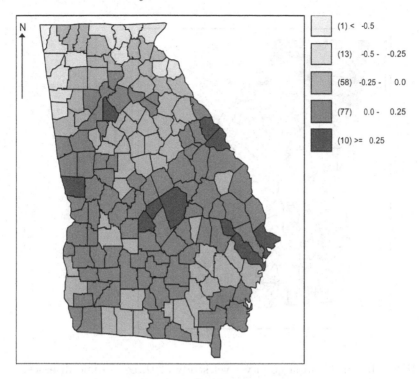

FIGURE 10.9: Georgia county level study of three chronic diseases, 2005: asthma: spatially-correlated random effect (u_1).

where $\phi_{l,k}$ is a binary inclusion variable with $\phi_{l,k} \sim Bern(p_l)$ essentially including w_l or not. $\delta_{l,k}$ are scaling paramaters measuring the effect of w_l on disease k. The w_l is the l th spatial field and assumed to have a Markov random field (MRF) prior distribution. L is unobserved in this formulation.

10.3.3 Multivariate Spatial Correlation and MCAR Models

10.3.3.1 Multivariate Gaussian Models

In general, once multiple diseases are admitted into an analysis there is a need to consider relations between the diseases. This can be done in a variety of ways. A basic approach to this is to consider cross-correlation between the diseases. There is a considerable literature on the specification of cross-correlation models for Gaussian processes (see e.g. Banerjee et al., 2014; Martinez-Beneito et al., 2017). In general, define an L-dimensional vector $\mathbf{Y}_i$ $i = 1, ..., m$ observed at a set of sites. For a multivariate Gaussian process, a common assumption would be that

$$\mathbf{Y} \sim \mathbf{MVN}_{mL}(\mu, \mathbf{A_Y})$$

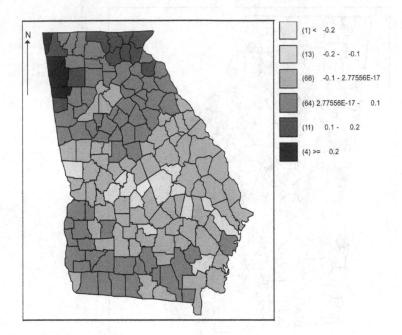

FIGURE 10.10: Georgia county level study of three chronic diseases, 2005: COPD: spatially-correlated random effect (u_2).

where μ is $m \times L$ and $\mathbf{A_Y}$ has dimension $mL \times mL$. It is convenient to consider a block representation of $\mathbf{A_Y}$ which stresses the covariance in cross-covariance form:

$$\mathbf{A_Y} = \left\{ \begin{array}{llll} A_{11} & A_{12} & . & A_{1L} \\ A_{21} & A_{22} & . & . \\ . & . & . & . \\ A_{L1} & & & A_{LL} \end{array} \right\}$$

where each of the diagonal block matrices are internal covariances within the given field whereas the off-diagonal block matrices define the cross-correlations between components. The dimension of the block matrices is $m \times m$ if all the fields are observed at the same m locations (sites), whereas if the different fields are measured at different numbers of sites then each diagonal matrix will be square and have different dimensions and the off-diagonals will not necessarily be square either.

Various models can be assumed for the overall covariance structure of a set of Gaussian fields. Often simple assumptions are made to allow for computation. Banerjee et al. (2014) discuss various examples of separable models and asymmetric cases (mainly for simple situations where each field is mea-

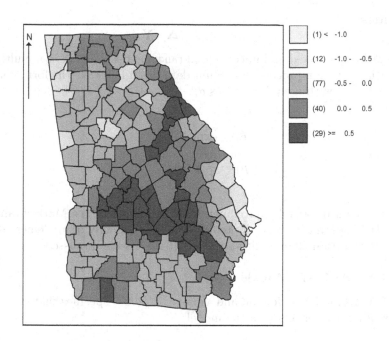

☐	(1) < -1.0
☐	(12) -1.0 - -0.5
▨	(77) -0.5 - 0.0
▩	(40) 0.0 - 0.5
■	(29) >= 0.5

FIGURE 10.11: Georgia county level study of three chronic diseases, 2005: angina: spatially-correlated random effect (u_3).

sured on the same grid). They also extend the analysis by considering the linear model for coregionalisation (LMC) which specified that a multivariate process is a linear function of *iid* spatial processes with zero mean, variance 1 and spatial covariance function $\rho(h)$ for distance h. More generally separate covariance functions $\rho_l(h)$ can be assumed so that the cross-covariance is defined as $A_{ll'} = \sum_{j=1}^{L} \rho_j(s - s')T_j$ for locations s and s', where T_j is the covariance matrix for the j th component. An alternative, computationally attractive, conditional specification was also proposed by Royle and Berliner (1999).

While in general full multivariate cross-correlation models could be employed for modeling continuous multivariate spatial processes, their implementation is not straightforward and in particular their computational demands often force the consideration of simpler formulations. Note that within a disease mapping context these models could form joint prior distributions for spatial random effects (rather than models for observed Gaussian fields), especially for case event models where continuous spatial effects are naturally favored. Hence for the i th case event we might be interested in the vector of intensities:

$$\lambda(s_i|\psi) = \exp{(\mathbf{\Delta}_i + \mathbf{Y}_i)}$$

where $\mathbf{\Delta}$ includes fixed and uncorrelated random effects and $\mathbf{Y}$ is a multivariate spatial Gaussian process. For count data this might take the form, for the i th small area with area denoted as a_i:

$$\theta_i = \exp(\mathbf{\Delta}_i + \mu_i)$$

$$\mu_i = \int_{a_i} \mathbf{Y}(u)du.$$

Often for count data and approximately for case event data a Markov random field (MRF) specification is adopted at least for simplicity of implementation. In the next section these multivariate CAR models are discussed.

10.3.3.2   MVCAR Models

The MVCAR model of Gelfand and Vounatsou (2003) specifies that the $m \times L$ matrix of random effects ϕ in the model

$$y_{il} \sim Poisson(e_{il} \exp[x_{il}'\beta_l + \phi_{il}])$$

is defined with a constraint that the spatial effects separate into non-spatial and spatially-structured effects:

$$\phi \sim \mathbf{N}_{mL}(\mathbf{0}, \mathbf{H}_1)$$

where $\mathbf{H}_1 = [\mathbf{\Lambda} \otimes (D - \alpha W)]^{-1}$ with $\otimes$ denoting Kronecker product , and D is a $m \times m$ diagonal matrix with elements which are the number of neighbors of the i th region and W is an adjacency matrix where $W_{ii} = 0$ and $W_{ij} = 1$ if the areas i, j are adjacent (i.e. $i \sim j$) and 0 otherwise.   Here $\mathbf{\Lambda}$ is a $L \times L$ positive definite matrix of non-spatial precisions, defining the relation between diseases and α is a common spatial autocorrelation parameter. This is denoted as the $MCAR\,(\alpha, \Lambda)$ model and it can be extended to allow for separate autocorrelation (smoothing) for each disease:

$$\phi \sim \mathbf{N}_{mL}(\mathbf{0}, \mathbf{H}_2)$$

where $\mathbf{H}_2 = [Q(\mathbf{\Lambda} \otimes I_{m \times m})Q']^{-1}$ and $Q = diag(R_1, ..., R_L)$ and $R_l = chol(D - \alpha_l W)$, $l = 1, ..., L$, where $chol()$ denotes the Cholesky decomposition. This has been termed the $MCAR(\alpha, \mathbf{\Lambda})$. Extensions and variants to these models have been proposed by Kim et al. (2001) and Jin et al. (2005). Restriction to the conditional ordering of the effects in the GMCAR model of Jin et al. (2005) led to a different approach.

10.3.3.3 Linear Model of Coregionalization

A classic approach to modeling cross-correlation between spatial fields is to adopt a simple model for the relation between selected fields. Within geostatistics, the linear model of coregionalisation (LMC) is commonly assumed for this purpose (Wackernagel, 2003; MacNab, 2016). In that model a set of random spatial functions $\{z_l(s); l = 1, ..., L\}$ is modeled via a linear combination of uncorrelated factors $(Y_u^l(s)$) and $u = 1, .., S$ components:

$$z_l(s) = \sum_{u=0}^{S} \sum_{l=1}^{L} a_u^l Y_u^l(s).$$

This idea has been used by Jin et al. (2008) in extending the multivariate models for disease mapping to allow order-free modeling. In their formulation the model at the likelihood level is

$$y_{il} \sim Poisson(e_{il} \exp[x_{il}'\beta_l + \phi_{il}])$$

where ϕ_{il} are random effects for each unit and disease. The joint distribution of ϕ is defined to be

$$\phi \sim \mathbf{N}_{mL}(\mathbf{0}, \mathbf{G})$$

where $\mathbf{G} = (A \otimes I_{m \times m})(I_{L \times L} \otimes D - B \otimes W)^{-1}(A \otimes I_{m \times m})'$ with $\otimes$ denoting Kronecker product and B includes smoothing parameters in the cross-covariances of the field, D is a $m \times m$ diagonal matrix with elements which are the number of neighbors of the i th region and W is an adjacency matrix where $W_{ii} = 0$ and $W_{ij} = 1$ if the areas i, j are adjacent (i.e. $i \sim j$) and 0 otherwise.

This is defined as a $MCAR(B, \Sigma)$ distribution , where the diagonal elements of B are correlated within a spatial process and the off-diagonals are the cross-correlations between any two processes. These are scalar quantities. Essentially, the difference with the $MCAR(\alpha, \mathbf{\Lambda})$ lies in the the elements of B: if $b_{jl} = 0$ and $b_{jj} = \alpha_j$ then the $MCAR(\alpha, \mathbf{\Lambda})$ results. Special prior distribution constructions must be examined to ensure that the eigenvalues of B lie in the correct range. Jin et al. (2008) give examples of its use and compare different formulations.

As with CAR models, the Leroux formulation can also be extended to the multivariate domain. Assume a spatial random effect for K diseases and m areas as ϕ_{ik} where $\phi = (\phi_1, ..., \phi_K)$ where $\phi_k = (\phi_{1k}, ..., \phi_{mk})$ and

$$\phi \sim \mathbf{N}(0, [\mathbf{Q}(\mathbf{W},\rho) \otimes C^{-1}]^{-1}). \qquad (10.5)$$

$\mathbf{Q}(\mathbf{W},\rho) = \rho[diag(\mathbf{W}1) - \mathbf{W}] + (1 - \rho)I$ is the precision matrix for the joint distribution and C a cross-variable covariance matrix. This can be programmed in Win/OpenBUGS and is available directly in CARBayes.

10.3.3.4   Model Fitting on WinBUGS

Currently, only the intrinsic (improper) version of the $MCAR\,(\alpha, \Lambda)$ model is available automatically on WinBUGS. This version forces the value of $\alpha = 1$, and this implies that it can be used as an prior distribution only, assuming that propriety of the posterior distribution can be assured. The command in WinBUGS for this is the mv.car distribution. It is also possible to fit a proper $MCAR\,(\alpha, \Lambda)$ if it is assumed, via the LMC, that

$$\phi = (A \otimes I_{m \times m})\mathbf{u}$$

where $\mathbf{u}_l\; l = 1, ..., L$ are assumed to have proper univariate CAR prior distributions (on WinBUGS: car.proper distribution) with common smoothing parameter α. This fixes A as it is the Cholesky decomposition of Λ although it may be preferred to allow a separate prior specification for Λ. Again by the LMC it is possible to extend this idea to fitting $MCAR(\alpha, \Lambda)$ models. In that case, as before, assign proper univariate CAR prior distributions to $\mathbf{u}_l$ $l = 1, ..., L$ but with separate smoothing parameters: $\alpha_l,\; l = 1, ..., L$. Assuming an inverse Wishart prior distribution for Λ determines A.

10.3.4   Georgia Chronic Ambulatory Care-Sensitive Example

In the example above concerning chronic ambulatory care-sensitive diseases, three diseases were examined: asthma, COPD, and angina. These were examined as counts for the year 2005 in Georgia counties. In Section 10.2.3, the analysis of these diseases was limited to two diseases only. Here all three are considered together in a multivariate framework. It is assumed that each disease has a log-linear link to a linear predictor which consists of random effect components. In particular it is assumed that two additive random effects are included in the form

$$\log(\mu_{l_i}) = \log(e_{l_i}) + \alpha_l + W_{l_i} + U_{l_i}$$

where

$$U_{l_i} \sim MVN(\mathbf{0}, \Sigma)$$
$$W_{l_i} \sim MCAR(1, \Omega).$$

The first effect is uncorrelated with zero mean and diagonal covariance matrix where $\Sigma = diag(\tau_1, ..., \tau_L)$. For the second term an intrinsic CAR model was assumed using the mv.car distribution. The 3×3 precision matrix has assigned to it a Wishart prior distribution with parameter matrix R, and the covariance matrix defined as Ω^{-1}. Additional assumptions about the model components were made. These include a Wishart prior distribution of the precisions of the uncorrelated effects (Σ^{-1}) and flat (uniform) priors for the intercept terms (α_l). The following code was used to specify the covariance priors and resulting standard deviations:

```
omega[1:3, 1:3] ~dwish(R[ , ], 3)
sigma2[1:3, 1:3] <- inverse(omega[ , ])
sigma[1] <- sqrt(sigma2[1, 1])
sigma[2] <- sqrt(sigma2[2, 2])
sigma[3]<-sqrt(sigma2[3,3])
```

The sigma[1:3] terms are conditional standard deviations of the disease components (1: asthma, 2: COPD, 3: angina). It is also possible to compute the correlations between spatial components in this formulation:

```
corr12<-sigma2[1,2]/(sigma[1]*sigma[2])    # asthma and COPD
corr13<-sigma2[1,3]/(sigma[1]*sigma[3])    # asthma and angina
corr23<-sigma2[2,3]/(sigma[2]*sigma[3])    # COPD and angina
```

The above analysis allows the computation of the correlations between the spatial random effects for each disease. This is useful but it is sometimes helpful to also consider the correlation of the relative risks directly. To compute the correlation between the relative risks it is possible to examine functionals of the posterior distribution. This can be easily implemented in WinBUGS with the commands below. RR1, RR2, RR3 are the relative risks for the asthma, COPD, and angina where within a for() loop with index i they are assigned as

```
RR1[i] <- exp(alpha[1] + S[1,i]+U[i,1])
RR2[i] <- exp(alpha[2] + S[2,i]+U[i,2])
RR3[i] <- exp(alpha[3] + S[3,i]+U[i,3])
```

where S is the MCAR and U represents (spatially) uncorrelated random effects. Then the empirical posterior correlation can be computed as:

```
mu1<-mean(RR1[])
mu2<-mean(RR2[])
mu3<-mean(RR3[])
sd1<-sd(RR1[])
sd2<-sd(RR2[])
sd3<-sd(RR3[])
mu12<-inprod(RR1[],RR2[])/N
mu13<-inprod(RR1[],RR3[])/N
mu23<-inprod(RR2[],RR3[])/N
CRR12<-(mu12-mu1*mu2)/(sd1*sd2)
CRR13<-(mu13-mu1*mu3)/(sd1*sd3)
CRR23<-(mu23-mu2*mu3)/(sd2*sd3)
```

The results of fitting this model to the three diseases yielded the following results. After convergence, the DIC was 3226.61 with pD as 392.92. This is

Disease	Asthma	COPD	Angina
Asthma		0.6617(0.128)	0.7536(0.105)
COPD	0.6004(0.033)		0.7825(0.099)
Angina	0.5417(0.044)	0.516(0.042)	

Table 10.2: Correlation of spatial random effects under MCAR model for Georgia chronic disease example: upper triangle, correlation of spatially structured effects; lower triangle, correlation of relative risks

ρ	Mean	95% Credible Interval
Asthma	0.9662	(0.8776, 0.9991)
COPD	0.9435	(0.8039, 0.9984)
Angina	0.9021	(0.6888, 0.9969)

Table 10.3: Posterior expected estimates for mixing parameters for multivariate Leroux model applied to Georgia 2005 data

slightly higher than the DICs for the separate analyses (DIC = 3225.5 and pD= 394.8). However this is not a large difference and so there appears to be little differentiating among models except that the MCAR model provided estimates of correlation and a lower pD of 392.9. From the point of view of parsimony, this might be preferred. The correlation estimates are given in Table 10.2. The top right hand triangle shows posterior expected correlation estimates from the posterior distribution for the spatially-structured random effects, while the lower triangle contains the empirical correlations calculated from the posterior sample for the estimated relative risks. It is interesting to note that the correlations between the spatially-structured effects are higher for angina versus COPD and asthma than those correlations for the relative risks. This effect might suggest more uncorrelated noise in the angina relation than in the asthma versus COPD relation, which remains relatively stable. In general, there does appear to be a strong correlation between the risks for these chronic diseases. The posterior mean relative risks are shown in Figure 10.12.

As a comparison of approaches, the Georgia example was also analysed using a multivariate version of the Leroux model (10.5). The DIC obtained for this model was 3203.64 with pD = 403.76. This is lower than the convolution model with MCAR spatial effect, although the pD is larger. A variant of the model (10.5) was employed whereby the correlation parameter is allowed to vary with disease:

$$\phi \sim \mathbf{N}(\mathbf{0}, [\mathbf{Q}(\mathbf{W}, \rho) \otimes C^{-1}]^{-1}).$$

Figure 10.13 displays the resulting posterior expected estimates of ϕ for the Georgia 2005 data. The posterior expected estimates of the mixing parameter ρ are given in Table 10.3

This suggests that most diseases are dominated by the spatially structured component, although angina is less well described.

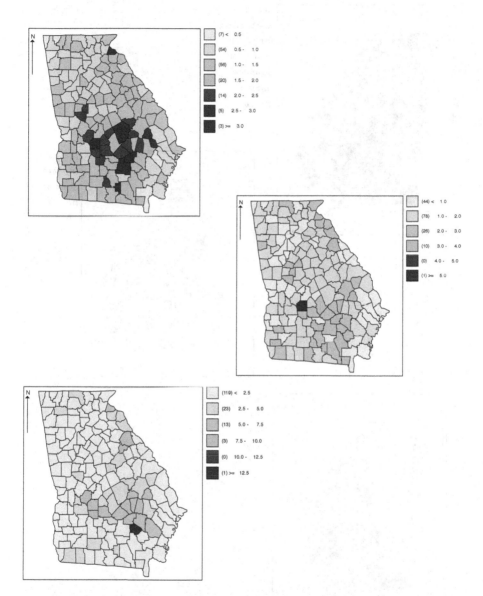

FIGURE 10.12: Georgia county level three-disease analysis, 2005: posterior expected relative risks under an additive UH and MVCAR model: top left: asthma, middle: COPD, bottom left: angina.

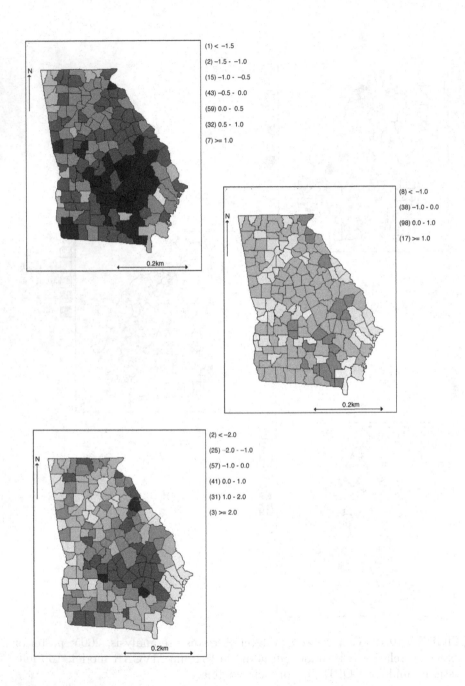

FIGURE 10.13: Georgia county level three-disease analysis, 2005: posterior expected estimate of spatial component under a multivariate Leroux model: top left: asthma, middle: COPD, bottom left: angina.

11

Spatial Survival and Longitudinal Analysis

11.1 General Issues

In many biostatistical applications there is a need to consider temporal variation. The most common examples are often found in clinical or behavioral intervention trials where a specific state can be reached by a patient. The time at which the state is reached could be of primary interest. The endpoint could be a vital outcome (death) disease remission, cure or cessation of a behavior. In all cases the time of the event is the important random variable. This is the typical scenario where *survival analysis* is employed. On the other hand, in some clinical or intervention studies, the variation of response over time is to be monitored. For example, cholesterol concentrations in blood might be monitored under different treatments, and the effects of these treatments over time examined. In an intervention trial for diet change, food intake might be repeatedly measured via self-report questionnaire. In these cases, the time of measurement is usually fixed and the measurement itself is the random variable.

In clinical trial applications there is not often a need, or interest, in examining the residential address of the patient or the even the neighborhood or county of residence (Savoca et al., 2017). As clinical trials are designed experiments and usually randomized, there should be less need to consider location or area of residence as a factor affecting outcome. Two factors should be considered, however, that could impact a decision to ignore spatial effects. First, at the design stage there could be a geographical bias in the areas displaying risk for the disease. This bias could manifest itself by differential recruitment. Any strong geographical differences might need to be represented in the study design if the study is to be representative of the population concerned. Even after design, in the analysis phase, there could be strong reasons to examine spatial effects. First, it could be important to know about any spatial effects such as confounding between space and intervention groups, in relation to outcome. Second, all studies whether designed or observational, have confounders which could be unknown or known but not included in the study. Hence if the researcher wants *a)* to explore factors affecting the study, and *b)* to make sure that confounding is allowed for in the analysis, then the use of spatial information can help with both these tasks. Clayton et al. (1993)

have stressed the usefulness of including spatial correlation terms in models to make allowance for confounder and ecological biases and so there are reasonably strong arguments for always including contextual terms that have spatial structure.

11.2   Spatial Survival Analysis

In survival analysis the time to endpoint (T) is the random variable. Assume that a sample of individuals has associated endpoint times. Denote this sample of times as $\{t_i\}$ $i = 1, ..., m$. For now assume these are all observed exactly. In addition to an endpoint time a geo-reference is also available. The geo-reference could be an address location, in which case it is denoted as s_i, or a contextual spatial effect, denoted as w_i. The contextual effect is simply a factor that may have spatial correlation so that areas close together have similar risk. In addition to these ingredients, each observation unit can have covariates associated and, for the i th unit/person these are denoted by the vector $\mathbf{x}_i$. These covariates could be individual, contextual or ecological (see e.g. Chapter 7.7). Reviews of Bayesian survival methods can be found in Gustafson (1998) and Ibrahaim et al. (2000). For Bayesian proportional hazard modeling see Carlin and Hodges (1999). A recent review of spatial survival is provided in Banerjee (2016). Early examples of the inclusion of spatial effects in survival models include Henderson et al. (2002), Banerjee et al. (2003), Carlin and Banerjee (2003), Cooner et al. (2006), Diva et al. (2008).

11.2.1   Endpoint Distributions

Often a failure time or endpoint distribution is specified for the time to endpoint and this is often chosen from distributions on the R^+ line. Common choices are among the Weibull or extreme value, log normal, or gamma families. Hence for a parametric survival model, at the first level of the hierarchy, the data model consists of an endpoint distribution. For flexibility, we will assume a Weibull distribution in what follows. The probability of an endpoint at time t_i under a Weibull distribution is specified by

$$f_T(t_i) \equiv f(t_i) = \rho \mu t_i^{\rho-1} \exp(-\mu t_i^\rho). \tag{11.1}$$

The survival and hazard functions derived from this specification are

$$S(t_i) = 1 - \int_0^{t_i} f(u)du = \exp(-\mu t_i^\rho)$$

$$h(t_i) = f(t_i)/S(t_i) = \rho \mu t_i^{\rho-1}.$$

The parameterisation emphasizes the modeling of a function of the mean of the distribution via μ. Note that this allows a straightforward interpretation of the model component for this distribution: covariates and contextual effects can be included within μ and the parameter ρ provides the shape of the distribution. Often modeling proceeds via the hazard function, rather than the density, and for the Weibull this decomposes into two components:

$$h(t_i) = h_0(t_i).h_1(t_i) = \rho t_i^{\rho-1}.\mu_i.$$

Here $h_0(t_i)$ is regarded as a baseline hazard, while $h_1(t_i)$ is a non-baseline component which is usually the focus of modeling. Usually it is assumed that a predictor term is linked to the parameter μ_i and each unit will have a different μ_i depending on covariates. A log-linear specification is often assumed. A non-spatial example could be

$$\log(\mu_i) = \mathbf{x}_i'\beta + v_i \tag{11.2}$$

where $\mathbf{x}_i'$ is a row vector of fixed covariates, β the corresponding parameter vector and v_i is an uncorrelated random effect. For the Weibull distribution, Carlin and Hodges (1999) suggested the formulation (11.2) in a non-spatial setting. This can be extended easily to include a spatial contextual effect. For example, we could specify

$$\log(\mu_i) = x_i'\beta + \Omega_i \tag{11.3}$$
$$\Omega_i = w_i + v_i$$

where Ω_i contains the unit specific random effect terms. These effects could be individual unit level or contextual.

While the Weibull is a flexible distribution there are many alternatives, some of which do not impose the constraints of proportionality of hazard. A wider class of model is the accelerated failure time (AFT) model. These models replace the t within the survival and hazard functions with a modulated function of covariates: $t \exp\{x'\beta\}$. This leads to a covariate acceleration or deceleration of risk. The Weibull is a special case of this general class. This leads to a linear model in the log of time:

$$\log T = \alpha + x'\beta + \epsilon$$

where ϵ is an error term independent of x (not necessarily zero centered) and α is an intercept. This model could also be extended with addition of random effects which are individual or contextual in the form of

$$\log T = \alpha + x'\beta + \Omega + \epsilon$$

where Ω is a random effect term (as in (11.3) above). Examples of the application of AFT models in spatial survial contexts can be found in Zhang and Lawson (2011), Lawson et al. (2014), and Onicescu et al. (2017a).

11.2.2 Censoring

Censoring is always an important issue in survival analysis as it is often the case that times are not observed exactly. For parametric models this can be treated via a survival function term product in the likelihood. For example for right censoring we can assume

$$L = \prod_u f(t_u) \prod_c S(t_c)$$

where u denotes uncensored and c denotes right censored. Note that this likelihood simplifies if you assume a censoring indicator γ which takes 0 for censored and 1 for uncensored, as

$$L = \prod_{all\ t} h(t)^\gamma S(t).$$

Other likelihood forms can similarly be derived for alternative censoring mechanisms.

11.2.3 Random Effect Specification

As most survival data is observed at the individual unit level there could be individual covariates, random effects or contextual effects relating to the individual. For example, the age of an individual could be a personal covariate and the location coordinates of the individual's address could be regarded as personal covariates also. In addition, there could be an individual level random effect which allows for frailty amongst individuals. This could be correlated spatially or uncorrelated. The residential address of the individual (unit) if known could be used directly. An uncorrelated frailty can easily be specified for each individual via

$$v_i \overset{iid}{\sim} N(0, \tau_v).$$

For a spatially-correlated term, w_i could be specified to have a full multivariate normal prior distribution with spatial correlation included within the covariance matrix, i.e.

$$\mathbf{w} \sim \mathbf{N}(\mathbf{0}, \mathbf{\Sigma}) \qquad (11.4)$$

where the elements of the covariance are functions of distance between locations: $\mathbf{\Sigma}_{ij} = f(d_{ij})$ where d_{ij} is the distance between the i th and j th location. Essentially this is a zero-mean spatial Gaussian process prior distribution. Note that it would also be possible to specify an intrinsic Gaussian prior distribution for $\mathbf{w}$ as long as a suitable neighborhood structure could be specified. A distance-threshold or Dirichlet tesselation neighborhood metric could be used to define neighbors. While it is commonplace to model uncorrelated frailty at the individual level, it is less common to find modeling of spatially-correlated effects at this level.

It is relatively straightforward to include spatial effects as contextual effects. For example, with cancer registry data, often individual outcomes include associated county, postal district, or ZIP code of residence. They often don't show an address location for confidentiality reasons. Hence contextual spatial information is often available for these data. This consists of aggregated spatial groupings or factors. For instance, in

$$\Omega_i = w_i + v_i$$

the two effects could be specified as spatially-contextual. For example, assume there are $j = 1, ..., J$ counties within which individuals can live. The county of residence for the i th individual is denoted $c_j \atop i \in j$ so that:

$$w_i = w(\underset{i \in j}{c_j}).$$

Hence, the effect for the i th person is assigned from the j th county where he or she lives. This also applies to any other contextual aggregate effects: for example, uncorrelated effects could also be defined in this way:

$$v_i = v(\underset{i \in j}{c_j}).$$

Henderson et al. (2002) modeled individual level frailty but resorted to modeling the spatial structure of leukemia survival data using district level (contextual) effects. In that work the covariance function $f(.)$ between the i th and j th *district* was modeled via the power exponential and Matérn covariance functions (as in Section 5.4.2). At this level of aggregation, it is natural to consider a CAR specification of the spatial effect at individual level:

$$w_i | w_{-i} \sim N(\overline{w}_{\delta_i}, \tau_w / n_{\delta_i})$$

where $w_i = w(\underset{i \in j}{c_j})$. Henderson et al. (2002) also compared their contextual models with fully specified covariance and CAR components and found that a Matérn covariance, specified with correlation function ($\rho_{ij} = \Sigma_{ij}/\tau$):

$$\rho_{ij} = (d_{ij}/\phi)^\kappa K_\kappa(d_{ij}/\phi)/[2^{\kappa-1}\Gamma(\kappa)] \tag{11.5}$$

where $K_\kappa(.)$ is a modified Bessel function of the third kind with $\kappa = 2$, gave the best fitting model based on the DIC goodness-of-fit criterion. Hence, in this case, the geostatistical model yielded a more appropriate model than the CAR model for district effects. This was not found to be the case in other comparisons of Gaussian prior distributions (e.g. see Best et al., 2005).

Banerjee et al. (2003) first reported the use of spatially-correlated frailty in application to linked birth-death individual level infant mortality data for the state of Minnesota. In their formulation a Weibull parametric model was assumed and the hazard was assumed to be

$$h(t_{ij}|\mathbf{x}) = \rho t_{ij}^{\rho-1} \exp\{\mathbf{x}_{ij}'\beta + w_i\}$$

where i th subject in the j th stratum $(i = 1, .., n_j)$, $\mathbf{x}'_{ij}$ is a row vector of covariate values and β the corresponding parameter vector. The random effect w_i was assumed to have either a CAR prior specification, an uncorrelated zero-mean Gaussian prior distribution, or a fully-specified multivariate normal prior specification as in Henderson et al. (2002). In fact, the fully specified model yielded the lowest DIC in model comparison in this case also.

11.2.4   General Hazard Model

Banerjee and Carlin (2003) proposed a relaxation of the Weibull model to allow a semi-parametric formulation whereby, with subject j, $(j = 1, ..., n_i)$ in the i th county:

$$h(t_{ij}|\mathbf{x}_{ij}) = h_{0i}(t_{ij}) \exp\{\mathbf{x}'_{ij}\beta + w_i\}$$

where h_{0i} denotes the county-specific baseline hazard. For an individual with censoring indicator γ_{ij} (0 if alive, and 1 if dead) the likelihood contribution is then

$$h(t_{ij}; \mathbf{x}_{ij})^{\gamma_{ij}} \exp\{-H_{0i}(t_{ij}) \exp\{\mathbf{x}'_{ij}\beta + w_i\}\}$$

where $H_{0i}(t_{ij}) = \int_0^{t_{ij}} h_{0i}(u)du$ is a county-specific cumulative baseline hazard, and the covariates are assumed to be not time-dependent. The baseline hazard appears in this likelihood and so must be estimated. Different approaches have been proposed for estimation of this baseline. One approach assumes a gamma process which is a function of a parametric cumulative hazard (see e.g. Ibrahaim et al., 2000). Another is the use of beta mixtures (Banerjee and Carlin, 2003). A related spatial model was developed by Bastos and Gamerman (2006) whereby they allowed time-dependent covariates which are fixed within small time periods and a CAR spatial frailty. They assume no separate baseline risk however.

11.2.5   Cox Model

The Cox proportional hazards model has been applied in a spatial context by Henderson et al. (2002). In the context of leukemia survival, the authors posited that the partial likelihood could be used without recourse to the estimation of the baseline. For ordered uncensored times $\{t_{(1)}, ..., t_{(m)}\}$, then the partial likelihood is given by

$$\prod_i [\exp\{\mathbf{x}'_{(i)}\beta\} / \sum_{j \in R_i} \exp\{\mathbf{x}'_j\beta\}]$$

where R_i is the set of those individuals at risk just before the i th event time. This can be used for the estimation of regression parameters. This is less parametric than models that include baseline components. Note that spatial effects can be included again as contextual effects by extending the specification of the intensity term $\exp\{\mathbf{x}'_{(i)}\beta\}$ to include random effects:

$$\exp\{\mathbf{x}'_{(i)}\beta + w_i + v_i\} \tag{11.6}$$

where w_i, v_i are random contextual effects which could be at an aggregate level such as district or county. In addition these effects could also be purely individual (in the sense of frailty rather than context).

11.2.6 Extensions

Extensions of the above approaches have been proposed into a variety of more complex applications. For example, spatial cure rate modeling has been proposed by Cooner et al. (2006) and joint spatial survival modeling of date of diagnosis and vital outcome from cancer registry data has been examined by Zhou et al. (2008b). In that approach, prostate cancer registry data was available for the state of South Carolina for 1997 through 2001. The endpoints are defined as T_{1ij} and T_{2ij} and their joint distribution is defined by conditioning as $f_{T_2}(t_{2ij}|t_{1ij})$ and $f_{T_1}(t_{1ij})$. In this formulation, flexible Weibull parametric models were assumed for both $f_{T_2}(t_{2ij}|t_{1ij})$ and $f_{T_1}(t_{1ij})$:

$$f_{T_1}(t_{1ij}) \sim Weib(\gamma_1, \lambda_{1ij})$$
$$f_{T_2}(t_{2ij}|t_{1ij}) \sim Weib(\gamma_2, \lambda_{2ij}|t_{1ij})$$

with log linear models for the scale parameters:

$$\log \lambda_{1ij} = \alpha_1 + \mathbf{x}'_{1ij}\beta_1 + W_{1ij}$$
$$\log \lambda_{2ij} = \alpha_2 + \mathbf{x}'_{2ij}\beta_2 + \beta_{2t}t_{1ij} + W_{2ij}.$$

In the second model, dependence on the the first endpoint is simply via a linear function dependent on t_{1ij}. Different forms of random effect combination were specified within a selection of models for W_{1ij} and W_{2ij}. Figure 11.1 displays the empirical Kaplan-Meyer (K-M) curves for the date of diagnosis (left panel) and vital outcome (right panel) endpoints. The data-based K-M curves are shown in solid line. The posterior predicted survival is shown on these curves (as dashed lines) and seems to approximate well the empirical behavior at least for ages less than 80. Eight models were fitted with different combinations of separate and common random effect structures. Of the eight models, a model with an uncorrelated effect (UH) for age and separate correlated effect (CH) for survival time had the lowest DIC (DIC = 115722, pD = 57.0). The second lowest DIC was for a model with separate UH effects only in each component (DIC = 115725, pD = 55.7). These were denoted model 6 and model 3 respectively.

Figure 11.2 displays various posterior average random effect maps for different fitted models for these data. The top row displays results for the best-fitting model 6. The left panel is the UH map for age-at-diagnosis while the right panel is the CH map for survival time. It is noticeable that the lowered

FIGURE 11.1: Nonparametric (Kaplan-Meier) survival curves compared to posterior predictive survival from the best fitting model based on DIC. Left: age at diagnosis, right: survival time after diagnosis. Solid line is data-based while the dashed line (with 95% credible limits) shows posterior predictive survival under the best-fitting model.

survival appears in the coastal and upstate areas of South Carolina whereas the UH map seems to display a relatively random pattern. The bottom panel highlights the similarity of the UH effect map for other models fitted for survival after diagnosis. It would appear that the CH component is absorbed by a single UH component under these models and so the result is similar to the CH map for the best fitting model.

Another recent extension has been to directly incorporate the spatial effect within the specification of the failure time density, hazard and survival functions. This avoids the assumption of a contextual spatial effect as it directly models the probabilty as a function of time and space. Some recent examples of this approach can be found in Onicescu et al. (2017b), and Onicescu et al. (2017a).

11.3 Spatial Longitudinal Analysis

There are many situations when variation in an individual response is time-dependent, and the focus is the monitoring over time of the associated outcome. The areas of application for these methods are manifold and in particular they are commonly applied in clinical trials and community-based behavioral intervention trials.

In many clinical settings, it is possible to conduct trials of new treatments. These trials are essentially designed experiments where patient outcomes are

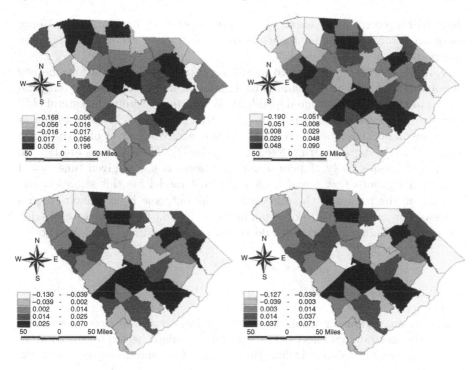

FIGURE 11.2: Posterior average maps of the random effects. Top left: UH effect for age-at-diagnosis for best model, top right: CH effect for survival after diagnosis for the best model; bottom left: UH effect joint model 3, bottom right: UH effect joint model 5, both for survival after diagnosis.

compared between treatment groups. Often these are monitored over time to establish whether a treatment has been effective or not. The simplest case, when only two time points are employed, leads to simple comparisons. However if more than two time points are monitored then considerable complication can arise. In what follows, I will emphasize a general framework for longitudinal analysis which incorporates spatial referencing. Recent general references to this area are Diggle et al. (2002), Verbeke and Molenberghs (2000), and Molenberghs and Verbeke (2005).

Define an outcome for the i th individual at a given time j as y_{ij}. Usually in designed studies the time period is fixed and measurements are made at these fixed times. Hence, the time label denotes a fixed time period. Further define $i = 1, ..., m$ individuals or individual observation units and $j = 1, ..., J$ time periods. The periods could be units of time, such as minutes, hours, days, weeks, years, etc., depending on the study design. Usually recruitment to the study will be randomized in a clinical or intervention trial setting. However in observational studies, where such control is not possible, it is likely that there will be considerable extra variation and potential for imbalance. Hence there are strong reasons for inclusion of confounder or random effects when observational designs are used.

A simple general approach is to consider a linear model formulation whereby the outcome of interest is assumed to depend on underlying parameters, and these parameters can be time-dependent. For example, serum cholesterol (ldl) levels could be measured in patients on a trial for a new cholesterol drug at two time points (baseline and 6 weeks). There are two groups: old drug, fixed dose (ODFD), and new drug, fixed dose (NDFD). These groups are labeled $l = 1$ and 2 respectively. The outcome of interest is y_{ilj} at given times $j = 1$ and 2, and groups $l = 1$ and 2. A statistical model for this situation will depend on the nature of the outcome. If the outcome is continuous then a Gaussian or gamma error model might be assumed. In the cholesterol trial, a continuous Gaussian variate could be assumed, as a first model, and so

$$y_{ilj} \sim N(\mu_{ilj}, \tau_y)$$

where μ_{ilj} is the mean for the i th person at time j in group l, and the τ_y is a variance.

The mean parameter would then be specified as a function of available covariates and random effects. For example, it might be important to allow for age of the individual. Define this as x_{1i}. One simple regression model might be

$$\mu_{ilj} = \alpha_l + \beta_1 x_{1i} + \beta_2 t_j$$

where t_j is the time of the j th period. Here each group has a separate intercept. The regression parameters would be commonly assumed to have zero-mean Gaussian prior distributions with small precisions. In this model a group effect is specified, with a covariate effect and constant linear effect over time. There is assumed to be no correlation in the error. More complex models could of course be envisaged. Addition of random effects can include frailty terms and a temporal dependence term (which can substitute for the regression on t_j):

$$\mu_{ilj} = \alpha_l + \beta_1 x_{1i} + v_i + \eta_j$$

with, possibly, a random walk prior dependence:

$$\eta_j \sim N(\eta_{j-1}, \tau_\eta), \ j > 1,$$

and a zero-mean Gaussian prior distribution with for the individual level frailty:

$$v_i \sim N(0, \tau_v).$$

If inference about spatial effects is important then a variety of possibilities exist. First, if individual level information is available then there are two approaches to the inclusion of space. First it may be possible to estimate an individual level effect directly for residential location data. Denote a spatially-structured effect as w_i. Also assume that $w_i \equiv w(s_i)$ where s_i is the i th residential address. A model could be assumed of the form

$$\mu_{ilj} = \alpha_l + \beta_1 x_{1i} + w_i + \eta_j$$

where w_i can be assumed to have a fully-specified multivariate Gaussian prior distribution and so

$$\mathbf{w} \sim \mathbf{N}(\mathbf{0}, \mathbf{\Gamma})$$

where the elements of $\mathbf{\Gamma}$ are defined as $\mathbf{\Gamma}_{ij} = \tau_w \rho(d_{ij})$ and $d_{ij} = ||s_i - s_j||$. Suitable forms of the correlation function $\rho(.)$ have been discussed elsewhere (Sections 5.4.2 and 11.2). Amongst these the power exponential and Matérn are common choices.

11.3.1 General Model

A general linear mixed model for a continuous outcome can be specified where the vector $\mathbf{y} : \{\mathbf{y}_1, ..., \mathbf{y}_m\}$ where $\mathbf{y}_i = (y_{i1}, ..., y_{iJ})$ is a realization of a random vector from a multivariate normal distribution of form:

$$\mathbf{y}_i = \mathbf{x}_i'\beta + \mathbf{z}_i'\gamma + \mathbf{e}_i \tag{11.7}$$

where $\mathbf{x}_i'\beta$ is a linear predictor (which can include a group indicator), $\mathbf{z}_i$ is a vector of g random effects for the i th individual, γ is a $g \times J$ unit matrix and the model for $\mathbf{e}_i$ is $\mathbf{e} \sim \mathbf{N}(\mathbf{0}, \tau\mathbf{\Sigma})$ where $N = mJ$, $\mathbf{x}'$ is an $N \times p$ matrix of covariates, β is a parameter vector, τ is a variance, and $\mathbf{\Sigma}$ is a block diagonal matrix with $J \times J$ blocks each representing the variance matrix of the vector measurements on a single subject. Various assumptions about the structure of the covariance can be made depending on the focus of the study. I do not pursue this here.

An extension of the approach can be easily made to generalized linear mixed models (GLMMs) whereby (11.7) is modified to allow different distributional assumptions at the data level and a link function is introduced to connect the mean to the linear predictor:

$$\mathbf{y}_i \sim f(\mu_i)$$
$$\text{where } E(\mathbf{y}_i) = \mu_i$$
$$\text{and } g(\mu_i) = \eta_i = \mathbf{x}_i'\beta + \mathbf{z}_i'\gamma.$$

Suitable models for $f(\mu_i)$ could be the Poisson, binomial, gamma, and others in the standard exponential family.

11.3.2 Seizure Data Example

As an example of the effect of spatial structure on the analysis of longitudinal data I have taken a famous data set which was examined by Breslow and

Clayton (1993): the epileptic seizure data (see also Diggle et al., 2002). This consists of a randomized clinical trial of anti-convulsive therapy to inhibit seizures in epilepsy. The trial obtained data from 59 patients, treated in two groups (0, 1: control, treatment) and the seizure count at four time points was monitored. There are three covariates: baseline seizure count (x_{1i}), treatment (x_{2i}), and age (x_{3i}) in years.

In Bayesian hierarchical modeling of longitudinal data there is no need to consider marginal models (Verbeke and Molenberghs, 2000) as a model for the full hierarchy is assumed and fitted. Hence our model will explicitly assume a full hierarchical model with random effects. The basic model is a GLMM with:

$$y_{ij} \sim Poiss(\mu_{ij}) \tag{11.8}$$
$$\log \mu_{ij} = \alpha_0 + \log(x_{1i}) + \alpha_1 x_{2i} + \alpha_2 x_{3i} + \gamma_j. \tag{11.9}$$

Here the baseline count is treated as a log (offset) with no estimable parameter while the treatment group is associated with parameter α_1. Age is also treated linearly with parameter α_2, while an overall intercept is also fitted. Finally the temporal effect in this model is defined by a time dependence. There is no regression on time but a random walk prior specification is used to model the variation. Hence the following is specified in the prior distributions:

$$\alpha_0 \sim U(-1000, 1000)$$
$$\alpha_1 \sim N(0, \tau_{\alpha_1}); \alpha_2 \sim N(0, \tau_{\alpha_2})$$
$$\gamma_1 \sim N(0, \tau_\gamma)$$
$$\gamma_j \sim N(\gamma_{j-1}, \tau_\gamma) \quad j > 1.$$

Regression parameters are given dispersed zero-mean Gaussian prior distributions with $\tau_{\alpha_1}, \tau_{\alpha_2}, \tau_\gamma \sim Ga(0.001, 0.001)$. Table 11.1, line 1 displays the overall goodness-of-fit results for the basic model with no spatial random effects. Under this model, the DIC was found to be 1775.4, with pD = 5.311. To demonstrate the effect of spatial referencing on a clinical trial data set, I have made a random allocation of patients to the 46 counties of South Carolina. While this single realization does not represent the possible spatial variation across counties, it does represent the type of variation one might expect if there were a set of contextual regions (of any size) within which the patients resided. While this initial assumption is crude in that it does not allow for population variation, it should demonstrate less spatial structure than a typically significant spatial confounder. I would then expect that there would be little or no advantage found in adding a spatial contextual effect at the county level. However, as will be seen below, this is in fact not the case, even for a blanket randomization such as this.

To examine the effects of spatial context I have fitted a sequence of simple spatial context random effect models with different combinations of UH effects (v_i) and CH effects (w_i). These are contextual effects and so they are defined

Model	pD	DIC
Basic model	5.311	1775.4
Model 1	33.93	1485.7
Model 2	38.24	1494.2
Model 3	36.0	1488.9

Table 11.1: Comparison of four models for seizure data: basic model and models 1 through 3

as

$$w_i = w(\underset{i \in j}{c_j})$$

$$v_i = v(\underset{i \in j}{c_j}).$$

They are defined as a conventional zero-mean Gaussian prior distribution for the $\{v_i\}$ and an improper CAR prior distribution for the $\{w_i\}$. The models fitted were as follows.

Model 1 is as (11.9) but with v_i added, model 2 is as (11.9) but with w_i added, and finally model 3 is a convolution model with v_i and w_i added to the basic model. The variance parameters for these distributions were assumed to have their distributions defined via standard deviations with $U(0,100)$ distributions (Gelman, 2006). The results of fitting these different models are found on Table 11.1. It is interesting to note that *all* models including spatially-referenced random effects (models 1, 2, 3) have a considerably lower DIC than the basic model. The best fitting model by the DIC criterion is model 1 with only the uncorrelated effect. Although the convolution model yields a slightly higher DIC, it has lower DIC and pD than a model with only the correlated spatial effect.

It is reassuring that the uncorrelated noise model has best fit as we did not present any overt spatial structure in the data. However the fact that any spatially-referenced model (whether correlated or not) leads to much reduced DIC is an important consideration. The reason for the closeness of models 1 and 2 is largely because the CH effect (w_i) will often mimic small scale variation found. Figure 11.3 displays the results of these different models. It is perhaps surprising how clustered even the UH component appears.

Of course one has to be careful of over-interpreting these results as they are based on a *single* realization of synthetic contextual labeling. On the other hand, it is commonly the case that a single geo-referenced sample would be available if the true spatial referencing were known. Finally it is interesting to consider how the parameter estimates for the non-spatial parameters performed under the different models. Table 11.2 displays the estimates for the final converged sample for the basic model and the "best" model (model 1). It is notable that under the random effect model the coefficient for the group

FIGURE 11.3: Posterior average effect maps for seizure example showing correlated and uncorrelated random effects. Top left: model 1 UH, top right: model 2 CH. Bottom row Left and right: model 3 UH, CH.

Model	α_0	α_1	α_2
Basic model	-6.188(0.623)	-0.079(0.046)	0.0101(0.003)
Model 1	-2.368(0.751)	-0.5028(0.074)	0.0207(0.006)

Table 11.2: Posterior average estimates of model parameters under four different seizure models

effect (α_1) becomes significant while under the basic model it is not well estimated. It seems therefore that by reducing the noise in the model the group effect becomes more apparent.

It is also interesting to compare standardized residuals for these models. I have computed the average standardized residual for each individual $\hat{r}_i = \frac{1}{4}\sum_j (y_{ij} - \hat{\mu}_{ij})/\sqrt{\hat{\mu}_{ij}}$ where $\hat{\mu}_{ij}$ is the posterior averaged value of μ_{ij}. Figures 11.4 and 11.5 display these residuals for the basic model and model 1. There is a noticeable reduction in variation in the residuals and a much better fit with model 1. Patients 16, 38, and 52 remain with large residuals but 10, 49, and 19 have been accommodated under model 1. Finally it is clear that random effect models are important here, and of course it would be useful in this modeling to consider *individual* frailty (besides contextual UH effects). Indeed this would often be the first step of extending a model, before considering any spatial effects. In this example an uncorrelated frailty term (constant over time) with $v_i \sim N(0, \tau)$ $i = 1, ..., 59$ leads to a substantial reduction in DIC (DIC =1277.180, pD =59.8) and also a significant group effect is found. Overall,

Model	γ_1	γ_2	γ_3	γ_4
Basic model	4.681(0.614)	4.636(0.614)	4.619(0.616)	4.530(0.615)
Best model (model 1)	0.607 (0.724)	0.559 (0.724)	0.543 (0.725)	0.455 (0.726)

Table 11.3: Parameter estimates for basic model compared to best fitting model (model 1)

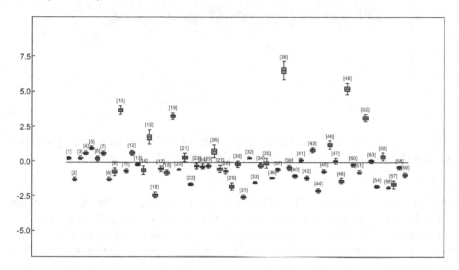

FIGURE 11.4: Standardised average residuals for the 59 individuals providing seizure data under the basic model.

in the "pretend" spatial data set it is clear that extra variation is present. Assuming the spatial referencing was correct, spatial contextual effects appear to be important. It is also clear that individual frailty could be even more important to consider in the context of longitudinal data analysis.

11.3.3 Missing Data

Missing data is a very important aspect of longitudinal studies. Apart from general forms of missing data found in all studies, missingness can often arise in longitudinal studies due to individuals failures to remain in the study. This *dropout*, as it is known, is common in clinical and intervention trials and often by the end of a study 20 to 30% of the participants may have left. There are different forms of missingness mechanisms. Verbeke and Molenberghs (2000) define three basic forms. While it may be important in given applications to consider which of these apply, here I will focus simply on the implications of geo-referencing on missingness. In particular the effect of geo-referencing on

FIGURE 11.5: Standardised average residuals for the 59 individuals providing seizure data under model 1 with spatial contextual random effect.

dropout will be discussed. Usually a geographical delimitation of a study area is made in the design of a trial. During the course of a trial it is possible for participants to move their residence. This is more likely to occur when longer term studies, such as behavioral intervention trials are considered. These trials can last up to 2 or 3 years.

Ultimately, the mobility of participants could lead to a move outside the study area. If there is a geographical limit on the study in terms of recruitment then those who remove from the study area would be regarded as dropouts. With intention-to-treat approaches to trials (Lachin, 2000) it may be important to use all data for those entered into the trial and so it may be important to consider this geographical dropout. Denote the outcome for the i th participant at time j as y_{ij}. Consider residential history defined as $s_i(t)$ where t is continuous and s denotes a spatial location (residence address). A study lasts for $j = 1, ..., J$ time points. Outcome measurements are made at the J time points. Often the residential history will be discretized into the study time periods so that s_{ij} may be censored between time points.

Besides location shifts, it would be useful to define a dropout indicator R whereby r_{ij} denotes presence (1) or absence (0) at the j measurement. While r_{ij} could be 0 for a range of reasons one such could be related to s_{ij}. Hence it might be important to consider (y_{ij}, r_{ij}, s_{ij}) jointly in modeling a given outcome. While drop-out can be modeled jointly with longitudinal outcomes (see e.g. Guo and Carlin, 2004 and Pike and Weissfeld, 2013,for longitudinal-survival and longitudinal-event-time examples), the geo-referencing of dropout appears to have been little explored.

11.4 Extensions to Repeated Events

There are many other application areas where spatial context can be useful. One such area that is closely related to longitudinal analysis is the analysis of repeated events. This is sometimes called repeated event analysis or event history analysis. For a recent review see, for example, Cook and Lawless (2007). The method is an extension of longitudinal and survival analysis where instead of making one measurement at different times or observing a single time-to-endpoint, the time period is fixed and within that period a sequence of events concerning an individual are observed. In the simplest case, the sequence could consist of just a single type of event and its repeated occurrence is observed. Hence for a single person this is a point process in time, assuming the event does not have finite duration. An example of a simple sequence would be a sequence of doctor visits where the time of visit is recorded. If we are only concerned about the times of the visits (and not their nature or duration) this can be considered a point process at the unit (patient) level.

At this point it is worth noting that there is an immediate connection between the linear or generalized linear modeling approach of longitudinal analysis and the time-based analysis for survival data. With repeated events, time is important but when fixed time periods are used and observations are collected within these time periods then the resulting counts of events can be considered within the framework of the conventional hierarchical model. For example for simple doctor visits the count of visits in time periods might be treated at the first level of hierarchy as a Poisson random variable whereas if the times are recorded directly then a (heterogeneous) Poisson process model might be appropriate. This, of course, mirrors the duality of the point process and count model when binning of events takes place in a spatial context.

11.4.1 Simple Repeated Events

Assume first that a patient resides at an address and makes repeated doctor visits. Denote the address as s_i and the sequence of visits as $\{t_{i,1}, ..., t_{i,n_i}\}$ where t denotes visit time within a study period (t_0, t_T) and n_i is the number of visits. If some basic assumptions are made concerning conditional independence given knowledge of all confounding and event history, independence of the events might be a reasonable starting model. If a modulated (heterogeneous) Poisson process (PP) were assumed for the event times then a conditional PP likelihood could be assumed whereby the first order intensity for the i th individual could be defined as:

$$\lambda_i(t) = \lambda_{i0}(t) \exp(\mathbf{x}_i'\beta + g_i(t)).$$

Note that here the covariates are included via the fixed vector x_i' and the $g_i(t)$ function could be a smooth function of time. The baseline function $\lambda_{i0}(t)$ is also unit specific. The associated unconditional likelihood is given by

$$\prod_{i=1}^{m} \prod_{j=1}^{n_i} \lambda_i(t_{ij}) \exp\{\Lambda_i\}. \tag{11.10}$$

where

$$\Lambda_i = \int_{t_0}^{t_T} \lambda_i(u) du. \tag{11.11}$$

For fixed covariates, then $\Lambda_i = \exp(x_i'\beta) \int_{t_0}^{t_T} \lambda_{i0}(u) \exp(g_i(u)) du$. Various methods can be used to include time-varying effects. Discretizing to allow piecewise linear terms in the baseline and $g_i(t)$ are possible, while semi-parametric models could also be assumed. Of course gamma or Dirichlet processes could also be used (Ibrahaim et al., 2000). Cook and Lawless (2007), in chapter 3, discuss various possibilities for Poisson process models.

11.4.2 More Complex Repeated Events

Two generalizations are immediate from the simple event case. First, multiple types of events could occur . Second, a feature associated with the event could be important and could vary with time or type of event. An example of the first situation could easily arise with the progression of a disease. Visits to doctors could be interspersed with hospital visits, or nurse visits. In fact, complex patterns of repeated events are more usual when making observational studies on disease progression, than in clinical settings. The observed data could then be of the form $\{t_{i,1}^s, ..., t_{i,n_i}^s\}$ where s denotes the event type. For a fixed number of event types (L) then s could be defined as $s = 1, ..., L$. The second situation arises when the event has a mark attached. For instance, a severity score or biomarker might be measured at a given visit. Alternatively the visit itself could have a duration.

In the first case, the mark or measurement could be jointly modeled with the event time and the correlation between the mark and time could be directly modeled. In this case the observed data for the i th individual would be $\{t_{i,1}, x_{i,1}...., t_{i,n_i}, x_{i,n_i}\}$, where x is the mark value. When duration is associated with event time, there can be a complication, as this type of mark directly affects the subsequent event times (as duration is time-based). In this case the observed data would be $\{t_{i,1}, d_{i,1}...., t_{i,n_i}, d_{i,n_i}\}$ where d is the event duration. This might be appropriate where hospital visits involve stays of different lengths.

Ultimately we might have a mixture of these situations where multi-type events also have marks or durations and so, for the mark case, we would observe $\{t_{i,1}^s, x_{i,1}...., t_{i,n_i}^s, x_{i,n_i}\}$.

Modeling approaches for these different situations depend on the observed data and also the study purpose. For example, if known times are observed it is often convenient to conditionally model the marks given the times so that we have the joint model

$$[x, t] = [x|t][t].$$ (11.12)

In this case, we consider the times to be governed by a point process model. In addition, the model for the marks could be a simple Gaussian distribution: $[x|t] \sim N(\mu(t), \tau)$. Here dependence on time could be made explicit in the mean parameterisation ($\mu(t)$). This mean function could be specified to include covariates (individual or contextual) as well as time dependence: e.g. $\mu(t) = x'\beta + \gamma(t)$. Note also that the point process model could also have covariate dependence and so it is debatable whether there should be multiple entries of the same covariate in each model. For example, patient age could affect both the mark (e.g. blood pressure) and doctor visit times.

On the other hand, if we do not observe directly the visit time but simply the number of visits within a fixed time period, then usually the mark would no longer be associated directly with the individual (as the information has been averaged). Either the mark would be based on the time period, or is an average mark for all events within the period. As the exact times have been lost the resulting data would consist of counts of visits and an average mark or the value of (say) a contextual variable pertaining to the time period. In this case the observed data would be y_{ij} where there are $j = 1, ..., J$ time periods. We might be interested in the joint distribution $[x, y] = [x|y][y]$. In effect, the model for the mark is conditioned on the count and a separate count model is specified. If the focus were on the counts per se then the alternative formulation of $[x, y] = [y|x][x]$ could be considered and often the visit frequency is modeled conditionally, i.e. via $[y|x]$ treating the mark as a covariate.

These formulations do not include any explicit spatial dependence. In the next section some proposals for how this dependence could be incorporated are proposed. As in Sections 11.2 and 11.3 the spatial effects can be regarded as contextual.

11.4.2.1 Known Times

11.4.2.1.1 Single Events As a first pass we could assume that the times follow a heterogeneous Poisson process (hPP) so that

$$f(t) = \lambda(t) \exp\left(-\int_{t_0}^{t_T} \lambda(u) du\right)$$

and conditional on n_i, the likelihood element for the i th individual would be

$$\prod_{i=1}^{m}\prod_{j=1}^{n_i}\lambda_i(t_{ij})\exp\{-\Lambda_i\}. \tag{11.13}$$

where $$\Lambda_i=\int_{t_0}^{t_T}\lambda_i(u)du$$

and $\lambda_i(t)=\lambda_{i0}(t)\exp(x_i'\beta+g_i(t))$. Note that it is possible to include within a Bayesian hierarchy the spatial effect, especially if it is not time-dependent. For example, a contextual effect at the individual level could be included

as $\Lambda_i=\exp(x_i'\beta+w_i)\int_{t_0}^{t_T}\lambda_{i0}(u)\exp(g_i(u))du$ where $w_i=w(c_j)$ as defined
$\underset{i\in j}{}$

in Section 11.2.3. More generally, the spatial location of the individual (s_i) could be incorporated in the PP model and a space-time formulation could be considered directly:

$$\lambda_i(s,t)=\lambda_{i0}(s,t)\exp(\mathbf{x}_i'\beta+g_i(t)+h_i(s)).$$

The resulting likelihood would be

$$\prod_{i=1}^{m}\prod_{j=1}^{n_i}\lambda_i(s_i,t_{ij})\exp\{-\Lambda_i\}$$

$$\text{where } \Lambda_i=\int_{t_0}^{t_T}\int_W\lambda_i(v,u)dvdu$$

where W is the study region area. In addition, variants of the specification could allow for nonparametric specification of the $g_i(t)$, $h_i(s)$ functions with a simple alternative being piecewise discretization in time and space. For space it may be possible to define a common surface, $h(s)$ say, and to form piecewise constant components from a tiling of the distribution of individuals. However if the spatial component is zero-centered then further approximations may be available (see e.g. Chapter 7.7). Of course integral approximations such as proposed by Berman and Turner (1992), could also be considered.

11.4.2.1.2 Multiple Event Types Alternatively, simpler intensity based methods can be pursued. Denote the intensity of the i th individual and j th type as $\lambda_{ij}(t|H_i(t))$, where $H_i(t)$ is the event history. A different period of observation is allowed for each individual: $[0,\tau_i]$. Denote the history of the i th subject as $H_i(t)=\{N_i(s),0<s<t\}$ where $N_{ij}(t)$ is the number of event of j th type occurring on the i th individual in the interval $[0,t]$ and $N_i(t)=\{N_{i1}(t),....,N_{iJ}(t)\}'$. Hence the history of the process concerns the

preceding count accumulation. The resulting likelihood is given for the times t_{ijk} , $k = 1, .., N_{ij}(t)$ so that

$$\prod_{j=1}^{J} \left\{ \prod_{k=1}^{n_{ij}} \lambda_{ij}(t_{ijk}|H_i(t_{ijk})) \exp(-\int_0^{\tau_i} \lambda_{ij}(u|H_i(u))du) \right\}. \tag{11.14}$$

This of course assumes that the distribution of events is functionally independent. In many situations this would of course not be reasonable. For example, a doctor visit might precipitate a hospital visit. When multiple events can arise with dependence over time then it is often useful to consider transition models for the occurrence of events of different types. These types of models specify the probability that an event of a given type will occur in an interval of time given the preceding event. Hence, such events are naturally specified conditioned on preceding events, their types and times or types of preceding events and their times. Alternatively a competing risk approach could be envisaged. I do not consider these methods further here.

Incorporation of random effects can be effected via redefinition of (11.14): $\lambda_{ij}(t_{ijk}|H_i(t_{ijk}), R_{ij}) = \lambda_{ij}(t_{ijk}|H_i(t_{ijk}), \exp\{W_{ij}\})$ where W_{ij} would represent the individual random effect for the j th event type. This could be decomposed into a number of individual specific or contextual random effects. For example, it would be possible to consider a set of L counties labelled c_l, $l = 1, ..., L$. Then we could consider as a first example, the hierarchical random effect model: $W_{ij} = (u_{ij} + v_i + w_i)$ where $w_i = w(\underset{i \in l}{c_l})$ and $v_i = v(\underset{i \in l}{c_l})$. These are spatial contextual effects and $u_{ij} = u_{1i} + u_{2j}$ where u_{1i} is a general individual frailty and u_{2j} is a type-specific effect. The prior distributions for the effects could be overdispersed zero-mean Gaussian for u_{1i}, u_{2j}, v_i while for w_i a CAR formulation would be possible. A variety of other formulations would be possible of course. As before it would be possible to consider the intensity as a function of both time and space and hence to include a specific spatial component in its definition. However it is probably simpler and more convenient to consider conditioning on the spatial component defined at a higher level of the hierarchy via a random effect.

11.4.3 Fixed Time Periods

When fixed time periods are observed, it is usual to collect events within the time periods into counts of events. Of course when this is done information about the time sequence of events is lost within the time period. Hence even if the residential location of individuals is known at a fine spatial resolution level, the resolution level in time is aggregated. In general, we denote the count within the j th time period as y_{ijl}, where i denotes the individual and $j = 1, ..., J$ denotes the time periods, and l denotes the event type ($l = 1, ..., L$).

11.4.3.1 Single Events

In the case of a single outcome then we observe $\{y_{ij}\}$ for a sequence of J times and $i = 1, ..., m$. The simplest modeling approach is to assume a generalized linear model for the counts with some form of time dependence. For example it could be assumed that $y_{ij} \sim Pois(\mu_{ij})$ and a log linear model could be assumed for the mean:

$$\log \mu_{ij} = \mathbf{x}'_i \beta + v_i + \gamma_j \tag{11.15}$$

where x'_i is a vector of individual fixed covariates, β is the corresponding parameter vector, v_i is an individual frailty and γ_j is a common temporal effect. The temporal effect could have a variety of specifications:

1. Uncorrelated prior distribution (for example $\gamma_j \sim N(0, \tau_\gamma)$)

2. Correlated prior distribution (for example a random walk: $\gamma_1 \sim N(0, \tau_\gamma)$, $\gamma_j \sim N(\gamma_{j-1}, \tau_\gamma)$ $j > 1$)

3. Trend regression on time $\gamma_j = \beta t_j$ (or a higher order polynomial) where t_j is the time of the j th period (start or end or middle by convention)

If the focus is on a parsimonious description of the overall behavior then option 2 may be favored as it is relatively non-parametric. However if a specific linear or polynomial estimate of trend is require then option 3 may be preferred. The individual frailty would usually be assumed to have an uncorrelated zero-mean Gaussian distribution:

$$v_i \sim N(0, \tau_v).$$

Incorporation of spatial contextual effects can follow as before by extending the model in (11.15) to include an individual contextual component:

$$\log(\mu_{ij}) = \mathbf{x}'_i \beta + v_i + w_i + \gamma_j \tag{11.16}$$

where $w_i = \underset{i \in l}{w(c_l)}$ and $w(c_l)|w(c_{-l}) \sim N(\overline{w}(c_{\delta_l}), \tau_w/n_{\delta_l})$ where $w(c_l)$ is the value of w for the l th county and $\overline{w}(c_{\delta_l})$ is the average value of w for the neighborhood (δ_l) of c_l. The number of counties in this neighborhood is given by n_{δ_l}. Note that v_i is an individual frailty effect here. An alternative specification could also assume an uncorrelated county effect, as for w_i except without correlation. Steele et al. (2004) describe essentially the model in (11.15) albeit with multiple events.

11.4.3.2 Multiple Event Types

In the case of multiple event types assume a count of the form y_{ijl} where l denotes the event type. Steele et al. (2004) describe a competing risk model where a multinomial form is assumed for the vector of events within a time period. The multinomial probability ratio (relative to the zero event case) was assumed to be defined by a logit link to a linear predictor with covariate random effect and trend components similar to (11.16) but without the spatial dependence. In general, one approach to these problems assumes that, conditionally on $N_{ij} = \sum_l y_{ijl}$

$$\mathbf{y}_{ij} \sim Mult(\mathbf{p}_{ij}, N_{ij}).$$

If it is important to consider the relative preference for visit types then this might be useful. Otherwise without conditioning it would be possible to consider

$$y_{ijl} \sim Pois(\mu_{ijl}).$$

A log linear link could be assumed whereby

$$\log(\mu_{ijl}) = \mathbf{x}_i'\beta + v_i + w_i + u_{il} + \gamma_{jl} \tag{11.17}$$

where v_i, w_i are contextual effects, as before, u_{il} is an individual level effect specific to the event type and γ_{jl} is a temporal effect specific to the event type. As before, the contextual random effects can be defined to depend on small area level geographies, while the temporal effects could be regression-based or random walk-based. In the multinomial model of Steele et al. (2004) the temporal component γ_{jl} was assumed to be defined by a quadratic regression in time and the individual random effect u_{il} was assumed to have a multivariate normal distribution (between event types only).

11.4.3.2.1 Asthma-Comorbidity Medicaid Example Sutton (2005) provides an example of the analysis of individual level outcomes with fixed time periods in a Bayesian setting. This work was based on Medicaid data on asthma (ICD-9 493) and congestive heart failure (CHF) (ICD-9 428, 402, 518.4), comorbidities for recipients between ages 50 to 64 in South Carolina for 1997 through 1999. Three groups were identified: asthma only, CHF only, and asthma and CHF. Of the 1857 individuals, 223 were in the comorbidity group. Recorded for each recipient were number of days between multiple dates of medical service; type of visit (inpatient, ER, outpatient, doctor's office) and recipient demographic information (age at first visit, gender, race, county of residence). As a brief guide, Figure 11.6 displays the event profiles summarized by time to second visit, for the asthma and CHF individuals separately.

An analysis was performed with the SC Medicaid data mentioned above where asthma and CHF were analyzed together. Here I only display the comorbidity group analysis. The analysis was carried out for 21 time periods (of 50 days each).

A model of the form

$$y_{ijl} \sim Poiss(\mu_{ijl})$$
$$\log(\mu_{ijl}) = \mathbf{x}_i'\beta + v_i + w_i + \gamma_{jl}$$

FIGURE 11.6: Distribution of number of days between first and second visits for recipients with asthma (top) and congestive heart failure (bottom).

was fitted to the data for the comorbidity group. Individual frailty was not required in this example, based on a variable selection criterion. Here the fixed covariates were β_0 (common intercept), β_1(age), factors for race and gender, and two spatial contextual effects at county level: $v_i = v(c_l)_{i \in l}$ with $v_i \sim N(0, \tau_v)$ and $w_i = w(c_l)_{i \in l}$ and $w(c_l)|w(c_{-l}) \sim N(\overline{w}(c_{\delta_l}), \tau_w/n_{\delta_l})$ where $w(c_l)$ is the value of w for the l th county and $\overline{w}(c_{\delta_l})$ is the average value of w for the neighborhood (δ_l) of c_l. All regression parameters had over-dispersed zero-mean Gaussian prior distributions while variances were assumed to have $Ga\,(0.05, 0.0005)$ distributions. The temporal effects for each type were as-

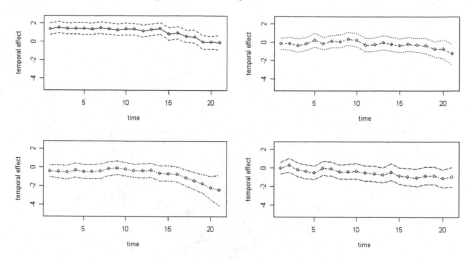

FIGURE 11.7: Posterior average temporal profiles for the effect γ_{jl}, for $l = 1, .., 4$. Mean profiles and 95% credible interval shown. Top left: $l = 1$, top right: $l = 2$, bottom left: $l = 3$, bottom right: $l = 4$.

sumed to have independent random walk prior distributions

$$\gamma_{jl} \sim N(\gamma_{j-1,l}, \tau_\gamma) \; \forall l.$$

The converged model fit yields the following results:

1. The posterior expected estimates (sds) of β_0 and β_1 respectively were -4.6 (0.9982) and -0.008553 (0.01118).

2. The posterior expected estimates (sds) of the race and gender effects were not significant for gender but showed a significance for the white versus African-American racial groups.

3. The temporal effects were estimated and shown in Figure 11.7. The sequence of visit types is l = 1, l = 2, l = 3, l = 4.

Finally the posterior average maps of the county-specific random contextual effects are shown in Figure 11.8. It is clear that there is some spatial effect (w map) displayed in the northeast of the state within largely rural areas, whereas the UH effect (v map) seems to be largely random. It should be noted however that analysis of discrete time events lacks a considerable amount of information due to the grouping within time periods and information about sequencing is lost. This constitutes an inevitable limitation to this form of analysis.

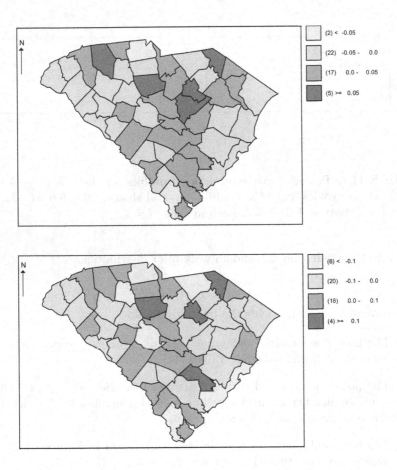

FIGURE 11.8: Posterior average maps of the county-level random effects (w_i and v_i) form the multiple event model for the Medicaid data.

12

Spatio-Temporal Disease Mapping

As in other application areas, it is possible to consider the analysis of disease maps which have an associated temporal dimension. The two most common formats for observations are:

(1) Geo-referenced case events which have associated times of diagnosis or registration or onset, i.e. we observe within a fixed time period J and fixed spatial window W, m cases at locations $\{s_i, t_i\}, i = 1, \ldots, m$.

(2) Counts of cases of disease within tracts are available for a sequence of J time periods, i.e. we observe a binning of case events within $m \times J$ space-time units: $y_{ij}, i = 1, \ldots, m, j = 1, \ldots, J$.

The analysis found for spatial data (see Chapters 5 and 6) can be extended into the time domain without significant difficulty.

12.1 Case Event Data

In the case event situation, few examples exist of mapping analysis. However, it is possible to specify a model to describe the first-order intensity of the space-time process (as in the spatial case). The intensity at time t can be specified as:

$$\lambda(s, t) = \rho g(s, t).f_1(s; \theta_x).f_2(t; \theta_t).f_3(s, t; \theta_{xt}), \tag{12.1}$$

where ρ is a constant background rate (in space $\times$ time units), $g(s, t)$ is a modulation function describing the spatio-temporal "at-risk" population background in the study region, f_k are appropriately defined functions of space, time and space-time, and $\theta_x, \theta_t, \theta_{xt}$ are parameter vectors relating to the spatial, temporal and spatio-temporal components of the model.

Here each component of the f_k can represent a *full* model for the component, i.e. f_1 can include spatial trend, covariate and covariance terms, and f_2 can contain similar terms for the temporal effects, while f_3 can contain *interaction* terms between the components in space and time. Note that this final term can include *separate* spatial structures relating to interactions which are not included in f_1 or f_2. The exact specification of each of these components will depend on the application, but the separation of these three components is helpful in the formulation of components.

The above intensity specification can be used as a basis for the development of Bayesian models for case events. If it can be assumed that the events form a modulated Poisson process in space-time, then a likelihood can be specified, as in the spatial case. For example, a parsimonious model could be proposed where a regression component and a random effect component is assumed:

$$\lambda(s,t) = \rho g(s,t) \exp\{\mathbf{P}(s,t)'\beta + T(s,t)\} \qquad (12.2)$$

where $\mathbf{P}(s,t)$ is a covariate vector, β a regression parameter vector and $T(s,t)$ is a random component representing extra variation in risk. The term $T(s,t)$ could be decomposed in a number of ways. For example, it could represent a spatio-temporal Gaussian process (Brix and Diggle, 2001). However, a simpler approach might be to consider: $T(s,t) = a(s) + b(t) + c(s,t)$ where a discretized version of the random fields could be envisaged so that any realization of the field $\{s_i, t_i\}$ has separable correlation structure and

$$a(\mathbf{s}) \sim MVN(\mathbf{0}, K_a(\tau_x, \phi)),$$
$$b(t) \sim N(f(\Delta t), \tau_b),$$
$$\text{and } c(s,t) \sim N(0, \tau_c I), \qquad (12.3)$$

where $K_a(\tau_x, \phi)$ is a parameterised spatial covariance matrix and I is an identity matrix, with variances τ_b, and τ_c, and Δt is a distance measure in space and time. In this approach the likelihood remains that of a conditionally modulated Poisson process.

This type of model can be included within a likelihood specification and a full Bayesian analysis can proceed using extensions to the analysis for purely spatial data. In these extensions, the integrated intensity of the process

$$\Lambda(\theta) = \int_W \int_0^J \lambda(u,v)\,dvdu$$

where $\theta = (\beta, \tau_x, \phi, \tau_b, \tau_c)$ is the parameter vector must be estimated or the background is concentrated out of the model by conditioning. In Lawson (2006b), an example of application of this model to a well known space-time case event data set was given (Burkitt's Lymphoma in the Western Nile district of Uganda for 1960 through 1975). In that data set the locations of cases and diagnosis dates (days from January 1st 1960) are known as well as the age of each case. There is no background population information in this example. While it might be possible to consider an approach where a population effect in $g(s,t)$ was estimated from the case event data, this is particularly assumption-dependent and was not pursued. The following analysis essentially assumes that the population is homogeneously distributed over space. This is of course a strong assumption.

The estimation of the $\Lambda(\theta)$ was adopted in this case. The intensity was integrated over space-time using Dirichlet tile approximations (Berman and Turner, 1992; Lawson, 2012). The model details were as follows. The basic form of the model assumed was

$$\lambda(s_i, t_i) = \exp\{\beta_0\}. \exp\{a(s_i) + b(t_i) + c(s_i, t_i)\},$$

where the prior distributions are defined as in (12.3) for the main components $(a(s_i), b(t_i), c(s_i, t_i))$. A zero-mean spatial Gaussian process was assumed for the spatial component with exponential covariance function $\tau_x \exp(-\phi d)$, where d is the distance between any two locations and with variance τ_x and covariance range ϕ. The temporal component is defined by $b(t_i) \sim N(a_t b(t_{i-1}), \tau_b)$ where a_t could take a variety of forms. This parameter could be constant or could be dependent on time differences, for example, $a_t = 1/\Delta t_i$ where $\Delta t_i = t_i - t_{i-1}$. The space-time component is a residual effect, namely $c(s_i, t_i) \sim N(0, \tau_c)$. The parameter prior distributions were assumed to be

$$\exp\{\beta_0\} \sim N(0, 0.001)$$
$$\tau_x \sim Ga(0.1, 0.1)$$
$$\phi \sim U(0, 2)$$
$$\tau_b \sim Ga(0.1, 0.1)$$
$$\tau_c \sim Ga(0.1, 0.1).$$

For the converged sample, the posterior estimate of ϕ was 0.0024 (sd: 0.0025) while those of $exp\{\beta_0\}$, τ_x, τ_b, τ_c were 11.77 (sd: 0.164), 8.481 (sd: 6.789), 22.68 (sd: 8.916), and 7.064 (sd: 5.087) respectively. All these parameters had positive lower and upper 95% credible limits. Figure 12.1 suggests a peak in the spatial component in the north and temporal variations with marked changes in the west of the area. However, the parameter estimates suggest that the overall rate and space-time component are well estimated but the spatial and temporal effects are not important in this example. Alternative formulations don't yield results of any great difference from this model. For example, a model including a covariate (age) was examined but the parameter for this covariate was found to have a credible interval crossing zero and so we have not reported this model here.

12.2 Count Data

Note that the above case event intensity specification can be applied in the space-time case where small-area counts are observed within fixed time periods

FIGURE 12.1: Burkitt's lymphoma: two displays of the components of a Bayesian model with a spatial Gaussian random field component, a temporal correlation component and an uncorrelated space-time component. The spatial and temporal components are displayed here.

$\{l_j\}$, $j = 1, \ldots, J$, by noting that

$$E\{y_{ij}\} = \int_{l_j} \int_{a_i} \lambda(u, v) \, \mathrm{d}u \, \mathrm{d}v,$$

where y_{ij} is the count in the i–j th unit, under the usual assumption of Poisson process regionalization. In addition, the counts are independent conditional on the intensity given, and this expectation can be used within a likelihood modeling framework or within Bayesian model extensions. In previous published work in this area cited above, the expected count is assumed to have constant risk within a given small-area or time unit, which is an approximation to the continuous intensity defined for the underlying case events. The appropriateness of such an approximation should be considered in any given application (see also Chapter 5). If such an approximation is valid, then it is straightforward to derive the minimal and maximal relative risk estimates under the Poisson likelihood model assuming $E\{y_{ij}\} = \lambda_{ij} = e_{ij}\theta_{ij}$, where e_{ij} is the expected rate in the required region or period. The maximal model estimate is $\widehat{\theta}_{ij} = y_{ij}/e_{ij}$, the space-time equivalent of the SMR/SIR, while the minimal model estimate is

$$\widehat{\theta} = \frac{\sum_i \sum_j y_{ij}}{\sum_i \sum_j e_{ij}}.$$

Smooth space-time maps, e.g. empirical Bayes or full Bayes relative risk estimates, will usually lie between these two extremes. If the full integral intensity is used, then these estimates have the sums in their denominators replaced by integrals over space-time units.

Development of count data modeling based on tract or period data has seen considerable development. In the context of a typical Poisson likelihood model where $y_{ij} \sim Pois(e_{ij}\theta_{ij})$, the log relative risk $(\log(\theta_{ij}))$ is usually the focus of modeling. Estimation of the expected rate or count over the regions and time periods also follows from the consideration of standardisation. For example, indirect standardisation based on a reference population could be used. In the case of a study region reference population, the expected rates could be computed as

$$e_{ij} = n_{ij}\frac{\sum_i \sum_j y_{ij}}{\sum_i \sum_j n_{ij}},$$

where n_{ij} is the population in the i–j th space-time unit. Other, more complex, standardisations could be pursued of course, especially if stratification of the population is to be represented in the expected rate calculation.

The first example of modeling space-time relative risk was by Bernardinelli et al. (1995). In their approach, they assumed a model for the log relative risk of the form

$$\log(\theta_{ij}) = \mu + \phi_i + \beta t_j + \delta_i t_j, \tag{12.4}$$

where μ is an intercept (overall rate), ϕ_i is an area (tract) random effect, βt_j is a linear trend term in time t_j and δ_i an interaction random effect between area and time. Suitable prior distributions were assumed for the parameters in this model and posterior sampling of the relevant parameters was performed via Gibbs sampling. Note in this formulation there is no spatial trend, only a simple linear time trend and no temporal random effect. The model in (12.4) above consists of spatial, temporal, and interaction terms. These could be extended in a number of ways. In general, we could consider three groups of components for $\log(\theta_{ij})$: $\log(\theta_{ij}) = \mu_0 + A_i + B_j + C_{ij}$ where A_i is the spatial group, B_j is the temporal group and C_{ij} is the space-time interaction group. In (12.4) above, $A_i = \phi_i$, $B_j = \beta t_j$, and $C_{ij} = \delta_i t_j$.

Waller et al. (1997) and Xia and Carlin (1998) (see also Carlin and Louis, 2000) subsequently proposed a different model where the log relative risk is parameterised as

$$\log(\theta_{ijkl}) = \phi_i^{(j)} + \delta_i^{(j)} + \text{fixed covariate terms } (kl),$$

where $\phi_i^{(j)}$ and $\delta_i^{(j)}$ are uncorrelated and correlated heterogeneity terms which can vary in time. This model was further developed and simplified by Xia and Carlin (1998), who also examined a smoking covariate that exhibited associated sampling error and spatial correlation. Their model was defined as

$$\log(\theta_{ijkl}) = \mu + \zeta t_j + \phi_{ij} + \rho p_i + \text{fixed covariate terms } (kl),$$

where an intercept term μ is included with a spatial random effect nested within time $\{\phi_{ij}\}$, a linear time trend ζt_j, and p_i is a smoking variable measured within the tract unit. In these model formulations no spatial trend is admitted and all time-based random effects are assumed to be subsumed within the ϕ_{ij} terms. In this formulation, $A_i = \rho p_i$, $B_j = \zeta t_j$, and $C_{ij} = \phi_{ij}$, and other covariate effects.

To allow for the possibility of time-dependent effects in the covariates included (race and age), Knorr-Held and Besag (1998) formulated a different model for the same data set (88-county Ohio lung cancer mortality, 1968 through 1988). Employing a binomial likelihood for the number at risk $\{n_{ijkl}\}$ with probability π_{ijkl}, for the counts, and using a logit link to the linear predictor, they proposed

$$\eta_{ijkl} = \ln\{\pi_{ijkl}/(1 - \pi_{ijkl})\},$$

where

$$\eta_{ijkl} = \alpha_j + \beta_{kj} + \gamma_{lj} + \delta z_i + \theta_i + \phi_i. \tag{12.5}$$

The terms defined are α_j, a time-based random intercept; β_{kj}, a k th age group effect at time j; γ_{lj}, a gender $\times$ race effect for combination l at the j th time; a fixed covariate effect term δz_i, where the z_i is an urbanization index; and θ_i, ϕ_i are correlated and uncorrelated heterogeneity terms which are not time-dependent. No time trend or spatial trend terms are used, and these effects will (partly) be subsumed within the heterogeneity terms and the $\alpha_j + \beta_{kj} + \gamma_{lj}$ terms. In this formulation, $A_i = \delta z_i + \theta_i + \phi_i$, $B_j = \alpha_j + \gamma_{lj} + \beta_{kj}$, including the covariate-time interactions and $C_{ij} = 0$.

More recent examples of spatio-temporal modeling include extensions of mixture models (Boehning et al., 2000), which examine time periods separately without interaction, and with the use of a variant of a full multivariate normal spatial prior distribution for the spatial random effects (Sun et al., 2000), and the extension of the Knorr-Held and Besag model to include different forms of random interaction terms (Knorr-Held, 2000). Although the more complex interaction terms proposed in the latter work did not fit the data example well, the simpler formulations seem to provide a parsimonious representation of space-time behavior in risk. For example, a log relative risk can be defined purely in terms of random effects via:

$$\log \theta_{ij} = \beta_0 + u_i + v_i + \tau_j + \psi_{ij} \tag{12.6}$$

where the correlated and uncorrelated spatial components (CH, UH) are defined to be constant in time (u_i, v_i). In addition, there is a separate temporal random effect (τ_j) and finally a space-time interaction term (ψ_{ij}). In this case, $A_i = u_i + v_i$, $B_j = \tau_j$, $C_{ij} = \psi_{ij}$. Often an autoregressive prior distribution can be used for τ_j: $\tau_j \sim N(\gamma \tau_{j-1}, \kappa_\tau)$. This allows for a type of non-parametric temporal effect (random walk when $\gamma = 1$). The prior distribution for the interaction term can be simply zero mean normal (i.e. $\psi_{ij} \sim N(0, \tau_\psi)$), but more complex prior distributions could be used. This model has also been applied within a surveillance context (Lawson (2004)). Extensions to non-separable space-time interaction can be made by different prior distribution specifications for ψ_{ij}. Denote ψ as the matrix of interaction terms $\{\psi_{ij}\}$. For example, Type II random walk interaction (Knorr-Held (2000)) is defined by the prior distribution

$$[\psi | \tau_\psi] \propto \exp(-\frac{\tau_\psi}{2} \sum_{i=1}^{m} \sum_{j=2}^{J} (\psi_{ij} - \psi_{i,j-1})^2),$$

whereas Type III interaction consists of time-averaged spatial correlation where

$$[\psi | \tau_\psi] \propto \exp(-\frac{\tau_\psi}{2} \sum_{j=2}^{J} \sum_{i \sim l} (\psi_{ij} - \psi_{lj})^2),$$

and finally Type IV interaction is fully space-time dependent and is defined as

$$[\psi | \tau_\psi] \propto \exp(-\frac{\tau_\psi}{2} \sum_{j=2}^{J} \sum_{i \sim l} (\psi_{ij} - \psi_{lj} - \psi_{i,j-1} + \psi_{l,j-1})^2).$$

These different prior distributions were fitted by Knorr-Held (2000) to the 21-year Ohio respiratory cancer data set (for white male counts only) but he found that Type II interaction was favored with Type I also offering a lower deviance than Type II or IV. However given that the interaction term could be regarded as a form of residual (made up of unobserved confounding unaccounted for by the main effects) and also the fact that identifiability of highly structured interaction (as in Type III and IV) from main effects may be doubtful, there may be a need to consider less structured priors in applications. While it is certainly true that in other spatial statistical applications non-separable space-time interaction could be important in making a parsimonious description of the variation (Gneiting et al., 2007), it may be the case that for epidemiological data where expected rates and covariates are often available, parsimonious description is possible without such assumptions.

In other developments for space-time count data, Zhu and Carlin (2000) have examined the use of covariates at different levels of aggregation within misaligned spatial regions. Misalignment is discussed in Section 9.2. In addition, there has been development of both descriptive and mechanistic models for space-time infectious disease modeling (Cressie and Mugglin, 2000, Knorr-Held and Richardson, 2003). This area is discussed in more detail in Chapter 14. Some recent examples of the use of spatio-temporal modeling can be found in Richardson et al. (2006), Abellan et al. (2008), Carroll et al. (2016), Lawson et al. (2017), Khan et al. (2017), Anderson and Ryan (2017).

Overall, there are a variety of forms which can be adopted for spatio-temporal parametrization of the log relative risk, and it is not clear as yet which of the models so far proposed will be most generally useful. Many of the above examples exclude spatial and/or temporal trend modeling, although some examples absorb these effects within more general random effects. Allowing for temporal trend via random walk intercept prior distributions provides a relatively non-parametric approach to temporal shifting, while it is clear that covariate interactions with time should also be incorporated. Interactions between purely spatial and temporal components of the models have not been examined to any extent, and this may provide a fruitful avenue for further developments. If the goal of the analysis of spatio-temporal disease variation is to provide a parsimonious *description* of the relative risk variation, then it would seem to be reasonable to include spatial and temporal trend components in any analysis (besides those defined via random effects).

Finally, it is relevant to note that there are many possible variants of the two basic data formats which may arise, partly due to mixtures of spatial aggregation levels, but also to changes in the temporal measurement units. For example, it may be possible that the spatial distribution of case event data is only available within fixed time periods, and so a hybrid form of analysis may be required where the evolution of case event maps is to be modeled. Equally, it may be the case that repeated measurements are made on case events over time so that attached to each case location is a covariate (possibly time-dependent) which is available over different time periods.

12.2.1   Georgia Low Birth Weight Example

An example of the application of a variety of space-time models to very low birth weight count data for the counties of Georgia for 1994 through 2004 is available. In this example the observed data is the count of births with very low birth weight (<1500 grams) in counties of Georgia for each year from 1994 to 2004. The total birth counts for the same periods and counties are publicly available from the Georgia Department of Health OASIS website (http//oasis.state.ga.us/). Figure 12.2 displays the crude rate ratios for this example. It is clear that some areas in rural Georgia have particularly high crude rates (>>0.0175).

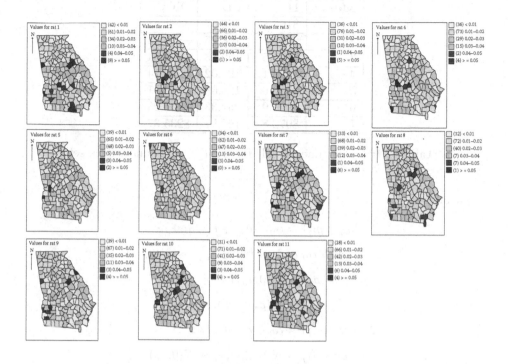

FIGURE 12.2: Georgia county level crude rate ratios for very low birth weight in relation to births from 1994 to 2004. Top row: from 1994 to 1997; middle row: 1998 to 2001; bottom row: 2002 to 2004.

In an analysis of very low birth weight, it may be important to provide a parsimonious description of the relative risk variation in space and time, and also to examine the data for possible risk anomalies. To this end I have examined a set of seven models that a priori may provide a parsimonious model. The choice is from a range of those discussed above and the models have been chosen to represent the different modeling approaches and also demonstrate features of the modeling process. This is in no way a comprehensive model fitting exercise, and alternative models could be hypothesized. The basic likelihood is assumed to be $y_{ij} \sim bin(p_{ij}, n_{ij})$ and the logit of the probability of very low birth weight is directly modeled. The overall crude rate ratio $(\sum \sum y_{ij} / \sum \sum n_{ij})$ for the counties of Georgia for this 11-year period is 0.0175. The first three models are simple separable models with no spatio-temporal interaction: a model with a random UH spatial and temporal term and an autoregressive temporal effect (g_j) (model 1); a model with no temporal dependence (model 2); and a model with only temporal trend and

Model	D̄	pD	DIC
1	8162.8	168.67	8331.68
2	8162.2	168.25	8330.43
3	8161.5	159.13	8320.64
4	8122.12	127.02	8249.14
5	8112.63	128.08	8240.71
6	7966.9	252.33	8219.23
7	8072.8	162.57	8235.37

Table 12.1: Space-time models for Georgia oral cancer data set

spatial UH.

1) $\log it(p_{ij}) = \alpha_0 + a_{1j} + v_i + g_j$

$\quad$ with $\alpha_0 \sim N(0, 0.0001),\ v_i \sim N(0, \tau_v)$

$\quad a_{1j} \sim N(0, \tau_{a1})$

$\quad g_j \sim N(g_{j-1}, \tau_g).$

2) $\log it(p_{ij}) = \alpha_0 + a_{1j} + v_i$

$\quad$ with $a_{1j} \sim N(0, \tau_{a1}), v_i \sim N(0, \tau_v).$

3) $\log it(p_{ij}) = \alpha_0 + a_1 t_j + v_i$

$\quad$ with $v_i \sim N(0, \tau_v),\ a_1 \sim N(0, \tau_{a1}).$

Table 12.1 displays the DIC results for models 1 through 7. Models 4 and 5 involve product interactions as per the orignal Bernardinelli model (12.4), with model 5 adding an uncorrelated random effect. Model 6 shows Type I ST interaction, and model 7 has a temporally dependent interaction prior distribution. It is clear that models 1 through 3 while parsimonious are far from the best models. The product interaction models (4 and 5) are more parsimonious, and the lowest among these is the original Bernardinelli et al. model with an added spatial UH component (model 5). It is also clear, however, that the models proposed by Knorr-Held yield the lowest DIC model. This is model 6 which exhibited Type I ST interaction and spatial CH and UH and temporal dependence. This model is less effective than the Type II interaction models although it is less parsimonious. Of course these results depend on prior specifications and in any particular applications, sensitivity to prior specification should be examined.

For the model with lowest DIC, various posterior summaries are available. Figure 12.3 displays the sequence of 11 years of exceedance probabilities for the lowest DIC model fitted to these data (model 6). These probabilities were

FIGURE 12.3: Georgia county level exceedance probability from a spatio-temporal model with Type I interaction. $\Pr(p_{ij} > 0.0175)$ is estimated as an average of posterior sample values of $I(p_{ij} > 0.0175)$. Top row: 1994 to 1997; middle row: 1998 to 2001; bottom row: 2002 to 2004.

estimated from $\widehat{\Pr}(p_{ij} > 0.0175) = \sum_{g=1}^{G} I(p_{ij}^g > 0.0175)/G$, where p_{ij}^g is the sampled value of p_{ij} from a posterior sample of size G. Given the caveats mentioned in Chapter 6 concerning the use of exceedance probabilities with inappropriate models, with the current "best" model we would expect reasonable reliability and stability in these estimates. It is notable that most counties where high exceedances are found are rural (Dougherty, Terrell, Marion, Baldwin, Handcock, Richmond, and Burke), although Richmond county includes Augusta. Periodically the counties within Atlanta also signal (DeKalb and Fulton). In general there appears to be a stable patterning of the very low birth weight in that the spatial clusters seem to persist over time, whereas space-time clusters appear periodically in Atlanta.

12.3 Alternative Models

As in the case of spatial disease modeling, there are a wide variety of model variants available in the space-time extension. For example, semi-parametric

models may be favored and it is straightforward to extend the spatial spline
models discussed in Section 5.7.2, to the spatio-temporal situation. Recent
examples of a form of spatio-temporal semi-parametric modeling is found in
Cai et al. (2012, 2013). This is not pursued here.

12.3.1   Autologistic Models

Another important variant that was examined in the spatial case, in Chapter
5, was the autologistic model. For binary data this is an attractive likelihood
variant. In Chapter 5, the ability of this model to capture some of the spatial
correlation effects was noted (see Section 5.7.1). Besag and Tantrum (2003)
proposed the use of autologistic models in a spatio-temporal setting. The use
of pseudolikelihood allows conditioning on the neighborhood counts which are
now time labelled. Define the binary outcome variable y_{ij} and assume that
$y_{ij} \sim Bern(p_{ij})$. A model for p_{ij} could be constructed as

$$\frac{\exp(y_{ij}.A_{ij}))}{1 + \exp(A_{ij})}$$

where A_{ij} is a function of the sum of neighboring areas and also a sum of
neighboring areas at previous times. For example, define the current sum
as $S_{\delta_i,j} = \sum_{l \in \delta_i} y_{lj}$ and the sum over the neighborhood at a previous time as
$S_{\delta_i,j-1} = \sum_{l \in \delta_i} y_{l,j-1}$. We can then consider a variety of models where space-
time dependence can be captured by different forms of $S_{\delta_i,j}$ and $S_{\delta_i,j-1}$. Table
12.2 displays the results of fitting a range of autologistic models to the 21-year
Ohio respiratory cancer county level data set. In Chapter 5, an analysis of one
year (1968) of this data was described. Here we examine the 21-year sequence
of data from 1968 through 1988. Once again, for the sake of exposition, we
threshold the i–j th value at 2:

$$y_{ij} = \begin{cases} 1 \text{ if } smr_{ij} > 2 \\ 0 \text{ otherwise} \end{cases}.$$

Then we consider $y_{ij} \sim Bern(p_{ij})$ with

$$p_{ij} = \frac{\exp(y_{ij}.A_{ij}))}{1 + \exp(A_{ij})}$$

with A_{ij} parameterised with a variety of covariates based on neighborhood
sums. We define two sets of neighbors. The simplest model is defined to
be a function of the sum of first order spatial neighbors, i.e. the neighbors
defined as the adjacent small areas (in this case I define adjacency as having
a common boundary). I also examine an extended neighborhood (second
order) where counties adjacent to the neighbors (excluding those already in
the first order neighborhood) are included. Hence the current sum of first

Model	DIC	pD	MSPE	DIC (added v_i, ψ_{ij})	MSPE
1	2488.49	42.78	0.4582	1833.71 (pD: 559.9)	0.2151
2	2386.66	61.5	0.4544	1765.17 (pD:548.6)	0.2266
3	2511.94	64.6	0.4534	1824.39 (pD: 565.5)	0.2107
4	2542.65	105.29	0.4407	2539.82 (pD: 129.3)	0.4348
5	1936.00	90.34	0.3344	-	-

Table 12.2: Autologistic space-time models: models 1 through 4 and convolution model (model 5)

order neighbors is $S_{\delta_{1i},j} = \sum_{l \in \delta_{1i}} y_{lj}$ while the second order is $S_{\delta_{2i},j} = \sum_{l \in \delta_{2i}} y_{lj}$. the sums at previous times are $S_{\delta_{1i},j-1}$ and $S_{\delta_{2i},j-1}$. The main autologistic models considered here are defined for the predictor A_{ij}:

1) $A_{ij} = \alpha_{1j} + \alpha_{2j} S_{\delta_{1i},j}$
2) $A_{ij} = \alpha_{1j} + \alpha_{2j} S_{\delta_{1i},j} + \alpha_{3j} S_{\delta_{1i},j-1}$
3) $A_{ij} = \alpha_{1j} + \alpha_{2j} S_{\delta_{1i},j} + \alpha_{3j} S_{\delta_{2i},j}$
4) $A_{ij} = \alpha_{1j} + \alpha_{2j} S_{\delta_{1i},j} + \alpha_{3j} S_{\delta_{1i},j-1} + \alpha_{4j} S_{\delta_{2i},j} + \alpha_{5j} S_{\delta_{2i},j-1}$.

Note that the regression parameters can be allowed to vary with time: there are no other random components in the model. For these models, α_{1j}, α_{2j}, α_{2j}, α_{3j}, α_{4j}, α_{5j} have been assumed to vary with time but there is no prior dependence, i.e. $\alpha_{*j} \sim N(0, \tau_{\alpha_*})$. Here we also compare a conventional convolution model with components $A_{ij} = \alpha_0 + \alpha_{1j} + u_i + v_i + \psi_{ij}$ and type I interaction (model 5) with $\alpha_0 \sim U(-a, a)$ with a large, $\alpha_{1j} \sim N(\alpha_{1j-1}, \tau_{\alpha_1})$, $u_i|u_{-i} \sim N(\overline{u}_{\delta_{1i}}, \tau_u/n_{\delta_{1i}})$, $v_i \sim N(0, \tau_v)$, $\psi_{ij} \sim N(0, \tau_\psi)$.

In this case, the random effect convolution model with Type I interaction appears to yield a relatively good model, based on DIC, compared to the autologistic model using a first order neighborhood and a single first order lagged neighborhood. To compare models with additional random effects it is reasonable to extend the autologistic models to include uncorrelated effects (v_i and ψ_{ij} where the same prior distributions are assumed as in the convolution model). For the four autologistic models this was carried out, and the resulting DICs are listed in the fifth column of Table 12.2. It is clear that the DICs are considerably lower than those of convolution model for the first three autologistic models.

While it is difficult to generalize from one data example, this result does suggest that autologistic models could be useful when modeling binary space-time health data, especially when added random effects are included. The added random effects included here are uncorrelated (v_i, ψ_{ij} Type I) and relatively simple to implement. Note that other goodness-of-fit measures can be examined, such as MSPE (see Section 4.1) and these may be useful when other features of the model such as predictive capabilities are important. The MSPE for each model was also calculated and in Table 12.2 the results are shown in columns 4 and 6. While the convolution model yields the lowest DIC compared to simple autologistic models, the autologistic models with added random effects yield lower MSPEs and DICs for most models. However, the model with lowest DIC is not the lowest MSPE model. In this case model 2, with a lagged neighborhood effect has the lowest DIC and model 3 has the lowest MSPE.

12.3.2   Spatio-Temporal Leroux Model

Other variants that can be considered in space-time involve the replacement of the spatial convolution term within the mean structure by the Leroux prior model (5.5.1). This would lead to

$$\log \theta_{ij} = \beta_0 + \phi_i + \gamma_j + \psi_{ij}$$
$$\phi_i|\phi_{-i}, W \sim N\left(\frac{\rho \Sigma_{k=1}^m w_{ki}\phi_k}{\rho \Sigma_{k=1}^m w_{ki} + 1 - \rho}, \frac{1}{\tau \rho \Sigma w_{ki} + 1 - \rho}\right)$$

where w_{ki} are neighborhood weights, ρ is a spatial correlation parameter, and τ is a precision. Further extensions, where different dependencies are assumed have been proposed. In particular the evolution of the spatial response surface can be modelled by replacing $\phi_i + \gamma_j + \psi_{ij}$ with a vector autoregressive process as per

$$\phi_t|\phi_{t-1} \sim N(\rho_t \phi_{t-1}, \tau^{-1}Q(W, \rho_s)^{-1}) \qquad (12.7)$$
$$\phi_1 \sim N(\mathbf{0}, \tau^{-1}Q(W, \rho_s)^{-1}).$$

This allows the temporal and spatial correlation to be jointly modeled (Rushworth et al., 2014). An adaptive extension of this has also been developed by Rushworth et al. (2017). These models and variants have been implemented using posterior sampling in CARBayes (CARBayesST: available from CRAN and URL: http://github.com/duncanplee/CARBayesST). Lee and Lawson (2016) further extend the models to allow for spatial clustering of time effects. For the Georgia VLBW example, I have fitted model (12.7) and it gave a DIC = 8242.352 with pD = 278.81, which is slightly higher than that for the Knorr-Held (Table 12.1, model 6: DIC: 8219.23, pD = 252.33). Hence, for these data and based on DIC, the Knorr-Held independent interaction model is preferred, although the average deviances are very close.

12.3.3   Latent Structure Spatial-Temporal (ST) Models

In Chapter 5 some approaches to spatial latent structure modeling were examined. Space-time data often provides a greater latitude for the examination of latent features, as inherently there is likely to be more possibility of complexity when dealing with three dimensions instead of two. Again count data is the focus although many of the proposals here could be applied in the case event situation. As in the spatial case, it is possible to extend the fixed convolution model to include a random mixture of effects. For example, one could propose a log-linear model,

$$\log(\theta_{ij}) = \alpha_0 + \sum_{k=1}^{K} w_{ij}\lambda_{jk}$$

where K is fixed and $\sum_{k=1}^{K} \lambda_{jk} = 1, \sum_{i} w_{ij} = 1$ and $w_{ij} > 0 \; \forall i, j$, with α_0 as overall intercept. In this formulation there is a weight for each space-time unit, which is normalized as a probability over space for a given time, while for identifiability a sum to unity constraint is placed on the temporal profiles. The weights here could be regarded as loadings but are not assigned to a particular component. The temporal profiles are labeled by component. An extension to this idea could be made where the number of components is allowed to be random and then the joint posterior distribution of $(K, \{\lambda_k\})$ would have to be sampled. This could be done via reversible jump MCMC (Green, 1995) or via variable selection approaches (Kuo and Mallick, 1998; Dellaportas et al., 2002; Petralias and Dellaportas, 2013).

Alternative formulations are possible and two of these are mentioned here. First it may be possible to use a principal component decomposition in this context. Bishop (2006) discusses how a special prior specification on the loading matrix W in a Gaussian latent variable formulation leads to parsimonious description. The columns of the loading matrix W span a linear subspace within the data space that corresponds to the principal subspace. Of course it might also be useful to consider time dependence in the specification of the component model. Extending the proposal of Wang and Wall (2003) it would be possible to consider a form such as

$$\mu_{ij} = e_{ij}\theta_{ij}$$
$$\log \theta_{ij} = \alpha_0 + \lambda_j f_i$$

where f_i is the spatially-referenced risk factor, with $\sum_i f_i = 0$ and λ_j is the temporally referenced loading. The factors could have a multivariate distribution such as

$$\mathbf{f} \sim \mathbf{MVN}(\mathbf{0}, \mathbf{C}(\zeta))$$

where $\mathbf{C}(\zeta)$ is a spatial covariance matrix. The spatial covariance could be fully specified with a distance dependence or could be a MRF so that

$$\mathbf{C}(\zeta) = \tau^{-2}(I - \rho W)^{-1},$$

where τ is a precision, ρ a spatial association parameter and W a neighborhood matrix.

Further it might be useful to consider an autoregressive prior distribution for the loading vector so that

$$\lambda_j \sim N(\lambda_{j-1}, \tau).$$

Further extension could be imagined (see, e.g. Tzala and Best, 2008).

An alternative to these approaches is to consider a mixture model extension of the spatial mixture models of Section 5.7.5.1. In this case the log of the relative risk is modeled via mixture product of separable components:

$$\log \theta_{ij} = \alpha_0 + \sum_{k=1}^{K} w_{i,k} \chi_{k,j}.$$

Here both $w_{i,k}$ and $\chi_{k,j}$ are unobserved but are separate functions of space and time. Identification is supported by the separation of the spatial loading weights and temporal profiles, although further conditions can be specified. There are K components, and constraints given by $\sum_i w_{i,k} = 1 \ \forall k, 0 < w_{i,k} < 1 \ \forall i, k$. In disease mapping studies it is reasonable to assume that underlying groupings of temporal risk profiles occur and these are region-based. Hence, we could easily be interested in finding spatial groupings of risk which are associated with specific temporal profiles.

Suitable prior distributions depend on the application. Some prescriptions are

$$w_{i,k} = w_{i,k}^* / \sum_l w_{l,k}^* \tag{12.8}$$

$$w_{i,k}^* \sim lN(\alpha_{i,k}, \tau_w), \ \forall i, k$$
$$\alpha \sim MCAR(1, \Lambda)$$
$$\chi_j \sim N(\chi_{j-1}, \Sigma)$$
$$\Sigma = \tau_\chi' I_0$$

where $\tau_\chi = \{\tau_{\chi 1}, ..., \tau_{\chi K}\}$, $\Lambda \sim Wish(\Phi, K)$, and fixed Φ.

I have examined asthma ambulatory-sensitive cases reported per year for <1 year age groups in counties of Georgia over an 8-year period (1999 to 2006). These data are publicly available from the Georgia Department of Health OASIS online health data system. The expected rates were computed from the overall rate (= total rate/total population <1year) times the local population count < 1 year. Figure 12.4 displays the SIR profiles for the 159 counties for the 8-year period. These incidence rates displayed considerable differences in their temporal behavior. It is of interest to examine whether there is any spatial clustering or aggregation in the temporal variation.

Figures 12.5 and 12.6 display the posterior averaged temporal profiles and weight maps for a converged run of model (12.8) (500,000 burn-in and 2000 sample size). It is notable that components 1 and 3 are well defined (and have decreasing and increasing trends respectively) while components 2 and 4 are

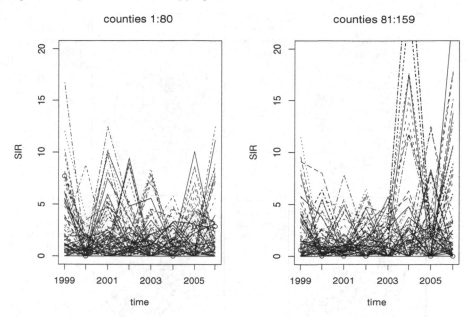

FIGURE 12.4: Georgia county level asthma ambulatory incidence for <1 year for 1999 through 2006. Standardised incidence ratios with expected rates were computed from the total period area rate adjusted for county × year population.

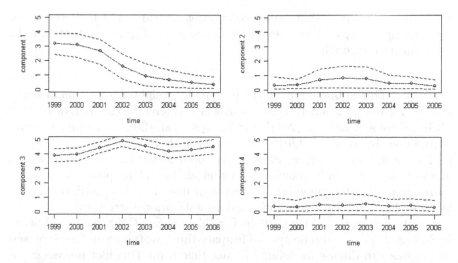

FIGURE 12.5: Space-time latent component mixture model with four fixed temporal components. Posterior expected temporal component profiles ($\chi_{j,l}$) for components 1 through N.

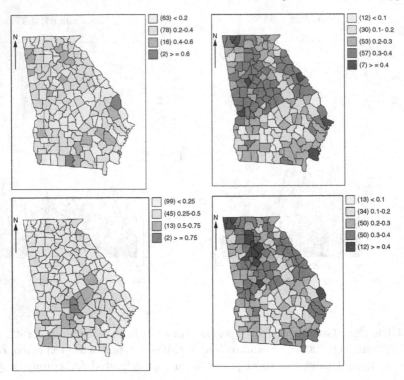

FIGURE 12.6: Space-time latent component mixture model. Posterior expected weight maps for four temporal components. Top left to bottom right: components 1 through N.

negligible. This is borne out by the probability weightings for components 1 and 3 compared to 2 and 4. These models are explored further in Lawson et al. (2010), Choi and Lawson (2011) and Lawson et al. (2012). A simpler variant is proposed by Li et al. (2012). Issues with identification of the temporal profiles under these mixture models have been avoided by the simplification of the temporal profile choice. Napier et al. (2017) proposed to limit the temporal profiles to simple profiles (linear incline or decline, peaked), thereby avoiding identifiablity issues with random walk prior distributions.

Finally, it is worth noting that the Dirichlet process (DP) mentioned briefly in Section 5.7.5, can also be applied in space-time. Kottas et al. (2008) present an approach to disease modeling in space-time using Dirichlet process extensions. See also Hossain et al. (2012) and Hossain and Lawson (2016) and Kottas (2016) for recent reviews.

13

Disease Map Surveillance

In many situations the retrospective analysis of disease maps is a major focus. In fact in previous chapters the analysis described has been applied to a retrospective collection of health outcomes. By *retrospective* I mean that the temporal labels of events are already known, i.e. we know the history of the process. There are situations where there is a need to consider how new data will be characterized and how well we can predict *new* events. In application to disease maps, we would be focused on the prediction of (say) counts of asthma in census tracts in 2013, when we only have counts up to 2012. This is in essence *prospective* analysis of disease outcomes.

13.1 Surveillance Concepts

The US Center for Disease Control (CDC) defines public health surveillance as:

> *The ongoing, systematic collection, analysis, and interpretation of health data essential to the planning, implementation, and evaluation of public health practice, closely integrated with the timely dissemination of these data to those who need to know. The final link of the surveillance chain is the application of these data to prevention and control. A surveillance system includes a functional capacity for data collection, analysis, and dissemination linked to public health programs.*

See Thacker and Berkelman, 1992; Thacker, 1994; Nsubuga et al., 2006.

This definition stresses that surveillance is essential in public health practice and that timeliness is a key component, as is dissemination to relevant parties. The ability to predict future disease outcomes is an important aspect of this activity. Timeliness can be in terms of minutes or hours or days for diseases that spread quickly (highly infectious agents), or in terms of months or years for less infectious or non-infectious diseases. Following the September 11th, 2001 World Trade Center attacks (9/11), there was raised awareness of the need for fast response to potential public health threats. This awareness has increased interest in being able to monitor and respond quickly to changes in disease behavior. This leads to the need to have efficient computational algorithms that can quickly process health data, and the need to make inference as early as possible.

Traditional public health surveillance was originally designed as a passive system whereby cases of disease (usually infectious) were reported individually as they arose, usually by primary care physicians. This individual case reporting allowed more detailed analysis of each occurrence. Prospective surveillance of data on disease by national or regional health agencies has historically been limited by the fear of data trawling and a high false positive rate. However, since 9/11 there is now a great emphasis on public health agencies' ability to react quickly and to detect unusual disease events within large databases. In fact data mining of multiple disease data streams has become important, partly because of the unfocused targeting of potential bioterrorist threats (Azarian et al., 2009). That is, it is unlikely that the targeting of a particular disease will be known and so a range of diseases must be considered. Added to this problem is the fact that rare diseases and transmission dynamics or degree of risk are often not well understood (for example, anthrax inhalation has hardly been studied in humans, see e.g. Brookmeyer (2006), Buckeridge et al. (2006)).

13.1.1 Syndromic Surveillance

The CDC (http://www.cdc.gov/ehrmeaningfuluse/Syndromic.html) defines syndromic surveillance:

Syndromic surveillance uses individual and population health indicators that are available before confirmed diagnoses or laboratory confirmation to identify outbreaks or health events and monitor the health status of a community. By automating public health data collection ..., syndromic surveillance provides public health information in near real time, often sooner than a laboratory test can even be completed. By getting more information more quickly, local, state, and federal public health can detect and respond to more outbreaks and health events more quickly.

Sosin (2003) noted that syndromic surveillance can be defined as the *systematic and ongoing collection, analysis, and interpretation of data that precede diagnosis (e.g., laboratory test requests, emergency department chief complaint, ambulance response logs, prescription drug purchases, school or work absenteeism, as well as signs and symptoms recorded during acute care visits) and that can signal a sufficient probability of an outbreak to warrant public health investigation.*

A recent review (Katz et al., 2011) stresses the range of definitions available but also stresses the usefulness of syndromic surveillance within the greater public health community. Statistical challenges facing early detection of disease outbreaks have been reviewed by Shmeuli and Burkom (2010), Fricker

Jr (2011) and Fricker Jr (2013). However only limited discussion has focused on Bayesian approaches or spatial and spatio-temporal approaches. An early volume that does address spatial issues is Lawson and Kleinman (2005). A recent review in this area can be found in Corberán-Vallet and Lawson (2016). In the following sections I consider some basic temporal surveillance ideas and then consider how Bayesian models of disease maps might be used or extended in the surveillance setting.

13.1.2 Process Control Ideas

In general it is possible to consider the monitoring of health outcomes as a form of process control, in that we want to consider action when some "unusual" disease event or event sequence occurs. In industrial process control usually an overall mean level for a process is specified and limits are set around the mean level. Action is taken when the process strays too far from the mean level and exceeds the limits specified. Control of variability of a process can also be considered. It is possible to consider variability and mean level jointly and also to extend this to multivariate applications where multiple processes are to be monitored. Extending these ideas to public health disease monitoring is possible but some important caveats must be considered:

- Disease events are discrete and at the finest resolution level form a point process.

- Disease events occur within a population that can vary in time and size but also in characteristics.

Figure 13.1 displays a sequence of estimated relative risks from counts simulated from a Bayesian model (aggregated within 100 arbitrary time periods). The counts were generated as a single 100-length realization from a Poisson distribution with mean $e_j\theta_j$ where the relative risk has a gamma prior distribution with parameters a, b where $a = b = 3$. The posterior mean (Bayes estimate) of θ_j is given by $(y_j + 3)/(e_j + 3)$. The posterior gamma distribution is used to set quantiles of risk for the estimates. In this case the 95% quantiles of $Ga(y + 3, e + 3)$ are shown (2.5%, 97.5% levels). It might be assumed that these levels could be used as threshold values for unusual disease events as they have already included the expected rates within the computation of the posterior distribution and allow discreteness of response. In this example the estimated risks fall within the limits defined. It would seem that monitoring relative risk estimates over time would therefore be a reasonable approach to temporal monitoring in a Bayesian setting.

However there are many caveats that complicate this scenario. First, the assumed count model may not be appropriate for the data. This could be due to event resolution, i.e. fineness of time measurement or correlation in the sequence. Second, there could be error in the recorded counts. Under-ascertainment is common in infectious disease monitoring for example. Third,

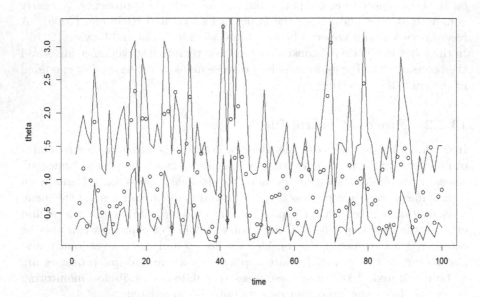

FIGURE 13.1: Estimated relative risks (posterior mean) from a simulated sequence of counts from a gamma-Poisson model with fixed expected rates (e_j) and relative risk (θ_j) generated from $Ga(3,3)$. The expected counts were generated from $N(5,1)$ distribution. 95% quantiles of $Ga(y+3, e+3)$ are solid lines.

it is not clear what threshold levels are appropriate for detection of unusual events. Fourth, unusual events can take different forms and may not be comparable to the drift or jumps found in industrial processes. These issues are discussed in more detail in the following sections.

The idea of risk adjustment on control charting implemented in a variety of settings has been proposed. These charts are based on summary statistics and use testing or confidence interval approaches to monitoring. Reviews of these approaches are given in Woodall (2006) and Woodall et al. (2010).

13.2   Temporal Surveillance

The most common form of disease surveillance, and one for which much has been written, is temporal surveillance. Some references to standard accounts of this area are given in Le Strat (2005), Sonesson and Bock (2003). In the simplest form a single stream of disease information (counts or event times) is monitored. Essentially this is either in the form of a point process (event times) or a time series of counts (in fixed time periods). If a sequence of times is monitored then in a Bayesian setting we would need to assume a data level model and also prior distributions for parameters. Define the time sequence of events as $\{t_1, \ldots\ldots t_J\}$ within a window of length T. We wish to model the event times and assume a Poisson point process model for these whereby $\lambda(t)$ is the first order intensity and the probability of an event at t^* conditional on T is $\lambda(t^*)/\Lambda(T)$, where $\Lambda(T) = \int_0^T \lambda(u)du$. A likelihood under this model can be set up and the model includes parameters within $\lambda(t)$. Some simple cases would be a trend in time (including a constant level): $\lambda(t) = \exp(\alpha + \beta t)$; an added random effect model to allow for extra variation: $\lambda(t) = \exp(\alpha + \beta t + v_t)$, where v_t is an uncorrelated component, or a double random effect model with uncorrelated and correlated variation $\lambda(t) = \exp(\alpha + v_t + u_t)$, where u_t replaces trend term to allow more non-monotonic trend in risk. Note that a fixed baseline risk of $\lambda_0 = \exp(\alpha)$ is assumed in these models. Proposals for analyzing such models in a surveillance context are few, though basic point process testing has been common (Mantel, 1967, Chen et al., 1993, Chen et al., 1997, Grigg and Farewell, 2004). Extension can be considered where the baseline is not constant or clustered so that $\lambda(t) = \lambda_0(t).\lambda_1(t)$.

In the case of aggregate counts within time periods, it is common to assume a Poisson data level model with observed counts for m time periods: $\{y_1, \ldots\ldots, y_m\}$ with associated expected counts $\{e_j\}$ and relative risks $\{\theta_j\}$ $j = 1, \ldots, m$. Hence,

$$y_j \sim Pois(e_j\theta_j).$$

Often a log link to a linear predictor is assumed and then models similar to those for the intensity of point process can be defined:

- Linear trend: $\log(\theta_j) = \alpha + \beta t_j$

- Additional random noise: $\log(\theta_j) = \alpha + \beta t_j + v_j$ where v_j is an uncorrelated noise term

- Random noise with correlation: $\log(\theta_j) = \alpha + v_j + u_j$ where u_j is a correlated noise term

Extensions to these models can include a variety of additional features: such as seasonal or cyclic effects or specific forms of background and foreground clustering and also non-linear or semi-parametric forms. A very useful review of temporal methods, albeit in the context of infectious diseases, can be found in Unkel et al. (2012). Implications of these models for surveillance methods will be discussed in the next sections.

13.2.1 Single Disease Sequence

Models for temporal effects can be derived and suitable prior distributions can be assumed. However in a surveillance context, there has to be consideration of how the data are to be analyzed when prospectively sampled. The fact that data arrives and decisions must be made based on the new data leads to the consideration of different approaches to model specification and use.

First, it is important to consider the focus of the analysis. If that focus changes to relative risk then we would be concerned about estimation of $\Delta\theta_t$, the change in risk between times $t - 1$ and t. This could be estimated from posterior quantities such as $\theta_t - \theta_{t-1}$ if we have data at t and include it in the model. However, it would be more usual in a prospective surveillance context to treat θ_t differently from θ_{t-1} if we want to monitor at time t. For example we may regard events up to time $t - 1$ as "normal" but at t we want to find out whether new data is consistent with the old data.

Second, there might be predictor information that could be used to enable early detection of changes (in risk). In the point process example we could assume a model, such as, $\lambda(t_i) = \lambda_0(t_i) \exp(\alpha_0 + x_i^T \alpha + \beta t_i)$, and estimate this with data including time t_i. New values for x_{i+1}^T and t_{i+1} could then be examined for changes in risk. Finally, a comparison of $\lambda_1(t_i)$ and $\lambda_1(t_{i+1})$ estimated with the new data might be appropriate.

Third a different approach might be assumed where predictive inference is employed. Under a Bayesian paradigm it is always possible to make a prediction from the predictive distribution for data given that already observed. For instance, for new data observation y^* and parameters θ and observed data y we can compute

$$p(y^* \mid y) = \int L(y^* | \theta) p(\theta | y) d\theta.$$

Hence we can compute a prediction of y^* given already observed data. One way to do this is to assume that current values of θ, say θ^t can be used to predict a new predicted value of y_{t+1}^*. It is then possible to compute a predictive residual from the observed value y_{t+1} and $y_{t+1}^* : r_{t+1}^* = y_{t+1} - y_{t+1}^*$. This residual could be examined for unusual or extreme departures from usual behavior. Note that there is a simpler possibility and that is to use the residual from a model fitted to all old and new data: $y_{t+1} - \widehat{y}_{t+1}$ where $\widehat{y}_{t+1}$ is the estimate under the full model. One problem that arises with such methods is that new data will be used to estimate the risk for the following period and so an adaptation can occur. This also requires refitting to the new complete data. This may not be a useful approach when continued surveillance is planned.

The use of residual-based methods has been criticized by Frisén and Sonesson (2005) as they do not consider any past evidence for changes in risk. That is, the history of the process is partially ignored. Optimal surveillance (Frisen, 1992; Frisen and Mare, 1991) seeks to evaluate the historical backdrop of a risk change.

Some alternative approaches to risk monitoring have been proposed. For example, Grigg and Spiegelhalter (2007) suggested the use of an exponentially weighted moving average estimator of the mean risk for count data time series in a Bayesian setting. They applied this to 30-day mortality after cardiac surgery in the UK over a 7-year period. Bayesian hidden Markov models have also been proposed for temporal surveillance data by Strat and Carrat (1999), but they only apply the approach to retrospective analysis. Switching models where a latent binary variable is estimated have been proposed by Martinez-Beneito et al. (2008).

13.2.2 Multiple Disease Sequences

Within a Bayesian formulation it is straightforward to extend temporal models into multiple time series. Define the K long vector outcomes at the j th time period as $\boldsymbol{y}_j$. At the data level a Poisson model could be appropriate where $y_{jk} \sim Pois(\mu_{jk})$ and $\mu_{jk} = e_{jk}\theta_{jk}$. A log link can be assumed so that $\log(\theta_{jk}) = \alpha_0 + \alpha_{1k} + \boldsymbol{x}_j^T\boldsymbol{\beta}_k + R_{jk}$ for covariates $(\boldsymbol{x}_j)$, and R_{jk} a set of time-dependent random effects. In this formulation it is possible to assume a correlation between parameters across diseases. For instance, $\boldsymbol{\beta} \sim N(\boldsymbol{\beta}^*, \Sigma)$ where Σ is a cross disease covariance matrix. Similar assumptions could be made for the correlation of R_{jk}. An alternative, syndromic view is that different diseases drive or presage other diseases and so a conditional model might be favored. For example, with two diseases (a, b) it might be that $\log(\theta_{jb}) = \alpha_0 + \alpha_{1b} + \boldsymbol{x}_j^T\boldsymbol{\beta}_b + R_{jb}$ and that $\log(\theta_{ja}|b) = \alpha_0 + \alpha_{1a} + \boldsymbol{x}_j^T\boldsymbol{\beta}_a + f(\theta_{jb}, \theta_{j-1,b}) + R_{ja}$. Examples of multivariate modeling in retrospective analysis are found in Paul et al. (2008). However, there are few examples of prospective analysis in this area (Paul and Held, 2011).

13.2.3 Infectious Disease Surveillance

In the case of infectious diseases, it is usually important to consider mechanisms of transmission in the formulation. Hence some transmission dynamic is usually included within models and transmission rates are estimated. A recent review of this area is given by Unkel et al. (2012). For influenza and meningococcal disease time series, Paul et al. (2008) examined likelihood models only, although Bayesian switching models have been proposed by Martinez-Beneito et al. (2008). Usually infectious disease models have mechanistic dependencies on previous levels or counts of disease so that $E(y_j) = f(y_{j-1})$ or $E(y_j) = f(y_{j-1}, y_{j-2}, y_{j-3}.....)$ or specified as a function of the mean level: $E(y_j) = f(\mu_j, \mu_{j-1}.......)$. Combinations of both forms have also been proposed.

An additional consideration in infectious disease surveillance is whether an endemic component should be modelled. Some diseases have background incidence which occurs outside of epidemic periods and so a two-component model is sometimes recommended: with endemic and epidemic parts (Held et al., 2006). Very rare diseases usually do not have such an endemic component (e.g. plague), but more common diseases may require one (e.g. sexually transmitted diseases). Influenza, for example, often displays limited background incidence outside of epidemic periods. Endemicity does depend on the temporal scale of the study as well. Longer time periods are more likely to find incidence of rare diseases.

13.3   Spatial and Spatio-Temporal Surveillance

Time is essential for surveillance. Changes over time are an explicit focus of this subject area and so it is perhaps inappropriate to consider *spatial* surveillance. In the same spirit, if spatial surveillance exists, it must consist of evaluation of a static mapped incidence of disease. Hence it is not clearly different from earlier spatial foci in this text. In the remainder of this section I specifically assume that surveillance is spatio-temporal.

The surveillance of disease maps can be viewed as an extension of temporal surveillance of single or multiple time series. One view is that a spatial map with fixed observation units is simply a multivariate time series with (spatial) correlation built into the adjacent series. Another view is that surveillance concerns map *evolutions* in that a spatial domain is changing over time. This in part depends on the units of observation. At a fundamental level, observation of disease in space and time usually consists of a location (s) and a date of diagnosis (t) and over a study period a sequence of disease occurrences is found at different locations and times. This sequence: $\{(s_1, t_1), (s_2, t_2), (s_3, t_3) \ldots \ldots\}$ forms a point process in space-time. In this situation, spatio-temporal point process models can be applied and there has been some application of these in a non-Bayesian setting (see e.g. Lawson and Leimich, 2000, Diggle et al., 2004, Diggle, 2005, Rowlingson et al., 2005, Diggle, 2007). A basic assumption of this approach is that a log Gaussian Cox process (LGCP) is assumed where

$$\lambda(s, t) = \lambda_0(s) \pi_0(t) R(s, t).$$

In the prospective setting it is assumed that the background population effect consists of two intensities (λ_0, π_0) that are separable and can be estimated from historical data. In Diggle et al. (2004) the background is estimated nonparametrically via kernel estimation. The risk function $R(s, t)$ is examined for exceedances $(\Pr(R(s, t) > c))$ as time progresses, while the background intensities were treated as determined (once estimated). MCMC methods were used to estimate relevant parameters but a fully Bayesian approach was not used.

Aggregation of events into m fixed spatial and J temporal units, yields counts of disease: y_{ij}, $i = 1, ..., m$; $j = 1, ..., J$. Usually for rare diseases we would assume $y_{ij} \sim Pois(e_{ij}\theta_{ij})$ as a data level model. Alternative, finite population models could be accommodated by a binomial distribution. For purposes here, I assume that θ_{ij} in a Poisson model is to be modeled and that e_{ij} is fixed and known. I assume a log link of the form

$$\log(\theta_{ij}) = P_{ij}$$

where P_{ij} is a predictor of some form which could include covariates (such as syndromic variables) and also smooth (random) effects.

Two major issue arise. First, how should components appear in P_{ij} to correctly describe "normal" health behavior and "unusual" behavior? Second, how can prospective analysis be carried out within a Bayesian paradigm?

13.3.1   Components of P_{ij}

In conventional spatio-temporal modeling (usually for non-infectious diseases) a separable decomposition is often assumed (see e.g. Chapter 12): that is $P_{ij} = \alpha + S_i + T_j + ST_{ij}$. This decomposition consists of main effects in space (S_i), time (T_j), and an interaction effect (ST_{ij}). This is similar to the decomposition used in the case event situation described above, albeit on the log scale. The question then arises: is this parameterisation appropriate for prospective surveillance? If the case event proposal of Diggle et al. (2004) is followed then S_i and T_j would be regarded as background and the space-time interaction component (ST_{ij}) would be monitored.

The questions then raised could be: how clearly identified are these components? and how can estimation proceed when differential risk changes occur? How can endemicity be accommodated?

13.3.1.1   Component Identification and Estimation Issues

Prospective surveillance requires that decisions be made about the health system as new observations arrive. Define the current count map at j th time as $\{y_{ij}\}$, $i = 1, ..., m$. The historical data is $\{y_{i,j-1}\}$, $\{y_{i,j-2}\}$, ... For prospective analysis there are a variety of estimation issues: 1) Should refitting of a model take place when new data arrives? 2) If refitting is not carried out then do we adjust the model when unusual events occur? 3) Should there be temporal components estimated under the background?

13.3.1.1.1   Refitting?   A fundamental feature of prospective analysis is that new data arrives and enlarges the data set available and also the parameter set may increase. To accommodate this it has often been proposed that

models should be refitted with new data (see e.g. Vidal-Rodeiro and Lawson, 2006b, Paul and Held, 2011). Note that if new parameters appear then the model actually changes its structure each time data appears. However refitting of models in prospective surveillance may lead to accommodation of data that is unusual, as the model attempts to adapt to the new data (Lawson, 2004, 2005). Clearly some comparison of the new data with model predictions at the new time point should be considered (before any refitting is considered). Following this comparison decisions about the state of the system should be made. This leads to estimation considerations.

With a set of spatial units, not all units will signal at any given time. Hence, some areas will no longer be in a background state. If areas have signalled, then a decision must be made concerning whether to "freeze" the background effect for those spatial units or not. The inclusion of temporal effects in normal background can also lead to estimation problems.

13.3.1.1.2 Background and Endemicity The issue of what should be included in the normal disease background is a major issue. Note that endemicity, a term synonymous with prevalence, represents the usual behavior of disease in an area. The correct modeling of this endemicity or background is important. Using historical data in the estimation of background has advantages, not least of which is the ability to self control area risk. However there is a disadvantage in that a choice of historical period that is stable could be misjudged. If the historical data includes unusual disease artifacts then false negatives could be found. Of course the use of expected counts could also be influenced by an inappropriate choice of reference population or period.

The modeling of the background is important in prospective surveillance. Primarily an efficient description of the spatial and temporal variation should be provided. To this end it would seem reasonable to include $S_i = v_i + u_i$, a spatial convolution model which is not time-dependent. The inclusion of a temporal component is more problematic. While in retrospective studies models with temporal trend, random walk structure and also interaction have been found to fit well, it is the focus of prospective surveillance to detect changes in time and so inclusion of adaptive time components is not likely to be a panacea.

While cyclic or seasonal effects may be important (see e.g. Knorr-Held and Richardson, 2003), it is less clear why temporal trend components or random effects should be included. Usually surveillance is carried out in real time and so a temporal trend is unlikely to be present (before any change or outbreak), whereas random effects, in particular non-stationary random walk effects are likely to track trends and other changes in risk. It has been found in variety of studies that removal of random temporal components from the background aids the identification of temporal changes in risk (Corberán-Vallet and Lawson, 2011, Corberán-Vallet, 2012). While temporal components are more controversial it is also the case that special consideration of the background spatial components may be needed. For example if it is known that a disease clusters under the background it may not be adequate to assume a CAR component prior distribution for u_i. Locations of clusters may be known a priori, such as for childhood acute lymphoblastic leukemias (ALL) and so the background should allow for these specific (known) clusters. In that case, specific cluster models may be needed.

Infectious diseases require additional consideration. First some rare diseases have no endemicity or background, and even some more common diseases only strike periodically and so aggregation in time and/or space is required to see any non-zero prevalence. For example, for influenza C+ laboratory notifications in South Carolina (Lawson and Song, 2010) there are many periods of zero counts interspersed by flu outbreaks. Of course these counts are under-ascertained, but there remain long periods of the year where there is little background evidence. This is particularly true also for some veterinary outbreaks and zoonotic diseases such as foot and mouth disease (FMD), avian influenza, and SARS. On the other hand, common sexually transmitted diseases (STDs) could show background prevalence and also outbreak characteristics. A decision to include a background in these diseases could be moot but important. The beginning and ending of an outbreak could fail to be detected if the wrong background model were chosen.

An example of an endemic-based model would be where

$$y_{ij} \sim Pois(\mu_{ij})$$
$$\mu_{ij} = e_{ij}\theta_{ij} + I_{ij}$$

(see e.g. Held et al., 2006, Heaton et al., 2012). The term $e_{ij}\theta_{ij}$ is regarded as the "usual" or endemic behavior while I_{ij} is an added effect due to outbreaks. The choice of an additive component independent of population is favored but an alternative would be to consider a form such as $e_{ij}[\theta_{ij} + I_{ij}]$.

13.3.2 Prospective Space-Time Analysis

In prospective analysis, it is assumed that we observe up to time j and want to consider the situation at $j + 1$. The following are different proposals for handling such a situation. It is possible to employ a Bayesian spatio-temporal model and to refit the model at each time point as new data arrive. In Chan et al. (2010), a Bayesian model-based approach does in fact perform well in the surveillance context, compared to cusum methodology for influenza-like illness outbreaks. In that paper the model assumed has a dependence on previous counts in the same spatial unit and the authors assess the risk using $\Pr(y_{i,j+1} > c | y_{i,1:j})$ where $y_{i,1:j}$ denotes all data up to j.

13.3.2.1   An EWMA Approach

Zhou and Lawson (2008) suggested using an exponentially weighted moving average (EWMA) approach to the estimation and updating of the relative risk θ_{ij}. In this approach the temporal change in the spatial vector of risk is updated at each time point via the EWMA:

$$\widetilde{\boldsymbol{\theta}}_{j+1} = \lambda \widehat{\boldsymbol{\theta}}_j + (1 - \lambda)\widetilde{\boldsymbol{\theta}}_j$$

where $\widetilde{\boldsymbol{\theta}}_{j+1} = [\widetilde{\theta}_{1,j+1}, ..., \widetilde{\theta}_{m,j+1}]^T$ is the one-step-ahead forecast of the map of risks, $\widehat{\boldsymbol{\theta}}_j = [\widehat{\theta}_{1,j}, ..., \widehat{\theta}_{m,j}]^T$ and λ is a smoothing constant. The issue of choice of λ is important as it can affect the degree to which new data affect estimates. Zhou and Lawson (2008) assume $\lambda = 0.2$. Now $\widehat{\theta}_{1,j}, ..., \widehat{\theta}_{m,j}$ are posterior mean estimates from a spatial convolution model at time j where $\log(\theta_{ij}) = \alpha + v_i + u_i$ with the conventional prior specifications. For monitoring, $Z_{i,j+1} = \widehat{\theta}_{i,j+1} - \widetilde{\theta}_{i,j+1}$ is computed along with a sample-based Monte Carlo p-value from

$$MCP_{i,j+1} = \frac{1}{G} \sum_{g=1}^{G} I(Z_{i,j+1}^g < 0).$$

The mean of $Z_{i,j+1}^g$ over the sample can also be used. A measure of percentage increase in risk (the PIR) was also defined as

$$PIR_{i,j+1} = \frac{\overline{Z}_{i,j+1}}{\frac{1}{G} \sum_{g=1}^{G} \widetilde{\theta}_{i,j+1}^g} \times 100.$$

The authors examined simulations and evaluated an average run length measure (ARL_0) and probability of false alarm $(PrFA)$ properties of the method. Figures 13.2, 13.3, and 13.4 display the temporal profiles for South Carolina county-level salmonellosis outbreaks for a range of months from 1995 to 2003 and a selection of mapped MCP and PIR estimates.

13.3.3   Sequential Conditional Predictive Ordinate Approach

Note that for a Bayesian model we can compute a Bayesian p-value from a predictive distribution which will determine whether a new observation is unusual:

$$p_{i,j+1} = \Pr(y_{i,j+1}^* > y_{i,j+1}|data).$$

This is slightly different from the model used by Chan et al. (2010) in that this directly compares predicted value to observed value. More recently, a surveillance approach based on conditional predictive ordinate evaluation has been proposed (Corberán-Vallet and Lawson, 2011). In this approach a sequential

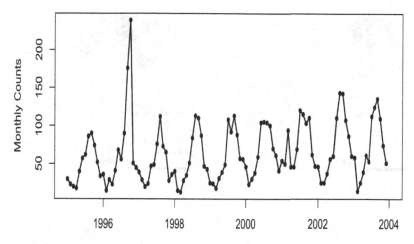

FIGURE 13.2: Time series plot of total monthly salmonella case counts for South Carolina from January 1 1995 to December 31 2003.

FIGURE 13.3: Monthly counts of reported salmonellosis cases in south carolina from 1995 to 2003 and (b) counts for October from 1995 to 2003.

conditional predictive ordinate (SCPO) is defined and used in evaluation of outbreaks or changes. The SCPO is defined by the relation

$$SCPO_{i,j+1} = f(y_{i,j+1}|y_{i,1:j})$$
$$= \int f(y_{i,j+1}|\theta_i, y_{i,1:j})\pi(\theta_i|y_{i,1:j})d\theta_i$$

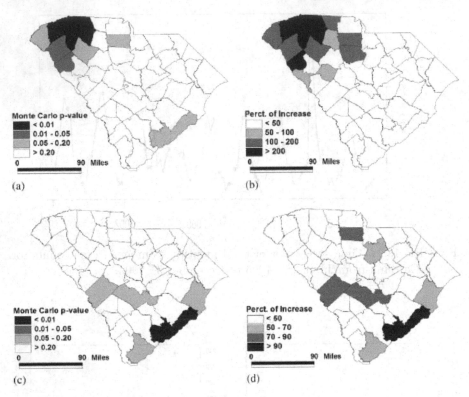

FIGURE 13.4: Spatial distribution of the Monte Carlo p and percent of increase for the salmonellosis case study data. (a) and (b): October 1996, (c) and (d): 2002.

where the notation $y_{i,1:j}$ means all the data up to time j. This can be approximated using an MCMC sample by

$$SCPO_{i,j+1} \approx \frac{1}{G} \sum_{g=1}^{G} Po(y_{i.j+1}|e_{i,j+1}\theta_{i,j}^{g}).$$

A case study was presented by Corberán-Vallet and Lawson (2011) which demonstrated the application of the SCPO. Figure 13.2 displays the time series plot of monthly salmonellosis case counts for the whole of South Carolina from 1995 to 2003. The counts were also observed within the 46 counties of the state. Figure 13.5 displays the resulting SCPO alarms for a selection of 6 month-periods.

This has also been extended to multivariate detection of spatio-temporal disease where shared components are allowed (Corberán-Vallet, 2012). In that

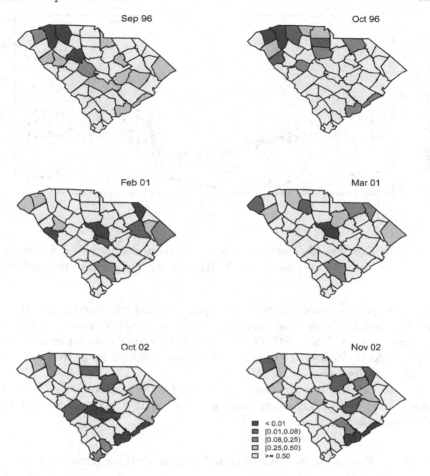

FIGURE 13.5: County-specific alarms detected using the sequential conditional predictive ordinate data for South Carolina salmonella data for a selection of 6-month periods.

case the multivariate SCPO (MSCPO) for p diseases is computed from

$$MSCPO_{i,j+1} = f(y_{i,j+1}^{k_1}, \dots y_{i,j+1}^{k_p} | \boldsymbol{y}_{i,1:j})$$
$$\approx \frac{1}{G} \sum_{g=1}^{G} Po(y_{i,j+1}^{k_1} | e_{i,j+1}^{k_1} \theta_{i,k_1}^g) \times Po(y_{i,j+1}^{k_2} | e_{i,j+1}^{k_2} \theta_{i,k_2}^g) \dots$$
$$\times Po(y_{i,j+1}^{k_p} | e_{i,j+1}^{k_p} \theta_{i,k_p}^g)$$

where the approximation is based on an MCMC sample of size G and it is assumed that $\theta_{i,*}^g$ is only spatially dependent. Corberán-Vallet (2012) applies this method to an example of weekly emergency room discharges for a group

FIGURE 13.6: Weekly counts of emergency room discharges for five diseases for South Carolina over one year. AURI = acute upper respiratory infection.

of five respiratory complaints (acute upper respiratory infection (AURI), influenza, bronchitis, asthma, pneumonia) for 46 South Carolina counties for 27 time periods. The MSCPO demonstrated its ability to detect a range of spatial effects. Figure 13.6 displays the disease variation for 27 weekly periods for the five diseases.

Figure 13.7 displays the alarm signalling for the 5 diseases for Charleston county over the weekly observation periods. Spatial maps of MSCPOs are also available of course.

13.3.4 Surveillance Kullback Leibler (SKL) Measure

Recently a more general measure has been developed which is complementary to the SCPO. Rotejanaprasert and Lawson (2016) suggested that unusual health events using the Kullback-Leibler (KL) divergence at location and time could be detected by

$$SKL(P_{ij}, P_{ij}^{en}) = \int ... \int P_{ij} \log(\frac{P_{ij}}{P_{ij}^{en}}) d\theta$$

where P_{ij} is the posterior distributon for the full data and P_{ij}^{en} is the posterior distribution $P(\theta | \mathbf{y}_{iT}^{en})$ where $\mathbf{y}_{iT}^{en} = \{y_{ij}^{en}\}_{i=1,...,m;\ j=1,...,T}$ generated from the underlying endemic state, so that $\mathbf{y}_{iT}^{en}$ is generated from the predictive distribution with a Poisson data likelihood: $P(\mathbf{y}_{iT}^{en} | \mathbf{y}) = \int ... \int P(\mathbf{y}_{iT}^{en} | \theta, \mathbf{y}) P(\theta | \mathbf{y}) d\theta$ where $P(\mathbf{y}_{iT}^{en} | \theta)$ is Poisson with mean μ_i^{en}. A comparison of SKL to SCPO demonstrated that each measure provides unique insight into the behavior of the spatio-temporal data: the SKL seems to be better at predicting change from endemic to epidemic, while the SCPO is good at detecting localised changes. Rotejanaprasert et al. (2016) also applied the SKL to real time case

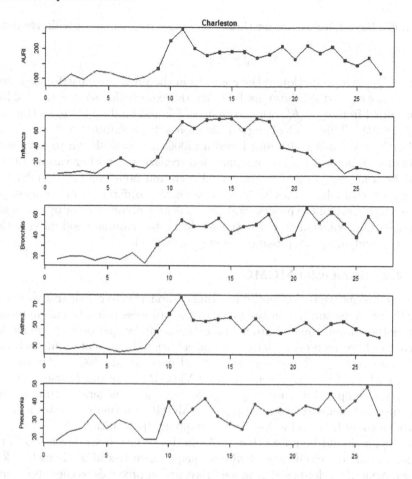

FIGURE 13.7: Time profiles for five diseases for Charleston County with detections superimposed on counts.

event surveillance of non-Hodgkin's lymphoma in South Carolina, and pointed out that SKL subsumes the SCPO within its formulation. A localised SKL was also derived, which can be used with the SCPO for local detection.

13.3.4.1 Residual-Based Approach

Residual-based methods are often used in monitoring. For spatio-temporal data it is possible to examine a Bayesian residual or a predictive residual. For instance, given data up to $j+1$ and an expected count at $j+1$ then we can compute

$$r_{i,j+1} = y_{i,j+1} - \frac{1}{G} \sum_{g=1}^{G} e_{i,j+1} \theta_{i,j+1}^{g}$$

which is a Bayesian residual for the current observations, or a predictive residual:

$$r^p_{i,j+1} = y_{i,j+1} - y^p_{i,j+1}$$

where $y^p_{i,j+1}$ is a prediction of the count from the predictive distribution. In a surveillance context it is often useful to lag the expectation so that the residual is computed from $e_{i,j+1}\theta^g_{i,j}$, the estimate of θ based on data for $y_{i,1:j}$ (Lawson et al., 2004). This kind of lagged result has been dubbed a "surveillance residual" by Vidal-Rodeiro and Lawson (2006a). It was shown to have good performance under a variety of simulation scenarios. Another variant of this could be a lagged predictive residual. Frisén and Sonesson (2005) criticized the use of residuals as they only address current differences or changes. Of course the history of the process will be to a degree accounted for by the model assumed. One solution to this issue would be to examine residual variation over time and/or space, possibly via cusum methods.

13.3.4.2   Sequential MCMC

Another alternative to sequential refitting of models is to consider a variant of MCMC where resampling of the history of the process is used. This approach reduces the computational burden in refitting complex spatio-temporal models which can have increasing data size and additional parameters at each new time point (see e.g. Berzuini et al., 1997, Doucet et al., 2001). An attempt to use such methods has been made by Vidal-Rodeiro and Lawson (2006b) in a spatio-temporal disease mapping context and considerable time savings were apparent compared to refitting using MCMC. Further exploration of this approach could be of value for disease mapping applications.

Finally, it should be noted that while much has now been developed in the toolkit of online surveillance of disease maps there is still a distinct lack of development of model-based or indeed Bayesian approaches to these problems. A Statistics in Medicine special issue (*Vol. 30, No. 5, 2011*) included a wide range of papers but none favored a model-based approach to biosurveillance in general, or indeed specifically for spatio-temporal surveillance, but instead all focused on hypothesis testing-based paradigms. This bias in the field will hopefully be reduced or corrected with future model-based developments.

14

Infectious Disease Modeling

The analysis of infectious disease has developed significantly since the seminal publication of Anderson and May (1992). While many basic models for spread of infection have been developed and applied to real data, in general these have been largely applied in the temporal domain only. For examples see Becker (1989), Mollison (1995), Isham and Medley (1996), Daley and Gani (1999), Keeling and Rohani (2007), Andersson and Britton (2013), and Becker (2015). An exception focusing on geographic issues is Sattenspiel and Lloyd (2009). The ability to accurately model how disease spreads, or has spread (retrospectively), also leads to consideration of surveillance (prospective detection of disease risk changes) and control strategies. In this chapter, the focus will be first on models for disease spread with emphasis on the spatial or geographical dimension. Secondly, I will review approaches to infectious disease surveillance.

Although much attention has been paid to the temporal evolution of infectious disease, there is an increasing awareness of the need to consider the spatial dimension of disease spread. The use of GIS systems and mobile GPS-enabled devices (mobile phones) in real time to monitor outbreaks (such as Ebola in West Africa in 2014; https://www.gislounge.com/empowering-gis-big-data/) is an example of the importance of spatial information in real time epidemic monitoring. The reason that spatial information is important is the need to assess both *where* an outbreak will spread and *when*, and to facilitate control measures and allocation of resources *spatially* in a timely manner. Not only is recording of spatial information important but the statistical analysis of the infectious disease data can provide understanding of the spatial variations in mode and intensity of transmission. Hence models that address the spatial and temporal structures of an epidemic spread should be useful adjuncts to monitoring and control.

14.0.1 Spatial Scale and Ascertainment

Spatial scale or level of resolution is an important factor in the analysis of infectious disease. Lee et al. (2016) consider the issue in some depth. Much emphasis has been placed on modeling of transmission at the level of the individual (see e.g. Deardon et al., 2010 for a review). A fundamental feature of this approach is that direct observation of the infection process between individuals should be recorded and hence the modeling of inter-individual transmission is the focus. There are very few examples of epidemics where the complete path of the epidemic process has been observed. One example was the Hagelloch measles epidemic in Germany in 1861 (see Becker, 1989, Lawson and Leimich, 2000, Neal and Roberts, 2004), where each person's infection status in an isolated village in Germany was recorded daily by a local priest. As it is not possible in most situations to observe fully the progression of infection, most epidemics are only partially observed. This leads to underascertainment (under-reporting) of infected cases.

Under-ascertainment can be considered a result of censoring, whereby a mechanism prohibits the observation of cases. A common example could be influenza reporting: most cases are not reported to doctors and so go unrecorded. In addition, when doctors encounter flu-like symptoms, a range of possible diagnostic endpoints could be possible for the patient's disease progression. Hence misdiagnosis could lead to ascertainment biases. Influenza-like illness (ILI) is in fact a diagnostic ICD code (ICD 10: J11.1). While primary care physicians record cases in health records for patients, only severe cases are confirmed via laboratory testing and so severity bias could be present. If the only information on case incidence comes from laboratory confirmation it could represent the 'tip of the iceberg'. It would then be important to estimate the unobserved case load. This would inevitably be assumption-based. Bailey et al. (2005) provide an example of estimation of under-ascertainment in a leprosy incidence study in Brazil.

14.0.1.1 Asymptomatic Cases

An additional concern with certain infections is that even when case enumeration is good, there can be asymptomatic infected people who do not present symptoms but could be infectious. Hence, for some diseases, there could be two infected states: symptomatic (observed) and asymptomatic (unobserved). For example, it is estimated that Zika virus has a 80% asymptomatic case load (https://www.cdc.gov/mmwr/volumes/65/wr/mm6502e1.htm).
Hence observed cases contribute only 20% of the infection potential at any time. One notorious asymptomatic typhoid carrier was Typhoid Mary, who was an asymptomatic chronic carrier of *Salmonella typhi*. As a cook in New York City and New Jersey in the early 1900s, she unintentionally infected many people until she was placed in isolation on an island, where she died 23 years later. Without any attention paid to asymptomatic sources of infection, the modeling of the infection process could be considerably skewed.

Usually, asymptomatic case load is estimated from a survey of an endemic population where the infectious status of a population sample can be obtained. This provides, at least, a static estimate of the asymptomatic proportion. Of course, if strata exist in the population, such as household clusters or age and gender splits, some differentiation or estimate of variability of the asymptomatic load could be made. From a modeling perspective, if the only information on asymptomatic cases is the population proportion, then a largely unobserved latent component should be included within the modeling framework. If such a component were to be ignored, the observed infection rates could be biased.

14.0.2 Individual Level Modeling

Deardon et al. (2010) provide a review of the basic modeling approaches. Assume that susceptible individuals have a location $(i = 1,..,m)$ and we consider discrete time periods indexed by $(j = 1,...,J)$. Each individual has associated a binary infection status and this can be defined as $y_{ij} \sim Bern(p_{ij})$ where p_{ij} is the probability that at the j th time period the individual at (i) becomes infected. A common model formulation for this probability assumes that $p_{ij} = 1 - f(\lambda_{ij})$ where $f(.)$ is a strictly positive function which maps the risk kernel (λ_{ij}) onto $(0, 1)$. Usually $f(\lambda_{ij}) = \exp(\lambda_{ij})$ is chosen, and $\lambda_{ij} = -\mu_{ij} \sum_{k \in I_{j-1}} \mu_k \, h(i,k) + \epsilon_{ij}$. Here, μ_{ij} is risk factors associated with the i th susceptible individual at the j th time, I_j is the infective set at the j th time, μ_k is the transmissibility of the k th infective, $h(i,k)$ is a kernel relating the i th person to the k th infective and ϵ_{ij} could represent extra infection potential.

The associated likelihood assuming a susceptible-infected-removed (SIR) model is

$$L(\mathbf{y}|\theta) = \prod_{j=0}^{T} \left[\prod_{i \in I_j \setminus I_{j-1}} p_{ij} \right] \left[\prod_{i \in S_j} (1 - p_{ij}) \right],$$

where S_j is the j th time susceptible population. The kernel $h(i,k)$ can act as a distance decay, e.g. $exp(-d_{ij})$ or d_{ij}^{-1}, or other contact propensity measure, while the final term ϵ_{ij} could consist of asymptomatic contribution to infection, external sources or random effects, such as $\epsilon_{ij} = \eta_i + \gamma_j$. Note however that the λ_{ij} must usually be negative overall so that $-\mu_{ij} \sum_{k \in I_{j-1}} \mu_k \, h(i,k) + \epsilon_{ij}$ cannot be positive; hence the correct scaling of ϵ_{ij} will be important. Note that individuals can transition beyond infection to recovery (or be removed from the pool of susceptibles). Deardon et al. (2010) provide an example of such a model applied to the FMD outbreak in 2001 in Cumbria, UK. In that case they assumed a risk kernel of the form

$$\lambda_{ij} = \exp(-\mu_{ij} \sum_{k \in I_{j-1}} \mu_k \, h(i,k) + \epsilon_{ij}) \qquad (14.1)$$

where $h(i,k) = d_{ik}^b$ with b estimated (posterior mean: -1.66), although a threshold change-point was also included. The extra term was assumed to be $\epsilon_{ij} = -\nu I_j$ as a function of the current infective count (ν estimate: -2.5×10^{-10}). The μ_{ij} was replaced by the susceptible vector times count of animal types on premises, while the μ_k is replaced by a transmission vector related to animal count in infected premises. The main spatial aspect of this model lies in the distance kernel expressed in kilometers, with a maximum range of 30 km.

14.0.3   Aggregate Level Modeling

Aggregation of this individual level model to counts in both spatial and temporal units is possible and in many cases data may only be available for aggregate areas and/or time periods. Hence it could be important to be able to develop methods that address this averaging of individual responses. If we aggregate into distinct non-overlapping regions within which the cases occur then for $m = 1, ..., M$ areas, and $j = 1, ..., T$ periods, we have

$$y_{mj} = \sum_{i \in m} y_{ij}$$

and we assume the counts are conditionally independent within the regions, so that y_{mj} has a Poisson-binomial distribution with $E(y_{mj}) = \sum_{i \in m} p_{ij}$ and $var(y_{mj}) = \sum_{i \in m} p_{ij}(1 - p_{ij})$. This is approximately Poisson but without mean-variance equality. It is possible to now specify a model for the counts with p_{ij} suitably parameterised. Assume the susceptibility of the i th subject at time $j - 1$ is $S_{i,j-1}$ and infection potential of the k th subject is τ_{kj} and $h(d_{ik}) = d_{ik}^b$. Then

$$p_{ij} = 1 - \exp(-\lambda_{ij}) \tag{14.2}$$

or

$$p_{ij} = 1 - \lambda_{ij}^{-1}) \tag{14.3}$$

where $\lambda_{ij} = S_{i,j-1} \sum_{k \in I_{j-1}} \tau_k h(d_{ik}) + \epsilon_{ij}$ and to ensure positivity of the final term, assume $\epsilon_{ij} = \exp(\eta_i + \gamma_j)$ where η_i, γ_j are random effects. Now the aggregation yields $E(y_{mj}) = \sum_{i \in m} [1 - \exp(-\lambda_{ij})]$ for model 1. The total number of subjects in m is n_{mj} of which there are y_{mj} 1s. This leads to

$$E(y_{mj}) = \begin{cases} n_{mj} - \sum_{i \in m} \exp(-\lambda_{ij}) & \text{model 1.} \\ n_{mj} - \sum_{i \in m} \lambda_{ij}^{-1} & \text{model 2.} \end{cases}$$

14.0.4   Poisson or Binomial Approximation

If we are prepared to make the Poisson approximation, $y_{mj} \sim Poiss(n_{mj} - \sum_{i \in m} \exp[-S_{i,j-1} \sum_{k \in I_{j-1}} \tau_k h(d_{ik}) + \epsilon_{ij}])$ for model 1. If the contact information is limited to constant rate across all subjects, as is susceptibility, and for simplicity we drop ϵ_{ij}, then $\lambda_{ij} = \exp(-S_{,j-1} \tau n_{I_{j-1}})$ where

τ is a constant transmission rate and $n_{I_{j-1}}$ is the number of infectives at $j - 1$ time period. For model 2 this is particularly simple as $y_{mj} \sim Pois(n_{mj} - \tau^{-1} \sum_{i \in m} \{S_{j-1} n_{I_{j-1}}\}^{-1}])$. This suggests that a binomial version approximation may also be relevant where $y_{mj} \sim bin(n_{mj}, p_{mj}^*)$ and *logit* $(p_{mj}^*) = \tau f_m(S_{j-1} n_{I_{j-1}})$. In later sections, 14.2.0.1 and 14.3.1, some variants of these models are discussed and applied to influenza data in South Carolina and parish level FMD data in Cumbria.

14.1 Descriptive Methods

Often space-time health data concerns infectious disease and its spread. Purely descriptive models that seek to mimic the behavior of the spread can be defined and have been proposed by Mugglin et al. (2002) for flu epidemics, and Knorr-Held and Richardson (2003) for meningococcal disease.

14.1.1 Case Event Data

In general it is possible to model the space-time labeling of infectious disease cases as in the non-infectious case. However, there are advantages to considering infectious disease case event modeling from a survival perspective. This is a simpler perspective than the one described in Section 14.0.2.

14.1.1.1 Partial Likelihood Formulation in Space-Time

An alternative approach is to assume that the observed process has only a time-dependent baseline, i.e. $\lambda_0(s, t) \equiv \lambda_0(t)$. This may be reasonable where the temporal progression of a disease is the main focus (such as in survival analysis). The set of observed space and time coordinates $\{s_i, t_i\}$ is conditioned upon, and a risk set (R_i) can be considered at any given time t_i. In the absence of censoring, $R_i = \{i, ..., n\}$. Then the probability that an event at (s_i, t_i) out of the current risk set is a case is just

$$P_i = \lambda(s_i, t_i) / \sum_{k \in R_i} \lambda(s_k, t_i).$$

This is an extension to the Cox proportional hazard model. Importantly in this formulation, when $\lambda_0(s, t) \equiv \lambda_0(t)$ the background hazard cancels from the model and the partial likelihood is given as

$$L = \sum_{i=1}^{n} [\log \lambda(s_i, t_i) - \log \sum_{k \in R_i} \lambda(s_k, t_i)].$$

Hence, this form enables relatively simple modeling of space-time progression of events. Lawson and Zhou (2005) use this approach to modeling progression of a foot-and-mouth epidemic, while it has also been used for a measles epidemic in a non-Bayesian context by Lawson and Leimich, 2000 (see also Neal and Roberts, 2004 for another measles modeling approach and Diggle, 2005).

14.1.2 Count Data

Often a descriptive approach would be considered first in the modeling of infection spread. By descriptive I mean using model elements to mimic the spread (without directly modeling the infection process). Mugglin et al. (2002) suggested a descriptive approach to flu space-time modeling in Scotland. They applied the model to weekly ER admissions for influenza in Scottish local government districts for 1989 and 1990. The model proposed for ER admission count y_{ij} in the ith district and jth time period was of the form

$$y_{ij} \sim Poisson(e_{ij} \exp(z_{ij}))$$

where e_{ij} is the number of cases expected under non-epidemic conditions, and z_{ij} is the log relative risk. Here z_{ij} is modelled as

$$z_{ij} = d_i'\alpha + s_{ij}$$

where $d_i'\alpha$ is a linear predictor including site dependent covariates, with d_i' the ith row of the $n \times p$ covariate design matrix and α a p-length parameter vector, and s_{ij} is defined by a vector autoregressive model $(\mathbf{s}_j : (s_{1j}, ...s_{mj})')$

$$\mathbf{s}_j = H\mathbf{s}_{j-1} + \epsilon_j.$$

Here, H is an $m \times m$ autoregressive coefficient matrix and ϵ_j is an epidemic forcing term. Spatial structure appears in both H and ϵ_j. The form of the epidemic curve is modeled by the Gaussian Markov random field prior distribution for ϵ_j:

$$\epsilon_j \sim MVN(\beta_{\rho(j)}1, \Sigma)$$

where β determines the type of behavior, $\rho(j)$ indicates the stage of the disease and Σ is a variance-covariance matrix. The model was completed with prior distributions specified for all parameters within a Bayesian model hierarchy. An alternative but somewhat simpler approach to descriptive modeling has been proposed by Knorr-Held and Richardson (2003). In their example, monthly counts of meningococcal disease cases in the departments of France were examined for 1985 through 1997. The model assumes the same likelihood as Mugglin et al. (2002) such that

$$y_{ij} \sim Poisson(e_{ij} \exp(z_{ij})).$$

At the second level they assume for the endemic disease process

$$z_{ij} = r_j + s_j + u_i$$

where r_j denotes temporal trend, s_j denotes a seasonal effect of 12 months and a CAR prior distribution for **u**. They assume no space-time interaction for the endemic disease. For the epidemic period an extra term is included:

$$z_{ij} = r_j + s_j + u_i + x_{ij} r_{ij}^T \beta$$

where x_{ij} is an unobserved temporal indicator (0/1) which is dependent in time (but not in space) and r_{ij} is a $p \times 1$ vector (a function of the vector of observed number of cases in period $j-1$) and β is a p-dimensional parameter vector. The authors propose six different models to describe the epidemic period depending on the specification of $r_{ij}^T \beta$. Whether an epidemic period is present completely depends on the value of x_{ij}. In this formulation the x_{ij} are essentially unobserved binary time series, one for each small area. Unlike the Mugglin et al. (2002) formulation, these series have to be estimated.

Both these approaches seem to have been successful in describing the retrospective epidemic data examined. It will be instructive to see whether they will be successful in the prospective surveillance of infectious disease.

14.2   Mechanistic Count Models

Mechanistic count models that address the infection mechanism have been proposed for the temporal spread of measles (Morton and Finkenstadt, 2005). These were based on susceptible-infected-removed (SIR) models where account is made of the numbers at each time point in each class. They also account for under-ascertainment in their model. For daily measles case reporting in London, Morton and Finkenstadt (2005) defined the true infective count for period j as I_j and the reported count as y_j linked by a binomial distribution to allow for under-ascertainment: $y_j \sim bin(\rho, I_j)$ where ρ is a reporting probability. The susceptible population at the $j+1$ th period is S_{j+1} while removal is D_j and additions B_{j+1}. Infectives and susceptibles are modeled as

$$I_{j+1} \sim f_1(r_j I_j^\alpha S_j, K_{j+1})$$
$$S_{j+1} \sim f_2(S_j + B_{j+1} - I_{j+1} - v D_{j+1})$$

where K_{j+1} is some underlying latent series of events, and f_1 and f_2 are suitable distributions. The distribution f_1 is called the transmission distribution. The term r_j is a proportionality constant that modulates the interaction term $I_j^\alpha S_j$ and can be regarded as an infection rate. The α term can also be estimated. Where the susceptible population is large compared to the infective population at each time period, the effect of $B_{j+1} - I_{j+1} - v D_{j+1}$ may be small and so simpler models could be conceived where $S_{j+1} \sim f_2(S_j)$. Of course, for finite small populations this could be a poor approximation.

Extending this to the spatial situation within a Bayesian hierarchical modeling framework is straightforward (see Lawson, 2006b, chapter 10). A space-time infection model could be proposed where there are $i = 1, ..., m$ small areas and $j = 1, .., J$ time periods. A simple form could be:

$$y_{ij} \sim bin(\rho, I_{ij}) \tag{14.4}$$
$$I_{ij} \sim Pois(\mu_{ij})$$
$$S_{ij+1} = S_{ij} - I_{ij} - R_{ij}$$
$$R_{ij} = \beta I_{ij}$$

where $\mu_{ij} = S_{ij} I_{ij-1} \exp\{\beta_0 + b_i\}$. The term $\exp\{\beta_0\}$ describes the overall rate of the infection process while a spatially correlated term b_i is assumed to have an ICAR prior distribution. The susceptible model is deterministic and with a fixed β the removal proportion is fixed. Many variants of these specifications could be considered. For example, we could specify $\mu_{ij} = S_{ij} f_*(I_{ij-1}, \{I\}_{\delta_{ij-1}}, \exp\{\psi_{ij}\})$ where dependence in f_* is on the previous count I_{ij-1}, on the counts in a predefined neighborhood $\delta_{ij-1}, \{I\}_{\delta_{ij-1}}$ say, and a linear predictor including both covariates and random effects which could be spatially or temporally correlated.

14.2.0.1 South Carolina Influenza Data

An example of the application of this model to publicly available flu culture positives (C+) from the 2004–2005 flu season reported for bi-weekly periods for the counties of South Carolina is given in Lawson and Song (2010), for the model specified in (14.4). In that case it was assumed that $\beta = 0.001$. Figure 14.1 displays the flu season count variations for a selection of four counties in South Carolina (Beaufort, Richland, Charleston, and Horry). Beaufort and Horry both have high older age group populations, while the main urban centers in the state are in Richland (city of Columbia) and Charleston (city of Charleston). Figure 14.2 displays the thematic maps of the counts for a selection of three time periods during the season.

Figure 14.3 displays the posterior average infection rates $(\widehat{\mu}_{ij})$ for a selection of four counties in the state, along with their 95% credible intervals. Interestingly, Horry county peaks much earlier than other areas (time periods 4 to 6) while Beaufort seems to display lag effects into periods 10 to 12. In fact there appears to be considerable spatial and temporal variation in the mean infection level.

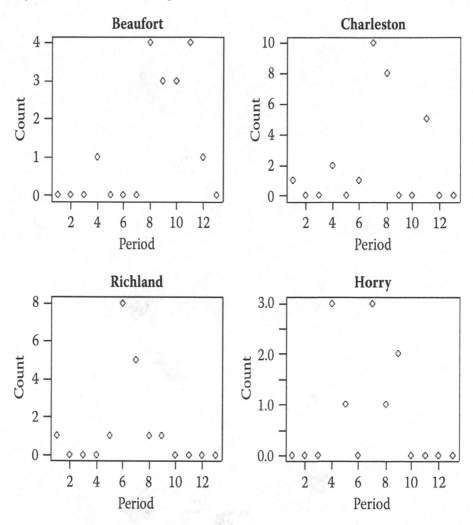

FIGURE 14.1: South Carolina influenza confirmed C-positive notifications. Count profiles for December 18 th 2004 to April 16 th 2005 for a selection of four counties.

14.3 Veterinary Disease Mapping

While most work in disease mapping has been targeted towards human health, there is a growing literature in the application of disease mapping to veterinary health. Veterinary health covers the analysis of managed animal populations, but may also cover wild populations, in particular where zoonosis (disease

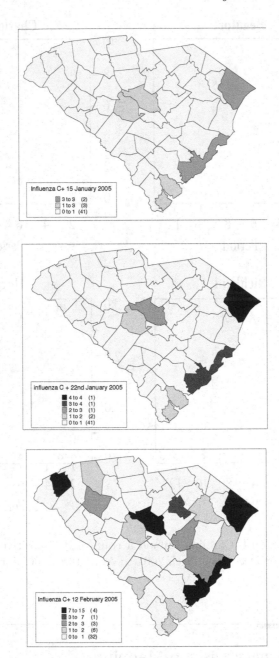

FIGURE 14.2: South Carolina influenza confirmed positive notifications. Count thematic maps for a selection of three time periods in the 2004-2005 season.

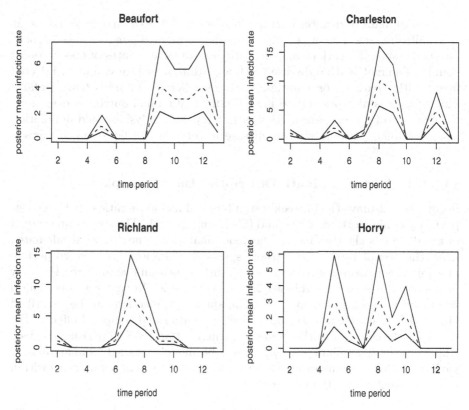

FIGURE 14.3: South Carolina influenza confirmed positive notifications. Posterior mean infection rate estimates for 13 time periods with credible 95% intervals for a selection of four urban counties: Beaufort, Charleston, Richland, and Horry.

transmission between species) is possible. This has arisen partly because of recent outbreaks of bovine spongiform encephalopathy in cattle, foot and mouth disease among sheep and cattle and the spread of SARS or avian flu and Ebola, and potential for evolution within the human population. Often in these examples, space-time variation is the most important feature to be modeled and so discussion of this topic here is justification.

The basic descriptive disease mapping techniques, such as the commonly used convolution models (Chapter 5) can of course be applied to veterinary data and there are now various examples in the literature (e.g. Stevenson et al. (2000), Stevenson et al. (2001), Stevenson et al. (2005), Durr et al. (2005)). Competing risk multivariate analysis has also been proposed (Diggle et al., 2005). An overview of GIS-based applications is found in Durr and Gatrell (2004).

Often the data from veterinary applications is akin to human health data. It is usually discrete and could be in the form of a marked point process in space and time (animal locations and their disease states and dates of observation), counts of animals with a disease (such as within farms), or counts of infected farms within parishes or counties. Within smaller spatial units (such as farm buildings) it is also possible to model individual animal outcomes over time. An example of a semi-parametric spatio-temporal analysis of wild population disease burden (chronic wasting disease in cervids) can be found in Lawson and Song (2009).

14.3.0.1   Foot and Mouth Descriptive Data Example

Figure 14.4 displays the bi-weekly standardized incidence ratios for FMD over parishes within Northwest England (Cumbria) in 2001. The space-time spread of the disease is clearly shown. The denominators for the SIR were calculated from the overall rate for the whole space-time window. In a retrospective analysis it is reasonable to standardise within such an overall rate. However in a surveillance context this would not be possible. In that case one option would be to use a historical rate. The spread of infection can be described via models that attempt to summarize the spatial and temporal effects. For example, the count of FMD premises (farms) within the i th parish at a given time period (j) in Cumbria (y_{ij}) could be modeled as a binomial random variable with various hierarchical elements. Define the number of farms within the i th parish as n_i, then assume:

$$y_{ij} \sim bin(p_{ij}, n_i)$$
$$\log it(p_{ij}) = A_i + v_i + \xi_j$$
$$\text{where } A_i = \beta_0 + \beta_1 x_i + \beta_2 y_i + \beta_3 x_i y_i.$$

The term A_i is purely a trend component in the spatial coordinates (x_i, y_i) of the parish, while v_i, ξ_j are random effects that are meant to capture the spatial and temporal random variation. This is the descriptive model reported by Lawson and Zhou (2005). The random terms are assumed to have prior distributions given by:

$$v_i \sim N(0, \tau_v)$$
$$\xi_j \sim N(\xi_{j-1}, \tau_\xi).$$

Hence temporal dependence is assumed to be modeled by an autoregressive term in the logit link. Suitable parameter prior distributions were assumed for β, τ_v, τ_ξ. There is no spatial correlation term, as it was felt that the dynamic nature of the risk would be better described via uncorrelated spatial risk and correlated temporal risk. This descriptive model was only partially successful in describing the variation. In this model the spatial trend component is fixed in time. Not considered by the authors was the possibility of making the regression parameters in the trend component time-dependent.

FIGURE 14.4: Foot and mouth disease (FMD) epidemic Northwest England for 2001. Bi-weekly maps of standardised incidence ratios for eight two-week period starting February 2nd (top left) through June 1st (bottom right).

This might be an attractive option in some cases as it would allow the spatial model to have a dynamic element. For example, it could be assumed that $\beta_j \sim \textbf{MVN}(\beta_{j-1}, \tau_\beta I_n)$, where I_n is a unit matrix where $n = 4$.

14.3.0.2   Mechanistic Infection Modeling

Infectious disease spread is particularly important in veterinary applications. There are few examples of mechanistic Bayesian modeling of such spread. Höhle et al. (2005) gives an example where swine fever within pig units is modeled spatially via a survival model where the hazard function is a function of the count of infected animals within the unit and also the count in neighboring units. Bayesian models for the United Kingdom (UK) FMD outbreak at farm level were also proposed where a survival model was assumed at the farm level (Weibull in this case) for the risk of infection and then a count model for the number infected conditional on the infection of the farm (Lawson and Zhou, 2005). This also included spatial dependence. These models are adequate where a relatively slow epidemic is apparent, but are likely to be inadequate when a full epidemic curve with peaking and recession is to be modeled.

14.3.0.3   Some Complicating Factors

There are a number of complicating factors that appear in veterinary examples that should be highlighted. First of all, it is often the case that for important infection epidemics, *intervention* by veterinary agencies will dramatically alter the progression of the disease (and also the ability to observe the progression). In the FMD outbreak in the UK in 2001, ring culling was introduced. This entailed slaughtering all farm animals within a fixed radius of a newly found case of FMD. This culling is an attempt to intervene in the spread of the disease. The effect is to introduce a particular form of spatio-temporal censoring during the epidemic and this can lead to considerable missing information that could affect model predictions. This also leads to the other important aspect of modeling veterinary disease: the surveillance or predictive capability of models. In the FMD outbreak in the UK in 2001, statistical models were used on a daily basis to predict the progression of the disease. The online surveillance of disease spread is very important and predictive capability is of course an important and natural ingredient of Bayesian modeling with recursive Bayesian updating an essential ingredient.

Finally, it is also important to note that there can be a major difference in data acquisition within veterinary health compared to human health. Animals usually do not report disease to vets! Hence they have to be sampled, and, unless registries of disease are set up with mandatory reporting, there is the possibility that *under-reporting* or *under-ascertainment* of cases could become a major problem. While this is less important for managed herds (such as on farms), it could be very important for wild populations.

For wild populations, animals are free to move and their mobility can make sampling very problematic. For wild populations, access to animals and the fact that they move around in space-time leads to extra complications. Distance sampling (Buckland et al., 2001) is one approach to assessing mobile population density. Remote sensing could be used for density estimation also.

However this does not usually allow the health of animals to be assessed. Hunter surveys and special culling have been used for deer population health (chronic wasting disease: Farnsworth et al., 2006; Lawson and Song, 2009). However these data are often prone to considerable biases due to the nature of hunting (choice of area, choice of animal, time of day, date of hunting) and it is not clear how representative these data are of the true population health. Faecal surveys may also be used.

Even with managed herds, the animals must be continually checked to find out whether they are diseased. Unless there is continual monitoring of uninfected animals over space and time, under-ascertainment is highly likely. Of course some modeling strategies are available to deal with this issue as noted above (Section 14.1.2).

14.3.1 FMD Mechanistic Count Modeling

The FMD outbreak in 2001 in Cumbria, England has been examined extensively at the individual infected premise (IP) level and at the parish level descriptively. An alternative approach at the parish level is to consider a mechanistic modeling of the disease progression where transmission dependence is included. Lawson et al. (2011) revisited the FMD epidemic, with updated data at the parish level and provided a modeling framework where counts of IPs within parishes (y_{ij}), for i th parish and j th time period (biweekly) are assumed to appear from the n_{ij} premises. In total there are $m = 138$ parishes and $J = 13$ time periods. While a binomial model could be assumed at the data level, it is convenient in this case to consider a Poisson model with

$$y_{ij} \sim Pois(\mu_{ij}) \tag{14.5}$$

$$\mu_{ij} = n_{ij}f(\{y_{l,j-1}\}_{l\in(i,\delta_i)}, \mathbf{x}, W_{ij}) \tag{14.6}$$

where $\mathbf{x}$ is a matrix of covariates, W_{ij} are confounders, and the dependence on previous counts both in the parish in question (i) and its neighborhood (δ_i) directly specified. The count is modulated by the total premises at a given time (n_{ij}). Note that n_{ij} can decrement over time as IPs are found in parishes.

14.3.1.1 Susceptible-Infected-Removed (SIR) Models

The core of SIR models is a transmission process or model. This can be identified in our case by the data level model given in Equations (14.5) and (14.6).

14.3.1.1.1 Transmission Models The specification of $f(\{y_{l,j-1}\}_{l\in(i,\delta_i)})$ defines the transmission processes assumed. These models can take a variety of forms with differing degrees of dependence. A simple model (M1) could be

$$\mu_{ij} = n_{ij}y_{i,j-1}^{\alpha_c} \cdot \exp\{ \alpha_0 + W_{ij} + \varepsilon_{ij}\} \tag{14.7}$$

$$\log(\mu_{ij}) = \log(n_{ij}) + \alpha_c \log(y_{i,j-1}) + \alpha_0 + W_{ij} + \varepsilon_{ij}$$

where the transmission rate is a function of α_0 and α_c and a residual space-time interaction effect is assumed as ε_{ij}. In this model new infected premises are generated from the count (n_{ij}) as well as the log of the previous infective number in that parish scaled by $exp(\alpha_o)$.

A variant of this model where the α_c is augmented by a local rate α_{ci} both with zero-mean Gaussian prior distributions always fitted better. This is a form of geographically adaptive regression (GAR). Also assumed was $W_{ij} = W_{1i} + W_{2i}$, purely spatial random effects with uncorrelated and correlated (ICAR) prior distributions respectively.

A second model (M2) was formulated where dependence on previous premises counts was also included:

$$\mu_{ij} = n_{ij}y_{i,j-1}^{(\alpha_c+\alpha_{ci})} \cdot [n_{i,j-1}]^{(\alpha_p+\alpha_{pi})} \cdot \exp\{ \alpha_0 + W_i + \varepsilon_{ij}\}. \tag{14.8}$$

Another extension added neighborhood dependence terms such as $\alpha_n \sum_{l \in \delta_i} y_{l,j-1}$ (M3). Composite forms of these models were also examined.

Within SIR models it is usual to define a removal process (R_{ij}) and an accounting equation as well as a transmission process. The accounting equation updates the premise count for a given time, by removing premises that are infected or were removed. Culling could remove premises, both infected or not (given that ring culls were used in the epidemic). Thus

$$n_{ij} = n_{i,j-1} - y_{i,j-1} - R_{i,j-1}. \tag{14.9}$$

In this case the final period count of premises is known after culling and so the complete vector of counts $\{n_{ij}\}$ is known for all times and does not need to be computed from (14.9).

Overall, it was found that the best fitting models based on the DIC criterion were M2 without the W_i effect and a model with lagged neighborhood and lagged infected count dependence. It is notable that in both models the previous total premises count is found to be important in generating new infectives.

14.3.1.2   Estimation of Termination

As part of the final analysis of the epidemic an attempt was made to estimate locally the final downturn of the outbreak. This was estimated by examining sequences of mean estimates over time and assessing whether there was a monotonic decline over a number of periods. It might be expected that waning conditions prevail when monotonicity is found. Specifically the difference

FIGURE 14.5: Posterior mean flag maps for time periods 4, 5, 10, 13 based on binary monotonicity indicator: $z_{ij} = 1$ if $\mu_{i,j-3} > \mu_{i,j-2} > \mu_{i,j-1} > \mu_{ij}$; else $z_{ij} = 0$.

functionals $\triangle \mu_{ij} = \mu_{ij} - \mu_{i,j-1}$ were monitored for consecutive times. A binary flag (z_{ij}) was activated when the posterior mean estimates were as

$$z_{ij} = \begin{cases} 1 & \text{if } \mu_{i,j-3} > \mu_{i,j-2} > \mu_{i,j-1} > \mu_{ij} \\ 0 & \text{otherwise} \end{cases}.$$

Figure 14.5 displays the result for model M2 over four time periods (4, 5, 10, and 13). For a selection of two parishes (8000, 8333), Figure 14.6 displays their location, while Figures 14.7 and 14.8 display the posterior mean profiles for both parishes (upper panel) and the binary flag profiles (lower panel).

14.4 Zoonoses

There is a growing awareness that diseases affecting animal species could affect the incidence of the same diseases in humans. Long recognized

FIGURE 14.6: Map highlighting the parishes 8000 and 8333.

FIGURE 14.7: Posterior mean profiles (with 95% credible intervals) for study period for parish 8000. Lower panel shows binary flag profile.

FIGURE 14.8: Posterior mean profiles (with 95% credible intervals) for study period for parish 8333. Lower panel shows binary flag profile.

zoonotic diseases include brucellosis (cows and pigs), anthrax (sheep), plague (rodents), trichinellosis and trichinosis (swine), tularemia (rabbits), and rabies (bats, raccoons, dogs, and other mammals). Zoonoses newly emergent in North America include West Nile encephalitis (birds), and monkeypox (prairie dogs). Many newly recognized infectious diseases in humans, including HIV/AIDS, Ebola infection and SARS, are thought to have emerged from animal hosts, although those hosts have not yet been identified (https://www.cdc.gov/ophss/csels/dsepd/ss1978/lesson1/section10.html).

14.4.1 Leishmaniasis

Leishmaniasis, which is a parasitic disease caused by sandfly bites, can affect animals as well as humans. Infection of humans could be directly from bites but also from contact with infected animals. In Brazil, dogs are particularly susceptible and it is estimated that visceral leishmaniasis uses, in the urban environment, domestic dogs as reservoirs (Corberan et al., 2017). Figure 14.9 displays the areas of coincident dog-human data available for 5564 Brazilian

FIGURE 14.9: Brazil Municipality data for dog and human leishmaniasis data for 2007 through 2011.

municipalities for leishmaniasis for 2007 through 2011. Figure 14.10 displays the human SIR high risk areas for 2007 through 2011. After fitting spatial models to both dog and human data for each year period of 2007 to 2011, posterior estimated exceedance probabilities were recorded for $\widehat{\Pr}(\theta_{ij}^d > 1)$ and $\widehat{\Pr}(\theta_{ij}^h > 1)$, with d and h denoting dogs and humans respectively. The coincidences of these high risk areas (where $\widehat{\Pr}(\theta_{ij}^d > 1) > 0.95$ and $\widehat{\Pr}(\theta_{ij}^h > 1) > 0.95$) for dogs and humans are found in Figure 14.11. In this example, there appeared to be no general overall correlation between dog incidence and human case load, but there is evidence of a cross-correlation which is spatially localised.

14.4.2 Tularemia

Tularemia (https://www.cdc.gov/tularemia/index.html) is an infectious disease caused by an intracellular bacterium, *Francisella tularensis*, which is tick- or deer fly-borne and zoonotic. The disease is endemic in North America and parts of Europe, with recurrent outbreaks in Sweden and Finland. *Francisella tularensis* has a wide range of hosts with transmission most commonly

FIGURE 14.10: Human Leishmaniasis standardised incidence ratio (SIR) for the period 2007 through 2011.

via arthropod vectors. Rodents could play a role in the zoonotic transmission of the disease after findings of a relationship between vole population cycles and human tularemia incidence in Sweden and Finland (Rossow et al., 2015). Specifically in Finland, rodent population dynamics displayed a spatiotemporal relationship with human tularemia cases, such that human case numbers appeared to peak one year after peak rodent densities. Rossow et al. (2015) reported the distribution of rodent (vole) incidence in Finland from 1995 to 2013. In addition they reported human case data recorded within 20 hospital districts. They also found a correlation between vole disease peaks and lagged human case numbers. Figure 14.12 displays the human case load variation in Finland with vole population cycles added over the period 1995-2013, for three areas of Finland.

FIGURE 14.11: Coincidence of probability of exceedance of relative risk ($\theta_{ij} >$ 1) for dogs and humans in Brazil estimated under a Bayesian spatio-temporal model.

In Rotejanaprasert et al. (2017), spatio-temporal models were fitted jointly to the human and rodent data. The rodent incidence was recorded as a binary (0/1: decline or increased/peak level) or three-level categorical variables (decline(1), increase(2), peak(3)). As the focus was on the relation between rodent and human incidence a space-time model was fitted to the human data of the form

$$y_{ij}^h \sim Pois(e_{ij}\theta_{ij})$$

with $\log(\theta_{ij}) = \alpha^h + u_i^h + v_i^h + \lambda_j^h + \delta_{ij}^h$ where $u_i^h + v_i^h$ is a spatial convolution term, λ_j^h had a random walk prior distribution, and the space-time interaction term δ_{ij}^h was modeled to be dependent on rodent previous ST interaction: $\delta_{ij}^h = \beta_i \delta_{i,j-1}^r$. This allows the localised clusters of human risk to be dependent on rodent space-time clustering. For the binary rodent model, a Bernoulli model was assumed with

$$y_{ij}^r \sim Bern(p_{ij})$$

with $\text{logit}(p_{ij}) = \alpha^r + u_i^r + v_i^r + (u_k + v_k)^{sr} + \lambda_j^r + \delta_{ij}^r$ where sr denotes regional level contextual effects. The rodent ST interaction term is given a zero mean Gaussian prior distribution. The coefficient β_i is given dependence

FIGURE 14.12: Vole population peaks and human case notifications for three regions within Finland for 1995 through 2013. Rossow H, Ollgren J, Hytönen J, Rissanen H, Huitu O, Henttonen H, Kuusi M, Vapalahti O. Incidence and seroprevalence of tularaemia in Finland, 1995 to 2013: regional epidemics with cyclic pattern. Euro Surveill. 2015;20(33):pii=21209. https://doi.org/10.2807/1560-7917. ES2015.20.33.21209.

on previous rodent number by $\beta_i = \beta_i^p y_{i,j-1}^r + \beta_i^n (1 - y_{i,j-1}^r)$ and a positive dependence measure can be computed: $DP_i = \frac{\exp(\beta_i^p)}{1+\exp(\beta_i^p)}$. This measure is close to 1 if human cases are high following an increase or peak in rodent cases in the preceding year. Figure 14.13 displays the posterior mean DP map under a binary rodent model for each hospital district. It is clear that the western district of Southern Ostrobothnia and part of Central Finland have high dependence on lagged rodent peaks but other areas are much less dependent.

FIGURE 14.13: Posterior mean DP measure under the binary rodent model for Finnish hospital districts.

14.4.3   Some Modeling Issues for Zoonoses and Infectious Diseases

The form of modeling to be undertaken for either host (human or animal) is important. First, should animal and human incidence be modelled jointly, or should animal incidence be treated as a conditioning predictor for human models? In the tularemia example above both rodent and human variation were modeled. If the animal variation is important to describe, joint modeling would usually be preferred. However, if the dependence on animal numbers is the focus, conditioning via a predictor would seem to be adequate. Second, whether mechanistic or descriptive models are chosen also depends on the focus. In the tularemia and leishmaniasis examples, descriptive models were assumed and functions of model parameters were the focus. In the FMD and flu examples, mechanistic models were chosen as the focus was on transmission. Third, this also relates to the idea that in infectious disease models dependence on previous incidence should be direct dependence and not be modeled via some mean level (see Equation (14.7)). The rationale for this is that transmission of infection should relate directly to those infected and where they are and are not on the mean level of infection. For zoonosis this means that animal counts (and previous human counts) should condition the current infections. A collorary to this is the fact that system-measurement (Kalman filter) models are not so useful in this context as the system part models the mean structure and the measurement part gives rise to the infection outcome, rather than a direct link to previous infections.

15

Computational Software Issues

Bayesian modeling relies on the evaluation of a posterior distribution and functionals of that distribution. Hence any methods that can accurately estimate posterior features can be useful in the disease mapping context. Monte Carlo sampling-based algorithms, such as Markov chain Monte Carlo variants such as Gibbs or Metropolis sampling, have been developed and have become commonly applied in many situations. Software such as WinBUGS, OpenBUGS, JAGS and CARBayes are all sampling-based and can be used for many of the model applications considered here. Alternative algorithms that can provide approximations to posterior distributions are also available and can provide fast computation.

In Chapter 3 the basic methods of posterior sampling and posterior approximation were reviewed. In this chapter the focus is on software implementation and comparison of available software. The following discussion focusses on R software and WinBUGS/OpenBUGS. In Appendix A, a basic R and WinBUGS/OpenBUGS review is presented. In what follows, some familiarity with R and basic WinBUGS coding is assumed. The focus here is on a review and comparison of software for a variety of spatial and spatio-temporal models.

15.1   Graphics on GeoBUGS and R

With R and Win/OpenBUGS it is possible to handle mapped data.

15.1.1   Mapping on R

On R the libraries maps and maptools provide facilities for the handling of polygon-based maps. In particular, maps allows the visualizing of polygonal data and with further programing allows the creation of thematic maps. The maptools package allows the importation of ArcView shape files to R (readShapePoly function). For example:

```
library(maptools)
polys<-readShapePoly("filename.shp")
```

```
plot(polys)
```

will generate a polygon object (polys) from a shapefile. In addition, Splus format export files from WinBUGS can be read into polygons on R. On GeoBUGS go to the adjacency tool and opt for 'Export Splus'. Then the export file:

```
polySC<-readSplus("SC_geobugsSPlus.txt")
plot(polySC)
```

will read the GeoBUGS export file 'SC_geobugsSPlus.txt' into a polygon object polySC. The package can also export maps to WinBUGS format via the function sp2WB. It should be noted that occasionally shapefiles hold extra polygon information and that they can be too complex to be handled by GeoBUGS. Sometimes the topology of the polygons must be simplified. This can be done on conventional GIS systems. On QGIS, for example, the option 'simplify geometries' is available.

Both spatstat and splancs have functionality for mapping point process data. For point referenced data where surface interpolation is needed, the general purpose akima R package is useful for converting coordinate systems for mapping routines. For high quality surface smoothing the multivariate B Spline package MBA can be used. For example, for a set of observations (z1: SIR for congenital deaths in 1990 in South Carolina) at county centroid locations (x1, y1: latitude, longitude) in dataframe asd,

```
library(MBA)
asd<-data.frame(x1,y1,z1)
mba.int <- mba.surf(asd, 400, 400, extend=FALSE)$xyz.est
image(mba.int, xaxs="r", yaxs="r",col=terrain.colors(20))
contour(mba.int,add=T)
```

produced Figure 15.1. In this case the smoothing depends on the evaluation points (400 in each direction).

15.1.2   Handling Polygon Objects

Once a polygon object is created it can be processed in a variety of ways. First, the polygon object can be used to create thematic maps from output data from model fits.

15.1.2.1   Thematic Maps

The R plot function can be used to produce basic thematic and choropleth maps (i.e. maps where discrete data classes are described by colors). In Appendix B.13 an R function fillmap is described which can be used and edited for special plots. However there is now a range of packages on R which can provide sophisticated mapping of polygon data.

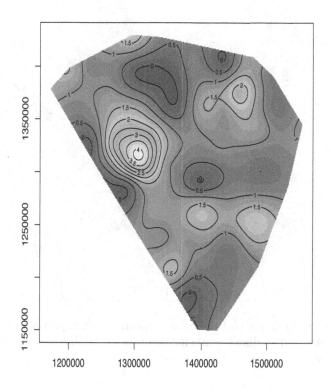

FIGURE 15.1: Multivariate B spline smooth of the SIR for congenital deaths in 1990 in South Carolina counties based on county centroids.

15.1.2.1.1 tmap The thematic mapping package tmap is relatively new but has sophisticated multi-map capabilities. qtm is a function within tmap which can provide fast plotting based on polygon objects and dataframes. An example of its use is given here based on the SCcongen90 dataframe which includes vectors: congential abnormality death count (obs), expected counts (expe), percentage below poverty (pov), and income (inc) in 46 counties of South Carolina for 1990. For plotting, a spatial polygon dataframe must be created from a dataframe and a polygon object. The dataframe should be organised so that it is in the order of the polygon object. The command is SpatialPolygonsDataFrame(Sr, data, match.ID = TRUE) where Sr is the polygon object and data is the dataframe. The matching is based on a common ID found in the polygon object. In the case of SCpoly, the ID is 'area' + 'areaID'. SCpoly and attr are combined in a SpatialPolygonsDataFrame called spg which is plotted using qtm:

```
SCcon<-data.frame(SCcongen90)
library(tmap)
```

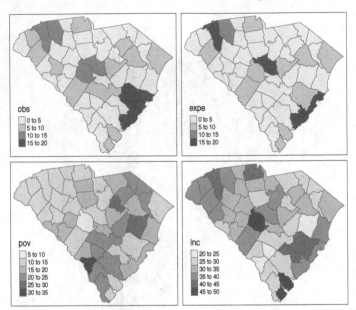

FIGURE 15.2: Multiple plot of count, expected count, percent of population below poverty, and median income for the South Carolina congenital mortality 1990 data set.

```
library(sp)
areaID<-as.character(seq(1:46))
area<-paste(c("area"),areaID,sep="")
attr<-data.frame(SCcon,row.names=area)
spg<-SpatialPolygonsDataFrame(SCpoly, attr, match.ID = TRUE)
qtm(spg,fill=c("obs","expe","pov","inc"),fill.palette="Blues",ncol=2)
```

This produces the multi-plot in Figure 15.2.

If input of the polygon object is from a shapefile then the combine.data.shapefile command in CARBayes can be used to combine the polygon object with dataframe. For the shapefile SC_county_alphasort.shp, the code for this could be:

```
SCcon<-data.frame(SCcongen90)
polySC<-readShapePoly("SC_county_alphasort.shp")
plot(polySC)
library(shapefiles)
library(CARBayes)
dbf <- read.dbf("SC_county_alphasort.dbf")
shp <- read.shp("SC_county_alphasort.shp")
dbf$dbf$AREA<-seq(1:46)
```

```
spdat<-combine.data.shapefile(SCcon,shp,dbf)
library(tmap)
qtm(spdat,fill=c("obs","expe","pov","inc"),fill.palette="Blues",ncol=2)
```

15.1.2.1.2 spplot An alternative multiple plotting function is available in the library **sp**. **spplot** is a lattice graphics function which can be used with a wide range of objects. It is particularly useful for plots where a common legend is to be used. For example, with spatio-temporal data, a sequence of time periods of relative risk might be the focus and so **spplot** can provide automatic plotting functionality for such a sequence. A simpler example, based on the **SCcongen90** data set, involves two variables (obs, expe) with common legend:

```
library(sp)
spplot(spdat,zcol=c("obs","expe"),col.regions=grey(seq(0.9,0.1,length=40)),
layout=c(2,1))
```

This is displayed in Figure 15.3.

For the very low birth weight (VLBW) example in Georgia counties as described in Chapter 12, a dataframe was set up for yearly counts for 1994 through 2004. The counts of VLBW in the 159 counties for a selection of years (1994, 1997, 2001, 2004) are mapped. The match ID for the Georgia export file is 'grid' + 'areaID' with the following code:

```
Gpoly<-readSplus("Georgia_Splus_export.txt")
plot(Gpoly)
library(sp)
areaID<-as.character(seq(1:159))
area<-paste(c("grid"),areaID,sep="")
attr<-data.frame(vlbw,row.names=area)
spg<-SpatialPolygonsDataFrame(Gpoly, attr, match.ID = TRUE)
spplot(spg,zcol=c("VL94","VL97","VL01","VL04"),
col.regions=grey(seq(0.9,0.1,length=40)),layout=c(4,1))
```

This code produced Figure 15.4.

15.1.2.1.3 ggplot2 Another popular plotting library is ggplot2 which is based on a GIS-like layer structure. The package can handle polygon objects and create sequences of thematic maps. The call structure has additive components, rather like layers in a GIS system. An example applied to the SCcongen90 data set where the SIR for congenital deaths is the focus, based on the SpatialPolygonDataFrame **spg** is given below:

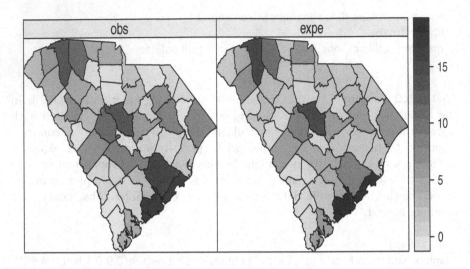

FIGURE 15.3: Observed and expected counts for SCcongen90 data set with common legend using spplot.

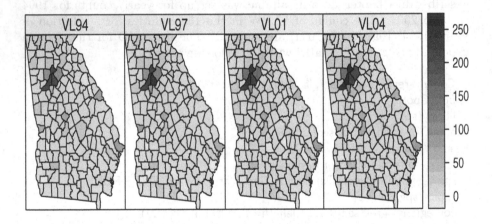

FIGURE 15.4: Sequence plot of VLBW counts for Georgia counties 1994, 1997, 2001, and 2004.

```
library(ggplot2)
require("rgdal")
require("plyr")
spg@data$id = rownames(spg@data)
spg.points = fortify(spg, region="id")
```

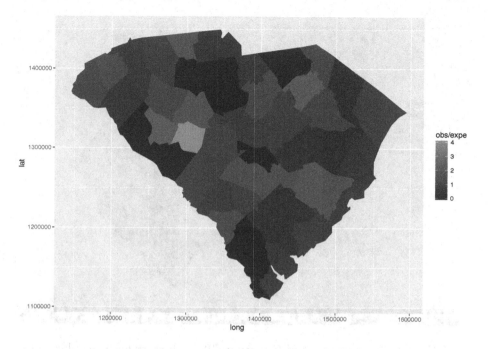

FIGURE 15.5: SIR thematic map for congential deaths in South Carolina counties in 1990 from GGPLOT

```
spg.df = join(spg.points, spg@data, by="id")
ggplot(spg.df)+aes(long,lat,group=group,fill=obs/expe) +geom_polygon()
```

This code produced Figure 15.5.

While ggplot2 provides a very sophisticated range of features for plot specialisation, it is not simple to use and possibly because of this many users resort to spplot or tmap for thematic mapping.

15.1.3   GeoBUGS and Quantum GIS (QGIS)

A direct conversion process to GeoBUGS format is useful and the GIS plug-in maps2WinBUGS is very useful in this regard. This plug-in can be downloaded free from *www.sourceforge.net*.   It can be embedded within Quantum GIS (QGIS) and Figure 15.6 displays the output from a shapefile layer in SPlus format.

The program allows the import of both MapInfo MIF or TAB files as well as ArcView shapefiles and exports GeoBUGS format map files. The GeoBUGS manual has conversion format recommendations for different GIS conversions to GeoBUGS format.

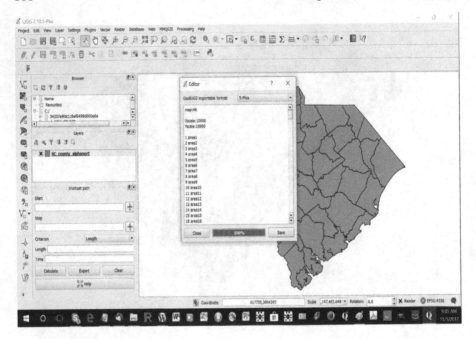

FIGURE 15.6: Screen shot of Quantum GIS with maps2winbugs plug-in displaying the SPlus output option.

15.2   Preparing Polygon Objects for WinBUGS, CAR-Bayes or INLA

15.2.1   Adjacencies and Weight Matrices

Polygon objects can also be used to create adjacency structures for fitting models using packages such as WinBUGS, OpenBUGS, INLA or CARBayes. For WinBUGS and OpenBUGS, if a *.map file is available then adjacencies are already available (adjacency tool) to the user. It is possible to also obtain adjacency information from polygon objects on R. WinBUGS and OpenBUGS use adjacencies to define neighborhoods for CAR models. Two vectors are used for this: an adjacency long vector (often called adj) and a vector that holds the number of neighbors (often called num). The library spdep has a facility to interrogate polygon objects to construct these vectors. The command poly2nb can be used to obtain a neighbours list. For the SCpoly object above we have

    Wnb<-poly2nb(SCpoly)

Then nb2WB outputs neighbors to WinBUGS format and so

```
adj2<-nb2WB(Wnb)
adj<-adj2$adj
num<-adj2$num
```

will create the vectors adj and num. In addition, an alternative for computing the num vector is

```
num<-card(Wnb)
```

where Wnb is a neighbors list.

The adj and num vectors can be combined in a dataframe, or copied into a WinBUGS or OpenBUGS .odc file directly. Note that while these vectors can be used on WinBUGS and OpenBUGS to fit spatial models, the associated GeoBUGS map is not available and it would still be necessary to import the associated .map file if one wanted to use GeoBUGS.

Finally, a note about preparing data for CARBayes and INLA. CARBayes is a R package that allows the fitting of a range of spatial models to aggregated count data. It is based on the traditional linear model specification style used by the lm and glm functions in R. Hence a formula is defined for the fixed effect part of the model and then the random effects are fitted next, incorporating the fixed effect formula via a suitable link function. A variety of spatial random effect models are available including ICAR, Leroux and adaptive variants (see Chapter 5). The package has also been extended for use with spatio-temporal aggregated count data (CARBayesST). Two basic features of CARBayes are that it requires the neighborhood specification of models to be given via a binary $0/1$ $m \times m$ matrix where each row represents an area and a 1 denotes a neighbor and 0 otherwise. This can be set up in R using the spdep function nb2mat which takes a neighbors list as argument. Style B yields a binary matrix:

```
W.mat <- nb2mat(Wnb, style="B")
```

In addition to the neighbor setup the package also has a facility for incorporating shapefiles into polygon objects and creating a SpatialPolygonDataframe which can be used directly in model fitting and by qtm and spplot. The command combine.data.shapefile(data,shp,dbf) allows the user to construct this SpatialPolygonDataframe.

INLA does not use posterior sampling and uses numerical integration to obtain posterior estimates. It is based on the Integrated Nested Laplace Approximation to a posterior distribution. The package has a range of facilities for general Bayesian hierarchical modeling and has spatial prior distributions such as ICAR, PCAR, and fully specified Matern covariance. The command structure is similar to CARBayes in that a model formula is specified (including random effects) and then the model is fitted. A Gaussian convolution model with unit level random effect for data in dataframe As, including y, x1, which are M length vectors is

```
ind<-seq(1:M);ind2<-ind
formula1<-y~1+x1+f(ind,model='iid',param=c(2,0.5))+
f(ind2,model='besag',graph="SCgraph.txt",param=c(2,0.5))
res1<-inla(formula1,family="gaussian",data=As)
```

This will fit a random effect in ind with precision having a gamma (2, 0.5) prior distribution, and an ICAR prior distribution with gamma (2, 0.5) precision prior in ind2 with the graph structure defined in SCgraph.txt, and regression parameters having default prior distributions.

INLA requires that instead of adj and num vectors, it uses a graph file which holds all the adjacency information. To this end the INLA package provides the command

```
inla.geobugs2inla(adj, num, graph.file="SCgraph.txt")
```

which converts the adj and num vectors into a INLA graph file for use in fitting spatial models. The head of this file looks like this:

```
46
1 5 33 30 24 23 4
2 5 41 38 32 19 6
3 4 25 15 6 5
4 5 39 37 30 23 1
5 5 38 25 15 6 3
6 4 38 5 3 2
7 3 27 25 15
```

The first entry is the number of areas (regions), the first column is an index to the areas, the second column is the number of neighbors, and remaining entries in a row are the adjacent regions. Hence region 1 has five neighbors and they are 33, 30, 24, 23, 4. The spdep library also has a command to produce an INLA graph file from a neighbors list:

```
nb2INLA("SCgraph.txt", Wnb).
```

15.3 Posterior Sampling Algorithms

15.3.1 Software

A range of different levels of sophistication in software is now available for posterior sampling. MCMC has permeated beyond the traditional confines of WinBUGS and OpenBUGS and can now be used on SAS (PROC MCMC), MATLAB® (http://helios.fmi.fi/~lainema/mcmc/), Stata (https://www.stata. com/features/bayesian-analysis) and MLwiN (Browne, 2012). In addition, WinBUGS and OpenBUGS can be called from R, MAT-LAB and Stata. Also, within the R platform, there are a range of packages that provide tools for fitting Bayesian models such as: BayesX, BayesMixSurv, spatsurv, Bayesvarselect, geoRglm, spBayes, spBayesSurv, spikeslabGAM, and JAGS. Additionally R has facilities for building posterior samplers (e.g. MCMCpack, mcmc). Recently the R package nimble has been developed to parse BUGS code and produce C++ versions which can be run using MCMC, MCEM or sequential MC.

From the viewpoint of spatial modeling and more specifically Bayesian disease mapping the most widely applicable sampling software remains Win-BUGS and OpenBUGS because, 1) these packages include Markov random field and spatial covariance prior distributions which are not available in many Bayesian packages, 2) they include a basic GIS module for mapping (GeoBUGS), and 3) they provide a flexible modeling language whereby non-standard models can be developed and fitted. Packages designed specifically for disease mapping are also available (R packages such as diseasemapping and SpatialEpi), and the main package reviewed here is CARBayes, which is a R package that can fit a range of predictor models with spatial random effects using MCMC.

15.3.1.1 WinBUGS and OpenBUGS

For many years the main software for Bayesian disease mapping models was WinBUGS and its more recent incarnation OpenBUGS. Models discussed in this work are based on WinBUGS and OpenBUGS and can be fitted on these platforms. OpenBUGS is the recent open-source implementation of WinBUGS (the last incarnation of WinBUGS was 1.4.3 from August 2007) and differs from WinBUGS mainly in the availability of different samplers that may be specified for different model components. Win-BUGS is available from the MRC Biostatistics Unit, Cambridge (www.mrc-bsu.cam.ac.uk/software/bugs/the-bugs-project-winbugs/), while OpenBUGS is available from the OpenBUGS project site (www.openbugs.net). The fundamental form of the package and its operation are the same. Both can run on Windows and other operating systems (WinBUGS via emulators in non-Windows systems).

WinBUGS is a windows package where models are specified via a model language and structured like R programs (including looping, assignments and logical statements). All models are specified as hierarchical with sets of stochastic nodes represented by cascades of parameter distributions. The data likelihood is the first level data model (stochastic node) which depends on parameters which usually have (prior) distributions and are hence stochastic nodes. A simple example of a model hierarchy for a disease mapping model is as follows:

$$[y_i|e_i, \theta_i] \sim Pois(e_i\theta_i)$$
$$\log(\theta_i) = \alpha + v_i$$
$$[\alpha|\tau_\alpha] \sim N(0, \tau_\alpha^{-1})$$
$$[v_i|\tau_v] \sim N(0, \tau_v^{-1})$$
$$\tau_* \sim Ga(2, 0.05).$$

This is a log linear Poisson model with uncorrelated random effect. Here the likelihoods $[y_i|e_i, \theta_i]$ and $[\alpha|\tau_\alpha]$, $[v_i|\tau_v]$ are stochastic nodes. WinBUGS model code for this is given below:

```
model{
for( i in 1:M){
y[i]~dpois(mu[i])
mu[i]<-e[i]*theta[i]
log(theta[i])<-alpha+v[i]
v[i]~dnorm(0,tauv)
}
alpha~dnorm(0,tau0)
tauv~dgamma(2,0.05)
tau0~dgamma(2,0.05)
}
```

Here the data y_i and expected count e_i and total number of units M would have to be read in, and α, v_i, τ_v and τ_0 would be estimated within the posterior sampler.

15.3.1.2 JAGS

JAGS (just another Gibbs sampler) is a variant of WinBUGS which can run BUGS model code and runs from R. It can handle a range of hierarchical models but does not have built in spatial prior distributions. With a recent version of JAGS downloaded from sourceforge and installed, the R package rjags can be used to run models. The example above was saved in a model text file: model1.txt, and data and inits in a list structure data1, inits1. The model was run with a single chain with adaptive phase of 2000 iterations. It was then updated for 5000 iterations. Sample summary is available in object asd:

```
modobj1<-jags.model(file="model1.txt",data=data1,inits=inits1,
n.chains=1,n.adapt=2000)
update(modobj1,n.iter=5000,by=10,progress.bar="gui")
asd<-coda.samples(modobj1,variable.names=c("alpha","tauv","tau0","v"),
 n.iter=5000, thin = 1)
summary(asd)
```

jags is a very efficient implementation but is limited for use in disease mapping due to the lack of provision for spatially structured prior distributions.

15.3.1.3 CARBayes

CARBayes is a R package that is designed to fit Bayesian disease mapping models using MCMC. The updates are coded in C++ and hence the process is relatively fast by comparison with WinBUGS. The samplers employed are a mixture of Metropolis-Hastings and Gibbs sampling. The package runs single chains of MCMC and provides convergence diagnostics in the form of Geweke single chain Z scores. Estimation is median-based rather than mean-based. Goodness-of-fit diagnostics are available (DIC, LMPL, and WAIC) and acceptance rates are also available. Recently the Metropolis steps for the random effects and the regression parameters have been re-engineered in version 4.7 compared to version 4.6 using the Metropolis-adjusted Langevin algorithm (MALA) rather than simple random walk Metropolis steps. For the regression parameters and random effects there is a choice between simple random walk Metropolis or MALA, with MALA being the default.

For a dataframe SCcongen90, which contains counts of deaths from congential abnormalities in South Carolina counties in 1990, the counts (obs) and expected counts (expe) are vectors, the polygon object is SCpoly, the weight matrix is W.mat and the code for fitting a BYM convolution model is given by

```
W.nb <- poly2nb(SCpoly)
W.mat <- nb2mat(W.nb, style="B")
form<-obs~1+offset(log(expe))
model.spatial<-S.CARbym(formula=form, family="poisson",data=SCcongen90,
W=W.mat, burnin=10000, n.sample=11000)
summary(model.spatial)
```

15.4 Alternative Samplers

15.4.1 Hamiltonian MC (HMC) and Langevin Sampling

A common approach to MH sampling is to use random walk transitions. These essentially ignore the nature of the posterior surface at the sampling points. To provide greater direction in sampling it is possible to augment the updating process with directional updates that take account of the local gradient of the posterior target. Hamiltonian Monte Carlo (Neal, 2011) sets up an auxiliary momentum variable r which helps to guide the sampler. The acceptance probability is given by

$$\alpha = \min\left\{1, \frac{\exp\{L(\widetilde{\theta}) - \frac{1}{2}\widetilde{r}.\widetilde{r}}{\exp\{L(\theta^{m-1}) - \frac{1}{2}r^0.r^0}\right\}$$

where $\widetilde{r} < -\widetilde{r} - (\epsilon/2)\nabla(\widetilde{\theta}), r^0 \sim N(0, I), \widetilde{\theta} < -\widetilde{\theta} + \epsilon\widetilde{r}$ (see e.g. Hoffman and Gelman, 2014). Metropolis-adjusted Langevin algorithm (MALA) is a special form of HMC. For the simpler MALA algorithm (as implimented in GeoRglm) the proposal distribution is multivariate Gaussian with mean $m(\theta) = \theta + (\frac{\delta}{2})\nabla(\theta)$ where $\nabla(\theta) = \frac{\partial}{\partial\theta}L(\theta)$ and variance δI_n.

15.4.2   Software

R-STAN provides an interface with the STAN C++ statistical modeling software package. STAN provides full Bayesian inference using the No-U-Turn sampler (NUTS), a variant of Hamiltonian Monte Carlo (HMC). The range of models available is wide and utilises efficient cross-validation metrics for model fitting. There is now some capability to fit CAR models within STAN also (see http://mc-stan.org/documentation/case-studies/IAR_Stan.html). GeoRglm was developed to fit generalised linear geostatistical models and provides a range of facilities where the continuous spatial process underlies the variation of disease risk. Often a spatial Gaussian process is the motivation for the assumed continuity of the risk. An example would be a Poisson model with log link:

$$y_i \sim Pois(\mu_i)$$
$$\log(\mu_i) = \alpha + S(x_i) + z_i$$

where it is assumed that the count y_i is measured at location x_i, and $S(x_i)$ is a zero mean spatial Guassian process with variance (σ^2) and correlation function $\rho(u)$. In this formulation a location is assumed to be a measurement site (fixed point), although as an approximation a centroid of a sub-region could suffice. A decision must be made whether the underlying process is continuous (such as for some environemental processes) or is better approximated by short range Markovian neighborhoods (as in CAR models). If the latter, then MRF models may be preferred. Note that for case event data where location of events is stochastic and can be considered to be governed by a log Gaussian Cox process, the R package lgcp could be used. This package fits a range of spatial and spatio-temporal point process models in a Bayesian framework using MCMC computation (MALA algorithms). The stochastic intensity of the process is assumed to be defined, for location s and time t by:

$$R(s, t) = \lambda(s).\mu(t).\exp\{\Upsilon(s, t)\}$$

with $\Upsilon(s, t)$ a spatio-temporal Gaussian process, and $\lambda(s)$ and $\mu(t)$ are separable spatial and temporal population backgrounds. These background effects are estimated as plug-ins to the package (i.e. estimated separately) and hence the estimation of the Gaussian process proceeds for a profile posterior distribution.

15.5 Approximate Bayesian Computation (ABC)

Approximate Bayesian computation relies on the idea that the likelihood is difficult to evaluate. Essentially, parameters are simulated from prior distributions, then outcomes simulated from likelihood and the simulated count is compared with the observed data via a loss function. An example of a likelihood-free MCMC sampler (see Section 3.7) applied to the county level South Carolina congenital mortality data 1990 for the gamma-Poisson model is coded below.

```
mu <-rep(0,46)
z<-rep(0,46)
theta<-rep(1,46);thetaOld<-rep(1,46)
thsamp<-matrix(0,nrow=46,ncol=10000)
thetaOld<-rgamma(46,1,1)
for (i in 1: 1000)
{repeat{
a<-rexp(1,1);b<-rexp(1,1)
theta<-rnorm(46,thetaOld,1);theta< -abs(theta)
mu<-diag(expe%*%t(theta))
z<-rpois(46,mu)
u1<-runif(1,0,1)
gamth<-dgamma(theta,a,b)*dnorm(thetaOld,theta,1)
gambot<-dgamma(thetaOld,a,b)*dnorm(theta,thetaOld,1)
rat<-gamth/gambot
rho=abs(mean(z)-mean(y))
print(rho)
if (u1<rat&abs(rho)<1)
{thetaOld<-theta;thsamp[,i]<-theta}
if (u1<rat&abs(rho)<1)
{break}
}
print(i)
}
head(thsamp[,1:10])
thetaM<-rowMeans(thsamp[,500:1000],dims=1)
```

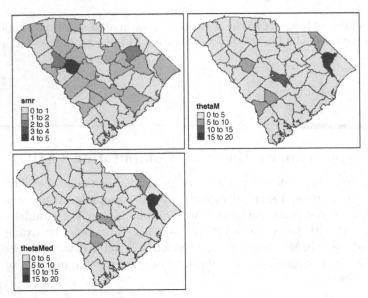

FIGURE 15.7: Comparison of relative risk estimates for the South Carolina congenital deaths data set. Top left: SMR. Top right: mean ABC estimate. Bottom: median ABC estimate.

```
library(robustbase)
thetaMed<-rowMedians(thsamp[,500:1000])
```

Figure 15.7 displays the SMR and mean and median estimates of the θ_i averaged from the final 500 sample iterations. It can be seen that while various features of the distributions are similar to that of the SIR there remain some significant differences. In fact, the fidelity of the estimation depends on the loss metric used and also the degree to which the prior distributions are non-informative. Making the sampling more MCMC-based and Markovian appears to improve the estimation. Without the MH transition kernel the sampling would be very inefficient.

15.6 Posterior Approximations

Posterior numerical approximation is an alternative to Monte Carlo integration as embodied in MCMC. A common approximation that has gained much interest recently is the Laplace approximation (Section 3.5.2.3) to the posterior distribution.

15.6.1 Software

INLA (www.r-inla.org) is the R package based on integrated nested Laplace approximation (see development in Section 3.5.2.3). This package has gained wide used in particular due to its ease of application (command-based as opposed to script-based) and its capability to handle very large data sets. The package replaces MCMC by numerical integration and hence does not require sampler convergence. Often computational efficiency alone supports its use. However at the time of writing there is no comprehensive user manual for the package. The web presence www.r-inla.org provides examples, definitions of model components, and a help forum, but no user guide. A basic guide to spatial and spatio-temporal models is provided by Blangiardo and Cameletti (2015).

15.7 Comparison of Software for Spatial and Spatio-Temporal Modeling

In this section a comparison of the platforms WinBUGS, CARBayes and INLA is presented for a range of spatial and spatio-temporal models commonly found in disease mapping. Three data sets will be analysed: 1) larynx cancer in Northwest England comparing geoRglm and OpenBUGS with a convolution model based on tesselation weights, 2) a spatial only data set consisting of incidence of oral cancer in Georgia counties for 2004, fitted with a UH log-normal model, a convolution model, and a Leroux model and reporting DIC, and MSPE, and LMPL, and 3) a spatio-temporal data set consisting of respiratory cancer mortality in the counties of Ohio for 1979 through 1988, with comparison based on deviance metrics.

15.7.1 Spatial Models

15.7.1.1 Larynx Cancer Data set, Northwest England

The analysis of these data was previously discussed in Chapter 8. Here a comparison of two computational platforms is demonstrated. These data consist of locations of larynx cancer cases and lung cancer controls in an area of Lancashire, Northwest England for 1973 through 1984. By conditioning on locations and assigning a binary case or control label to each location, the label data (y_i) can be modeled with a Bernoulli distribution at the data level:

$$y_i \sim Bern(p_i)$$
$$\log\mathrm{it}(p_i) = \alpha_0 + x_i^t \alpha + W_i$$

where a logistic regression arises as a function of a linear predictor with fixed effects (in this case, distance to incinerator and age of patient), and random effects (W_i). If it is assumed that the background confounding is continuous (even when the individual risk modeling should address some spatial structured variation), a generalised linear geostatistical model (GLGM) could be applied. The package geoRglm is designed to fit such models where the random effect is assumed continuous and described by a zero mean Gaussian process (S). Here $W_i = S_i$ and $\{S_i\}$ is assumed to be described by zero mean multivariate Gaussian distribution with spatial covariance structure. In the example below, an exponential covariance form $(\sigma^2 \exp(-d/\phi))$ was assumed. geoRglm was run with distance from source (dis) and age (age) as covariates, and default prior specification for σ^2 and β parameters:

```
geodata2<-as.geodata(gorb, coords.col = 1:2, data.col = 3,
data.names = NULL, covar.col = 5:6,
covar.names = "obj.names",units.m.col = units.m)
model<-model.glm.control(trend.d = ~dis+age,
trend.l = "cte", cov.model="exponential",aniso.pars = NULL)
mcmc2 <- mcmc.control(S.scale = 0.09, n.iter = 30000,
 burn.in=20000, thin=10,phi.scale = 0.2, phi.start = 2.0)
prior2 <- prior.glm.control(phi.discrete = seq(0.2,3,0.2))
asd<-binom.krige.bayes(geodata=geodata2,
 coords=geodata2$coords, data = geodata2$data,
 units.m = geodata2$units.m,model=model, prior=prior2,
mcmc.input=mcmc2)
```

The resulting converged model fit gave posterior means of -6.163 (intercept), -58.043 (distance), and 0.0104 (age) with standard deviations, respectively, of 1200.54, 3724.43, 7.562383e-06 . Clearly the age effect is positive and well estimated but the distance effect, though negative, is not well estimated. (Note that an alternative to using geoRglm for fitting GLGMs is to employ spBayes with the function spGLM, which also fits smooth continuous spatial effects using a binomial or Poisson data model.)

The equivalent model was fitted on OpenBUGS, under the assumption that a convolution model could be fitted instead of a continuous spatial random effect. This can be acheived by using tesselation neighbors (Delauney or Dirichlet); neighborhoods are defined by neighboring tiles or triangles. deldir was used to derive such neighbors. Once neighbors are obtained then an adjacency vector is available. Essentially the model fitted was

$$y_i \sim Bern(p_i)$$
$$\log it(p_i) = \alpha_0 + x_i^t \alpha + W_i$$
$$W_i = v_i + u_i$$

which includes both uncorrelated (v_i) and correlated (ICAR: u_i) components to allow for different forms of confounding. The converged sampler based on the code, with higher level prior distributions suppressed, is given below:

```
model {
for (i in 1:N){
ind[i]~dbern(p[i])
indP[i]~dbern(p[i])
logit(p[i])<-gam0+gam1*dis[i]+gam2*age[i]+v[i]+W[i]
v[i]~dnorm(0,tauv)
log(L[i])<-ind[i]*log(p[i])+(1-ind[i])*log(1-p[i])
invL[i]<-1/L[i]
ppl[i]<-ind[i]-indP[i]
ppl2[i]<-pow(ppl[i],2)
x1[i]<-x[i]
y1[i]<-y[i]}
mspe<-mean(ppl2[])
for(k in 1:sumNumNeigh){wei[k]<-1}
W[1:N] ~car.normal(adj[],wei[],num[],tauW)

.}
```

This code yielded parameter estimates for the fixed effects of: intercept: -3.125 (0.467), distance: -0.5101 (1.201) and age 0.00385 (0.00672). Hence, there are some differences in the estimation particularly for the intercept term, which in this case was well estimated. The distance effect is not well estimated, but as before age is well estimated and positive.

Goodness-of-fit was assessed for each model fit by mean square predictive error (MSPE), log marginal penalised likelihood (LMPL) and DIC. These quantities have to be specially programmed for R and/or WinBUGS and OpenBUGS. The MSPE was 0.10579 (0.00949) for the convolution model and 0.0514 (0.00791) for the GLGM. The LMPL results yield -137.92 for the GLGM and -224.36 for the convolution model. The DIC was 518.37 for the GLGM and 518.39 for the convolution model. The DIC results are close but the LMPL and MSPE results seem to favor the GLGM. The parameter estimates from both approaches yield poor estimation of the distance effect but better estimation of the age effect. Differences between these approaches are clear, but it is not certain whether exactly comparable prior settings or sampling regimes could lead to the differences, or whether the differences are substantive model improvements.

15.7.1.2 Georgia County Level Oral Cancer Incidence, 2004

These data were discussed in Chapter 5 in connection with fitting convolution models. Here a comparison is presented between OpenBUGS, CARBayes and

INLA in the analysis of these data. First a random intercept (frailty) model
is considered:

$$y_i \sim Pois(e_i\theta_i)$$
$$\log(\theta_i) = \alpha_0 + v_i$$
$$\alpha_0 \sim U(-10000, 10000)$$
$$v_i \sim N(0, \tau^{-1})$$
$$\tau \sim Ga(1, 0.01).$$

The OpenBUGS code is

```
model
{for( i in 1 : m ) {
 y[i] ~dpois(mu[i])
 mu[i] <- e[i] * theta[i]
 log(theta[i]) <- al0 +v[i]
v[i]~dnorm(0,tVac)
ypred[i] ~dpois(mu[i])
log(L[i])<-y[i]*log(mu[i])-mu[i]-logfact(y[i])
invL[i]<-1/L[i]
PPL[i] <- pow(ypred[i]-y[i],2)}
mspe <- mean(PPL[])
tVac~dgamma(1,0.01)
al0 ~dflat()}.
```

The CARBayes code for data in dataframe GOrCa and spatial polygon object
Gpoly is

```
library(spdep)
library(maptools)
library(CARBayes)
Gpoly<-readSplus("Georgia_Splus_export.txt")
W.nb <- poly2nb(Gpoly)
W.mat <- nb2mat(W.nb, style="B")
## UH model
form<-y~1+offset(log(expe))
model.spatial<-S.CARleroux(form,family="poisson",data=GOrCa, W=W.mat,
   fix.rho=TRUE,rho=0,burnin=10000,
n.sample=11000, thin=1,MALA=TRUE)
```

The INLA code for the uncorrelated effect model is:

```
formulaUH = y~f(region, model = "iid",param=c(1,0.01))
resUH = inla(formulaUH,family="poisson",
data=GOrCa,control.compute=list(dic=TRUE,cpo=TRUE),E=expe)
```

A convolution model with UH and CH (ICAR) components is fitted next.

The Win/OpenBUGS code is

```
model
{for( i in 1 : m ) {
y[i] ~dpois(mu[i])
mu[i] <- e[i] * theta[i]
log(theta[i]) <- al0 +v[i]+W[i]
v[i]~dnorm(0,tVac)
ypred[i] ~dpois(mu[i])
log(L[i])<-y[i]*log(mu[i])-mu[i]-logfact(y[i])
invL[i]<-1/L[i]
PPL[i] <- pow(ypred[i]-y[i],2)}
W[1:m]~car.normal(adj[],wei[],num[],tauW)
for(j in 1:sumNumNeigh){wei[j]<-1}
mspe <- mean(PPL[])
tVac~dgamma(1,0.01)
.
.
.}
```

The CARBayes code is given below:

```
library(spdep)
library(maptools)
library(CARBayes)
Gpoly<-readSplus("Georgia_Splus_export.txt")
W.nb <- poly2nb(Gpoly)
W.mat <- nb2mat(W.nb, style="B")
form<-y~1+offset(log(expe))
model.spatial<-S.CARbym(form,family="poisson",data=GOrCa, W=W.mat,
burnin=10000, n.sample=11000, thin=1,MALA=TRUE)
```

The INLA code is, using adj and num vectors:

```
inla.geobugs2inla(adj, num, graph.file="Georgia.txt")
formCONV = y~f(region, model="iid",param=c(1,0.01))+
f(region2,model="besag",param=c(1,0.01),graph="Georgia.txt")
resCONV = inla(formCONV,family="poisson",
data=GOrCa,control.compute=list(dic=TRUE,cpo=TRUE),E=expe)
```

Finally the Leroux model is fitted and correlation parameter is estimated. The OpenBUGS code for this is

```
model
{for (i in 1:m){y[i]~dpois(mu[i])
ypred[i]~dpois(mu[i])
p.sig[i]<-step(s[i])
log(mu[i])<-log(e[i])+log(theta[i])
log(theta[i])<-alph+s[i]
s[i]~dnorm(S[i],tau[i])
tau[i]<-inv.delta*(1-lam+lam*num[i])
S[i]<-(lam/(1-lam+lam*num[i]))*sum(Ws[cum[i]+1:cum[i+1]])
log(L[i])<-y[i]*log(mu[i])-mu[i]-logfact(y[i])
invL[i]<-1/L[i]
pres[i]<-y[i]-ypred[i]
PP[i]<-pow(pres[i],2)
PP2[i]<-abs(pres[i])}
for (i in 1:NumNeig){Ws[i]<-s[map[i]]}
mape<-mean(PP2[])
mspe<-mean(PP[])
inv.delta~dgamma(1,0.0001)
alph~dflat()
lam~dunif(0,1)}
```

The CARBayes code using the MALA option is

```
form<-y~1+offset(log(expe))
model.spatial<-S.CARleroux(form,family="poisson",data=GOrCa, W=W.mat,
burnin=10000, n.sample=11000, thin=1,MALA=TRUE)
```

The Leroux model is not available currently on INLA but it can be fitted using Bayesian model averaging in the INLABMA package (using the leroux.inla function). A mixture form close to the Leroux model specification is available in the testing version of INLA (bym2). The code below fits a Leroux model using INLABMA by using different (fixed) lambda values:

```
rlambda <- seq(0.03, 0.8, length.out = 20)
errorhyper <- list(prec = list(prior = "loggamma",
param = c(1, 0.01), initial = log(1), fixed = FALSE))
form2 <- y ~1+offset(log(expe))
lerouxmodels <- mclapply(rlambda, function(lambda) {
leroux.inla(form2, d =GOrCa, W = W.mat,
lambda = lambda, improve = TRUE,
family = "poisson",
control.predictor = list(compute = TRUE),
control.compute = list(dic = TRUE, cpo = TRUE))})
```

	DIC	pD	MSPE (SD)	LMPL
OpenBUGS				
UH	420.5	33.8	2.9 (0.8)	-216.17
Convolution	410.5	12.2	3.0 (0.8)	-216.44
Leroux	446.2	1.0	4.1 (0.9)	-223.62
$\hat{\rho} = 0.558$				
CARBayes				
UH	419.24	24.04	2.94 (5.95)	-187.42
Convolution	418.48	23.55	2.88 (5.10)	-187.43
Leroux	418.77	24.28	3.072 (6.13)	-192.49
$\hat{\rho} = 0.430$				
INLA				
UH	422.76	31.13	7.96 (43.15)	-266.92
Convolution	422.76	31.13	8.17 (43.52)	-232.62
Leroux	436.54	20.8	8.89 (48.26)	-241.67
$\hat{\rho} = 0.440$				

Table 15.1: Goodness-of-fit statistics computed on three packages with three different spatial models

```
resLER<- INLABMA(lerouxmodels, rlambda, 0, impacts = FALSE)
```

The resLER object can be examined as an INLA object.

All the above models were fitted to the Georgia oral cancer data for 2004 and Table 15.1 displays the results of applying goodness-of-fit criteria under each package. The DIC and pD were automatically produced on all packages. The LMPL is available directly on CARBayes and INLA (via calculation of $\sum_i \log(cpo_i)$) but the cpo calculation has to be completed on R when using WinBUGS or OpenBUGS. On the other hand, the MSPE can be computed easily on WinBUGS or OpenBUGS as simulation from the predictive distribution is available for any data model. For CARBayes and INLA, sampled fitted values must be used to generate predicted outcomes and hence the posterior predictive loss once the model is fitted.

Some notable differences appear between and within the packages' results. Assuming that the LMPL is computed in the same manner across packages, it is noticeable that LMPLs for CARBayes are higher (less negative) than for the other packages. As they seem quite different between packages, it is perhaps wise to only consider the between model comparison for each package. In all cases, the packages seem to favor, for these data, the convolution or UH model, showing least negative LMPLs for OpenBUGS and CARBayes, but the INLA result suggests that the convolution model is better. Basing results on DIC there is less difference between packages, but some marked differences in estimated pDs. CARBayes shows little difference between the models, whereas INLA and OpenBUGS do not favor the Leroux model. Both CARBayes and INLABMA yield ρ estimates which are very close (0.43 and 0.439) for the mixing parameter of the Leroux model, whereas the OpenBUGS yields a slightly higher estimate.

15.7.2　Spatio-Temporal Models

All three packages provide facilities for fitting spatio-temporal risk models such as those discussed in Chapter 12. Here, two basic models with associated Poisson data models will be demonstrated. I assume fixed time and spatial units. Assume that y_{ij} is the count outcome in the i th area and j th time period, $i = 1, ..., m$; $j = 1, ..., J$. Likewise e_{ij} and θ_{ij} are the expected counts and relative risks respectively. Assume a separable model so that

$$y_{ij} \sim Poiss(e_{ij}\theta_{ij})$$
$$\log(\theta_{ij}) = \alpha_0 + S_i + T_j + ST_{ij}$$

where S_i is spatial, T_j is temporal, and ST_{ij} a space-time interaction term. Here two basic models for spatio-temporal risk variation are examined:

$$1) \quad \log(\theta_{ij}) = \alpha_0 + v_i + u_i + \gamma_j \tag{15.1}$$
$$2) \quad \log(\theta_{ij}) = \alpha_0 + v_i + u_i + \gamma_j + \psi_{ij}. \tag{15.2}$$

Model 1 assumes no space-time interaction and a spatial convolution with random walk in time:

$$v_i \sim N(0, \tau_v^{-1}); \ u_i \sim ICAR(\tau_u^{-1}); \gamma_j \sim N(\gamma_{j-1}, \tau_\gamma^{-1}).$$

Model 2 is the same as model 1) but with an ST interaction term ψ_{ij}. This is assumed to have a independent zero mean Gaussian prior distribution: $\psi_{ij} \sim N(0, \tau_\psi^{-1})$. Note that for the CARBayes models the $v_i + u_i$ is replaced by a Leroux prior model.

The data to be modeled is the Ohio county respiratory cancer mortality data for 1979 through 1988. For OpenBUGS code, the data is read in as matrices $y[i, k], e[i, k]$ where there are m regions (counties) and T time periods (years). Code for model 1) is below:

```
model
{for (i in 1:m)
{for (k in 1:T)
    {y[i,k]~dpois(mu[i,k])
 log(mu[i,k])<-log(e[i,k])+log(theta[i,k])
 log(theta[i,k])<-a0+g[k]+u[i]+v[i]
    }
 v[i]~dnorm(0,tauv)}
```

```
g[1]~dnorm(0,taug)
for (k in 2:T){
g[k]~dnorm(g[k-1],taug)}
u[1:m]~car.normal(adj[],weights[],num[],tauu)
for(k in 1:sumNumNeigh) {weights[k] <- 1}

........
........}
```

For this model the DIC and pD are 5782.5 and 100.54 respectively on Open-BUGS.

For the ST interaction model the code is extended by addition of the ψ_{ij} (psi[i,k]) term:

```
model
{for (i in 1:m)
{for (k in 1:T)
    {y[i,k]~dpois(mu[i,k])
 log(mu[i,k])<-log(e[i,k])+log(theta[i,k])
 log(theta[i,k])<-a0+g[k]+u[i]+v[i]+psi[i,k]
 psi[i,k]~dnorm(0,taupsi)}
 v[i]~dnorm(0,tauv)}
 g[1]~dnorm(0,taug)
 for (k in 2:T){
 g[k]~dnorm(g[k-1],taug)}
 u[1:m]~car.normal(adj[],weights[],num[],tauu)
 for(k in 1:sumNumNeigh) {weights[k] <- 1}
  .............

}.
```

The CARBayes code involves the use of CARBayesST which has a range of spatio-temporal model formulations. Note that for CARBayesST the data must be in $m \times T$ long vectors, where the first m entries are for time 1, the second m for time 2 etc. For the dataframe dataOhio created with vectors Yobs, EXPECT, and YEAR:

```
dataOhio<-cbind(matE,matY,YEAR);names(dataOhio)<-c("EXPECT",
 "Yobs","YEAR")
```

Model 1 : no interaction

```
form1<-Yobs~offset(log(EXPECT))
model1<-ST.CARanova(formula=form1,family="poisson",data=dataOhio,
```

```
W=W.mat,interaction=FALSE,burnin=10000,n.sample=12000,thin=1)
```

Model 2 : ST interaction

```
form2<-Yobs~offset(log(EXPECT))
model2<-ST.CARanova(formula=form2,family="poisson",data=dataOhio,
W=W.mat,interaction=TRUE,burnin=10000,n.sample=12000,thin=1)
```

The INLA code for this example assumes that the input of $T \times m$ length vectors where the first T entries are for region 1, the next T for region 2 etc:

```
year<-rep(1:10,len=880)
region<-rep(1:88,each=10)
region2<-region
ind2<-rep(1:880)
data<-data.frame(yL,eL,year,region,region2,ind2)
```

Model 1: no interaction

```
formula1<-yL~1+f(region,model="iid",param=c(1,0.01))+
f(region2,model="besag",graph="OHIO_map.txt",param=c(1,0.01))+
f(year,model="rw1",param=c(1,0.01))
result1<-inla(formula1,family="poisson",data=data,
E=eL,control.compute=list(dic=TRUE,cpo=TRUE))
```

Model 2: ST interaction

```
formula2<-yL~1+f(region,model="iid",param=c(1,0.01))+
f(region2,model="besag",graph="OHIO_map.txt",param=c(1,0.01))+
f(year,model="rw1",param=c(1,0.01))+f(ind2,model="iid",param=c(1,0.01))
result2<-inla(formula2,family="poisson",data=data,
E=eL,control.compute=list(dic=TRUE,cpo=TRUE))
```

Table 15.2 was compiled from running the code for each package in turn for models 1 and 2. Note that CARBayesST models employ a Leroux model, instead of a convolution model which was assumed for the OpenBUGS and INLA models. Differences between the package goodness-of-fit measures are apparent. First it should be noted that CARBayes uses posterior median estimates instead of mean estimates and so it could be expected that deviance estimates may vary. Differences in mean deviance do occur between packages, but the larger differences appear in the pD estimation. In particular, for the ST interaction model (model 2) the pD is markedly different between INLA

	Mean Deviance	pD	DIC
Model 1			
OpenBUGS	5682.0	100.54	5782.5
CARBayes*	5828.81	112.79	5941.6
INLA	5686.70	86.07	5772.7
Model 2			
OpenBUGS	5627.0	413.56	6033.6
CARBayes*	5751.41	366.71	6118.1
INLA	5640.1	113.65	5753.7

*CARBayes uses Leroux model for spatial effects.

Table 15.2: Comparison of goodness-of fit for two spatio-temporal models using OpenBUGS, CARBayes and INLA

and the MCMC packages. There is also a similar difference, though less marked, for model 1. The pD estimation on INLA is based on the projection matrix while for MCMC its is based on differences between the sample average deviance and deviance computed at the posterior mean.

Finally it is not easy to make comparisons directly between packages as there could be assumptions made that are not completely transparent to the end-user. For example, while OpenBUGS allows some choice of samplers, CARBayes only allows the choice to use MALA or not. Choice of sampler should not be critical if a converged MCMC sample is found; however it could impact the speed with which convergence arrives. INLA on the other hand, does not provide clear guidance concerning choice of prior distributions, as no user guide is currently available. For example, for the MSPE calculation for spatial models (see Table 15.1), sampled fitted values were used to generate predictions of the outcomes for CARBayes and INLA. However the resulting MSPEs are markedly different. The CARBayes results are close to the OpenBUGS results but INLA is substantially larger.

15.8 Pros and Cons of Software Packages

While the majority of models fitted for this work were programmed in Win-BUGS and OpenBUGS, there is now a greater availability of efficient packages which could be readily used without having to resort to writing model code. The ease with which both INLA and CARBayes can be set up and the speed of execution are major factors in the wider application of Bayesian disease mapping methods.

Table 15.3 is a compilation of the facilities available within the packages, at the current time of writing (November 2017; R 3.4.2). The table is meant to allow a novice user to assess whether using a particular package provides the support and/or facilities needed to perform analysis on a project.

As noted above in Section 15.7, there are differences between the three packages in terms of computation of goodness-of-fit measures. These differences suggest concern about which package is better in terms of accuracy. Some accuracy assessment has been made previously for INLA and MCMC for log Gaussian Cox processes (LGCPs) and also with OpenBUGS for disease mapping.

Taylor and Diggle (2014) performed a simulated evaluation of MCMC and INLA for a spatial LGCP with constant expectation and parameters treated as known values. In this restricted scenario, they found that MCMC yielded more accurate estimates of predictive probabilities than INLA. Further, Teng et al. (2017) compared a variety of INLA specifications (including SPDE) to HMC and variational approximations for LGCPs. They found that hyperparameters were more accurately estimated via HMC with associated increased computation time.

While INLA is computationally very efficient, it does not provide accurate estimation of predictive distributions and further SPDE tends to over-smooth the latent Gaussian field if mesh size is too crude. Numerical instability can arise in SPDE when step size is not chosen carefully for the Newton Raphson algorithm. Posterior variability, which relates to hyperparameter estimation accuracy, can also be problematic for the variants of Laplace approximation (including SPDE) evaluated.

In a simulated evaluation of OpenBUGS and INLA in the context of disease mapping, Carroll et al. (2015) found that OpenBUGS provides less biased estimates of spatial random effects than INLA, but also found that comparable estimation accuracy could be found. Default precision prior specification in INLA produced extremely large precision estimates compared to ground truth. Hence it would appear that INLA requires the use of weakly informative precision prior distributions (such as Ga (1, 0.5)) to obtain accurate precision estimates. However, OpenBUGS and INLA both show increased bias in estimation of uncorrelated precisions when spatial effects are accurately estimated under Ga (1, 0.5), in a convolution model. The overall conclusion appears to be that the choice of prior specification for precisions of both correlated and uncorrelated effects is crucial, and in particular for INLA, where the default prior specification overestimates significantly.

In conclusion, it is clear that any package chosen should be used exclusively because mixing of results from different packages is not sensible. INLA appears to provide accurate estimation of regression parameters, but has issues with estimation of precisions. If missingness is a major issue (particularly in ITT situations) then it would be difficult to see how INLA or CARBayes could be used (without extensive extra programming). On the other hand, if large data sets are to be modeled ($m > 5000$, say) then efficient computa-

Criterion	Win/OpenBUGS	INLA	CARBayes
Ease of use	Must write program	Simple command-based but no user guide	Simple command-based
Big data	No limit (but slow speed)	No limit specified but large models can reach computational limits	No limit but computational speed
Missingness	Yes: both outcome and predictors	Can set up to handle outcome missingness	Outcome missingness only
Computation speed	Slow with large data and parameter sets	Fast	Fast (single chain used)
GIS	Yes: useful for quick mapping	Must use R	Must use R.
Special spatial models	Yes	Spatial and ST, but no multivariate models	Some spatial, ST and multivariate models
Flexibility	Very flexible modeling environment	Some flexibility but joint modeling limited	Flexible within range of models
Intention to treat (ITT)	Yes: all data can be used for analysis	No: complete case only if there is missingness in predictors	No: only outcome available is missingness
Special models:			
MVCAR	Yes	No	MVLeroux only
Zero-inflated models	Yes using mixture likelihood	Yes: special programmed, general mixtures not available	No
Structural equation	Yes	No	No
Measurement error	Yes	Yes	No
User-defined likelihood	Yes	No	No
User-defined prior	Yes	Yes	No
Bayesian model averaging	Yes	in INLABMA	No
Stochastic partial differential equations	No	Yes (SPDE)	No
Infectious disease modeling	Yes	Possible but difficult to set up	No
Geostatistical	Yes but not flexible	Yes	No

Table 15.3: Comparison table for software choice

tional alternatives to Win/OpenBUGS should be sought. As alternatives in the future, STAN and NIMBLE (https://r-nimble.org/) may be preferred as they allow faster MCMC. Nimble 0.6-8 has now incorporated improper and proper CAR models within its prior specifications (dcar_normal, dcar_proper).

Appendices

Appendices

Appendix A: Basic R, WinBUGS/OpenBUGS

It is useful to be able to manipulate data, design models, and analyze output from posterior sampling with suitable tools. The package R, which is freely available (*www.r-project.org*), is a very useful tool for pre- and post-analysis of Bayesian models. Not only is R readily available, it also includes state-of-the-art procedures for manipulating and analyzing data and has very sophisticated graphics capabilities. It also has functionality for interacting with MCMC programs and in particular has functions that can process MCMC outputs (CODA, BOA).

A.1   Basic R Usage

R is an object-oriented language which is platform-independent and command-driven. This latter feature seems regressive given the common use of graphical user interfaces (such as in S-Plus). On the other hand, this allows wide availability across platforms. There is no doubt that this feature does frustrate the occasional user, particularly when data input must be command-based.

Review of basic R features is found in Maindonald and Braun (2003) and more extensive use in modeling is covered by Faraway (2006). We assume some basic familiarity with R.

A.1.1   Data

Most often data can be processed as vectors or matrices within R. For the South Carolina congenital anomaly mortality example the data, consisting of county-based observed counts and expected rates, are read into a dataframe adat:

```
adat<-list(m=46,
y=c(0,7,1,5,1,1,5,16,0,17,4,0,0,1,1,7,1,3,0,0,8,2,13,7,0,8,0,3,2,4,1,
11,0,1,2,3,3,8,6,14,3,11,6,0,1,5),
e=c(1.129778827,6.667008775,0.650279674,6.988864371,0.95571406,
1.123210345,5.908349156,8.539026017,0.601016062,18.92051111,
2.272694617,1.73736337,2.019808077,1.688099759,1.747216093,
3.221840201,1.835890594,5.221942834,0.978703751,1.254579976,
```

6.407553754,2.676656232,16.57884744,3.077333607,1.087083697,
7.606301637,1.018114641,2.15774619,2.844152512,2.955816698,
0.985272233,9.22871658,0.38097193,1.855596038,1.579719813,
1.579719813,2.647098065,4.791707292,4.144711859,15.70852363,
0.765228101,11.32077795,6.256478678,1.500898035,2.085492893,
7.297583004)).

The expected rates are computed from state-wide age-gender stratified rates and applied to county age-gender group populations and then summed.

Summarization of the basic data can be achieved via various commands, one of which is displayed here.

```
smr<-adat$y/adat$e
summary(smr)
```

This produces a numerical summary of the vector smr:

Minimum	First Qtr.	Median	Mean	Third Qtr.	Maximum
0.0000	0.5404	0.8908	0.9282	1.2620	3.9200

A.1.2 Graphics

For non-spatial data a variety of summarization graphics are available. For example, histogram or density estimates are available. The commands

```
par(mfrow=c(1,2))
hist(adat$y,breaks=50,xlab="count",main = "")
plot(density(adat$y),main="")
```

produced Figure A.1. We can also easily get crude relative risk estimates by computation of the standardized mortality ratio for the data which is just given by plot(density(smr),main="").
Figure A.2 displays the density estimate for the SMR for this count data.

Another display demonstrates some further range of measures available.

```
par(mfrow=c(2,2))
hist(adat$y,xlab="count",main="")
hist(smr,xlab="standaridised incidence ratio",main="")
plot(adat$y,adat$e,xlab="expected",ylab="count",main="")
boxplot(smr,xlab="standaridised incidence ratio")
```

These commands led to the display in Figure A.3.

A variety of functions are available for graphical representation of spatial data on R. A common practice with spatial data is to display a pixellated image of a smoothed (estimated) surface. In addition, contouring is useful and

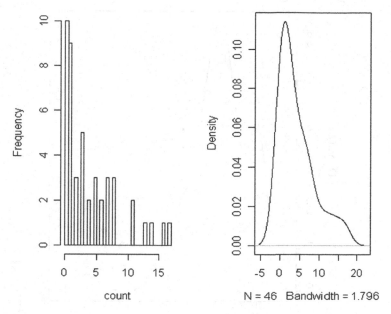

FIGURE A.1: Histogram and density estimate of the 46 county level counts for the South Carolina congenital anomaly mortality example.

can be added easily to such a plot. Although perspective plotting is possible and visually attractive, these plots are not particularly informative and are not discussed here. Until recently, the AKIMA and SM packages provided the simplest functionality for such surface interpolation and plotting. In AKIMA, the interp function provides a mathematical interpolator which allows the input of unordered x,y,z vectors:

```
library(akima)
asd1<-interp(xcen, ycen, smr).
```

The output is a list with three elements which are x and y grid points and associated interpolated values (z), These can be displayed via image or contour functions:

```
image(asd1,col=gray(20:0/20),xlab="x coordinate",ylab="y coordinate")
contour(asd1,add=T)
```

Figure A.4 displays the results of using interp for the smr data with county centroids (xcen,ycen).

The above commands yield a grey scale pixellated image map with a contour map overlain with default contour levels. An alternative package that can be used for such surface interpolation and plotting is the SM package.

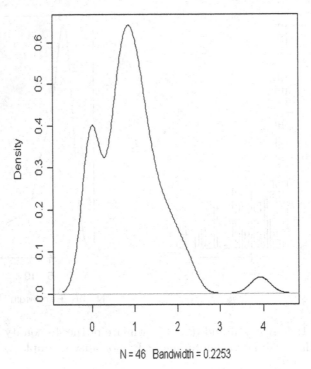

N = 46   Bandwidth = 0.2253

FIGURE A.2: Density estimate of the SMR for South Carolina anomaly mortality data.

This package was developed to provide nonparametric density and regression estimation capability. The function sm.regression provides a nonparametric regression estimator for x, y, z data, based on the work described in Bowman and Azzalini (1997):

```
xyMAT1<-cbind(xcen,ycen)
asd2<-sm.regression(xyMAT1,smr,h=c(3000,3000),display="image")
```

and will display a contour surface for the interpolated smr with a fixed smoothing (specified by h). An evaluation grid can be specified as well as different smoothing procedures.

More recently a more sophisticated smoothing and interpolation package has been developed. This package is called MBA and implements multilevel B spline surface interpolation. The main function is mba.surf and relies on a multilevel B spline approximation (MBA) algorithm. The R implementation of the code was developed by Andrew Finley and Sudipto Banerjee at University of Minnesota. The mba.surf function has considerable flexibility in the specification of the local smoothness and flexibility of the surface. An

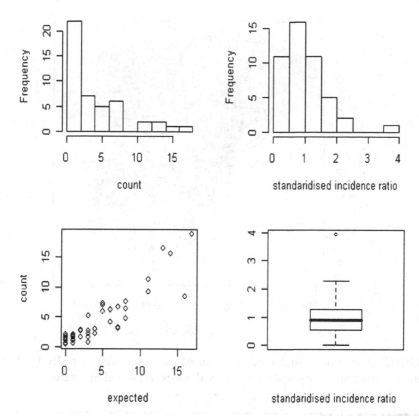

FIGURE A.3: Display of four basic plots of the observed data. Clockwise from top left: histogram of the observed counts; histogram of the SIR; boxplot of the SIR; scatterplot of observed versus expected count.

example of its application follows where we examine the smr surface for the South Carolina congenital anomaly data:

```
xyMat2 <-cbind(xcen,ycen,smr)
asd2<-mba.surf(xyMat2,50,50)$xyz.est
  image(asd2,col=gray(20:0/20),xlab="x coordinate",ylab="y coordinate")
  contour(asd2,add=T)
points(xcen,ycen)
```

Figure A.5 displays the image and contour map for the smr example using the multilevel B spline code function mba.surf .

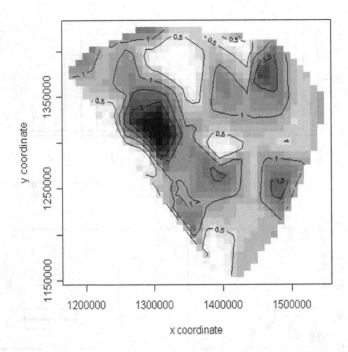

FIGURE A.4: Image and contour map of the interpolated SMR for South Carolina congenital anomaly mortality data example using interp.

A.2   Use of R in Bayesian Modeling

R is a useful package for data manipulation and post MCMC processing as it can a) pass information to WinBUGS, b) run WinBUGS. In addition to this specific purpose it can be useful in simpler scenarios also. R is used extensively in Bayesian course teaching (see for example www.mrc-bsu.cam.ac.uk/bugs/weblinks/webresource.shtml for extensive examples). The simulation facilities in R allow a wide range of simple Bayesian modeling to take place. With simpler Bayesian models there could be two approaches to posterior inference. First, the exact form of the posterior distribution could be a known distribution and so summarization of the posterior information can made directly from the distribution. For instance, the beta posterior distribution that arises from the binomial likelihood and beta prior distribution can be viewed via the use of dbeta function (the distribution function for the beta distribution):

```
theta<-seq(0.3,0.6,0.001)
w=1;alpha=1;beta=1;
```

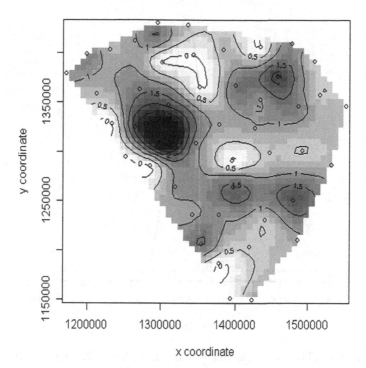

FIGURE A.5: The SMR image and contour map for South Carolina congenital anomaly mortality example using the MBA package.

```
plot(theta,dbeta(theta,436,543),type="l",xlab="theta",
ylab="",xaxs="i",yaxs="i",yaxt="n",bty="n",cex=2)    # likelihood of θ
 lines(theta,dbeta(theta,alpha,beta),lty=2)    # prior distribution
 lines(theta,dbeta(theta,437+alpha,543+beta),lty=2)    # posterior distribution
```

This sequence of commands plots the likelihood, beta (1, 1) prior distribution (dotted line) and the posterior distribution, and resulted in Figure A.6.

Second, we may know the posterior distribution but may find it easier to *sample* from the posterior distribution. On R it is possible simulate from a wide range of distributions and so if the form of the posterior distribution is known then we can take a posterior sample. For example a sample of 1000 values from a beta distribution with parameters 437 + alpha, 543 + beta could be obtained in the vector samp by:

```
samp<-rbeta(1000,437+alpha,543+beta)
summary(samp)
```

FIGURE A.6: The binomial likelihood, (flat) beta (1, 1) prior distribution and the posterior distribution from R commands.

The final statement provides a summary of the sample values in samp.

Alternatively, sometimes within simulation studies, we would be interested in generating synthetic data from given parameter values for parametric distributions. For example, it is commonly assumed that incident counts of disease (y_i) within small areas (census tracts, counties, ZIP codes, etc.) follow a Poisson distribution with mean $e_i\theta_i$ where e_i is a known expected rate and θ_i is a relative risk. Under simulation it would be usual to assume a particular form for the risk, and then to simulate form the given Poisson distribution. Hence if for example $\theta_i = \exp(\alpha + v_i)$ where $\alpha = -.2$ and $v_i \sim N(0, 1)$ then we could simulate this straightforwardly within R. For a sample size of m, the following code will leave a vector of m counts in ysim:

```
v<-rnorm(m,0,1)
theta<-exp(-.2+v)
mu<-e*theta
ysim<-rpois(m,mu).
```

A.3   WinBUGS and OpenBUGS

While it is not the intention here to provide a detailed review of WinBUGS usage in disease mapping, some basic features of the package should be highlighted here to motivate further use. OpenBUGS is an open source version of WinBUGS and while some facilities in the package are unique the main program structures and calls are identical. WinBUGS is the focus from now on. More advanced features will be introduced when they arise in later chapters. The reader is directed to Lawson et al. (2003) for a more detailed introduction to a selection of topics in this area. Example code for examples in that work are available from http://academicdepartments.musc.edu/phs/research/ lawson/data.htm.

The WinBUGS online help facilities include a range of examples that are also very useful to follow. The examples can be executed within the package and so can be seen in operation. The MRC Biostatistics WinBUGS website (www.mrc-bsu.cam.ac.uk/bugs/welcome/shtml) is also a very useful resource.

A.3.1   Simulation

It is possible to use WinBUGS for simulation as it is a feature of the package that simulation-based methods are kernels of the methodology. WinBUGS is based on the idea that Bayesian models can be represented as hierarchies with parameter nodes. Each parameter in the model can be represented as either a stochastic node (in which case it has a distribution) a constant, in which case its fixed, or a logical node, in which case it is the result of an assignment operation based on an expression. Given below is an example of simple simulation code for a Poisson distributed variable y (y) with mean $\mu = e\theta$ (mu $< -\exp * $ theta)   where a gamma distribution is assumed for θ at the next level of the hierarchy.

```
model;
{
 y ~dpois(mu)
 mu <- exp * theta
 theta ~dgamma(0.05,0.05)
}
```

In this example, $\theta \sim Gamma(0.05, 0.05)$, the expected rate ($e = 1$) is loaded as data

```
list(exp=1)
```
while all stochastic nodes are initialized before the simulation starts:

FIGURE A.7: Time series of simulated counts from WinBUGS under a simple gamma-Poisson model.

list(theta=0.5,y=2).

Following the initialization, the number of values of y simulated is set in the sample monitor. Figure A.7 displays the series of counts obtained with 10000 iterations with the above model.

More sophisticated simulation is possible within WinBUGS of course, but it is often more convenient to exploit the interactive capabilities of R in simulation of data.

A.3.2 Model Code

The previous example demonstrated some features of WinBUGS model code, albeit in a simple simulation. When data is included within a model then a data likelihood must be specified. WinBUGS does not distinguish between a parameter node that has associated data and a stochastic node. The only difference is when data are loaded and a stochastic node with data is not simulated. Hence, both data and parameters are treated alike. WinBUGS assumes that if data are not assigned to any stochastic node then the node must be initialized and updated. The model code below is an example where data are included.

In this case, again we use the simple example of count data within small areas. Assume that there are m regions, the count in the i th region is y_i, the expected rate for the same region is e_i and the relative risk is θ_i. Assume also that $[y_i|e_i, \theta_i] \sim Poisson(\mu_i)$, and $\mu_i = e_i\theta_i$. At the next level of the hierarchy assume that $\theta_i \sim gamma(a, b)$. Hence this model assumes that the relative risks are exchangeable samples from a gamma distribution. Finally, at the next level of the hierarchy, the gamma parameters: a and b, have exponential distributions with fixed parameters $\lambda = 1/mean = 0.1$. This leads to a large variance and mean of 10. With these hyper-prior distributions the expected value of θ_i is 1 with a relatively small variance. For this model the WinBUGS code is given below.

```
model{for (i in 1:m)
{     # Poisson likelihood for observed counts
      y[i]~dpois(mu[i])
      mu[i]<-e[i]*theta[i]
      # Relative Risk
      theta[i]~dgamma(a,b)}
# Prior distributions for "population" parameters
a~dexp(0.1)
b~dexp(0.1)
# Population mean and population variance
mean<-a/b
var<-a/pow(b,2)}
```

In the above specification the for loop extends over all subscripted variables (y[],mu[],e[],theta[]). Note that the for specification has the same loop format as that found in R or S-Plus. The prior distributions are specified outside the main loop for parameters which are not subscripted. Note that theta[i] is subscripted and so its distribution must be specified inside a for loop.

A variant of this model defines a log linear model for the relative risk:

```
model{for (i in 1:m)
{     # Poisson likelihood for observed counts
      y[i]~dpois(mu[i])
      mu[i]<-e[i]*theta[i]
      # Relative Risk
      log(theta[i])<-a+v[i]
v[i]~dnorm(0,precv)}
# Prior distributions for "population" parameters
a~dnorm(0,preca)
siga<-1/preca
sigv<-1/precv
preca~dgamma(0.5,0.0005)
precv~dgamma(0.5,0.0005)}
```

This model can be run with the data from the adat list (m,y,e) and initialization of the stochastic parameters (preca,precv,a,v) can be made using:

```
list(preca=0.1,precv=0.1,a=0.1,
v=c(0,0,0,0,0,0,0,0,0,0,0,0,0,0,0,
0,0,0,0,0,0,0,0,0,
0,0,0,0,0,0,0,0,0,0,0,0,
```

FIGURE A.8: Display of the acceptance rate monitor for the metropolis-Hastings sampler adopted by WinBUGS, for the Poisson log linear frailty relative risk model for 20,000 iterations.

node	mean	sd	MC error	2.50%	median	97.50%
a	-0.00171	0.03814	0.001706	-0.08869	-3.20E-04	0.0733
preca	1258	1490	49.17	17.76	732.6	5429
precv	2183	1596	215.5	151.7	1947	6.00E+03
siga	0.01844	0.4321	0.009608	1.84E-04	0.001368	0.06029
sigv	0.001223	0.001814	2.47E-04	1.67E-04	5.14E-04	0.00663

FIGURE A.9: Display of summary of posterior parameter estimates as provided by WinBUGS for a sample of 2000 after a burn-in of 20,000 iterations for the log linear Poisson relative risk model with simple frailty random effect.

0,0,0,0,0,0,0,0,0,0,0,0)).

The Metropolis acceptance rate monitor from this program on WinBUGS for a single chain of 20,000 iterations is shown in Figure A.8. Metropolis-Hastings updating was chosen by WinBUGS in this case. This monitor checks for a reasonable acceptance rate. In general, the rate should lie between 0.25 and 0.5 (see Section 3.3.7.1). Hence, in this case the sampler seems to have reasonable acceptance rate.

The sample Monitor on WinBUGS also provides a wealth of information concerning parameters that have been monitored, the choice of monitored parameters being left to the user. In Figure A.9, the sample monitor has been set to record the parameters a,preca,precv, siga,sigv and the display shows part of the stats option on the Sample Monitor. Displayed are the posterior summaries for each parameter for the chosen sample size (mean, standard deviations (sd), MC error, 2.5% and 97.5% percentiles, and the median). These are the default summaries, but can be altered to produced, e.g. alternative percentiles. All parameters can be monitored and summarized in this way. For those that are indexed by region (such as theta[], mu[], v[]), then a complete list of summaries for each region is produced.

Figure A.10 displays the thematic map of the posterior expected spatial distribution of the theta[] parameter (relative risk) for the Poisson log linear model. This map is easily obtained from the GeoBUGS menu option within

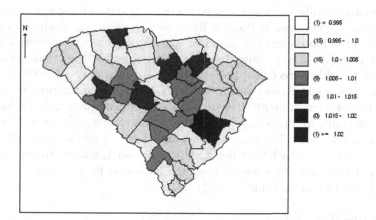

FIGURE A.10: Display of the GeoBUGS posterior expected relative risk map
(theta[]) for the 46 counties of South Carolina for a 2000 sample size and the
Poisson log linear frailty model.

WinBUGS. The region geographies (in this case county boundaries) must be
available as a GIS-format file (ArcView, Epimap, Splus, ArcInfo) and this can
be imported to WinBUGS for use with GeoBUGS. Once the map geographies
are input, any parameter indexed by the region index can be plotted via the
Map Tool. The order of regions must be the same as the parameter vectors
in the model of course. In the Poisson log linear model above, there are three
parameter vectors that can be plotted: theta[], mu[], and v[]. Figure A.11
displays the map for the posterior mean uncorrelated frailty term v[].

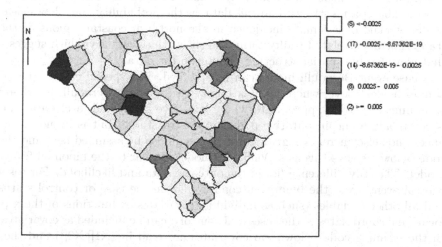

FIGURE A.11: Display of the posterior mean uncorrelated frailty term (v[],)
for a sample of size 2000, for the Poisson log linear relative risk model with
uncorrelated frailty term.

Alternative models can be specified for different data formats of course. The above example assumed a Poisson likelihood for count data within counties. Another situation commonly found in spatial analysis of disease risk is that a finite population is measured within an area and within that population a binary variable is recorded on each unit. For example, an example would be total brain cancer cases recorded within counties and the number of male cases. Here we examine another South Carolina example where we have the total case count for 1996 through 2000 of brain cancer by county. The male count is also known. The sex difference may have a spatial expression. The model code below can be applied to this case and is similar to the previous Poisson model except we have a binomial likelihood for the male count data (malc[i]) out of a total count (totc[i]).

```
model{for (i in 1:m)
{      # binomial likelihood for observed counts
      malc[i]~dbin(p[i],totc[i])
      logit(p[i])<-a+v[i]
v[i]~dnorm(0,precv)}
# Prior distributions for "population" parameters
a~dnorm(0,preca)
preca<-1/siga
precv<-1/sigv
siga~dgamma(0.001,0.001)
sigv~dgamma(0.001,0.001)}
```

The mapped results of this analysis after 10,000 iterations are displayed for a sample size of 2000, in Figures A.12 and A.13. It is clear that there is considerable spatial structure in this data, as the probability map shows clear peaks of male over female incidence in the northern coastal regions of the state. The uncorrelated frailty term also shows this tendency which suggests that considerable spatial structure remains in the data.

A case event Bernoulli model can be specified, as a special case of the binomial model above, where y[i] is a binary vector and a Bernoulli likelihood is assumed. This is appropriate where you have individual level data, and is commonly found in putative pollution hazard studies but is by no means uncommon elsewhere. A logistic linear link can also be assumed here and the code below displays this as a WinBUGS model, similar to the binomial frailty model. The only difference lies in the outcome data and likelihood. For case-control event data, the binary outcome would be the case or control status and all other variables (such as individual covariates or functions of the x ,y locational coordinates of the cases and controls) can be included as covariates. In the example code below, two covariates are read in (X1[],X2[]) and they have regression parameters B1 and B2 with zero-mean Gaussian prior distributions. The precision parameters have diffuse Gamma prior distributions.

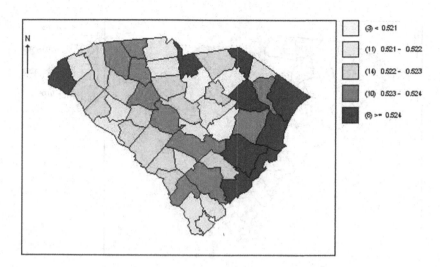

FIGURE A.12: Display of the posterior mean probability of male cancer incidence by county in South Carolina for the five-year brain cancer case example (1996 to 2000).

This a binary random effect logistic linear model.

```
model{for (i in 1:m)
{      # Bernoulli likelihood for observed counts
    y[i]~dbern(p[i])
     logit(p[i])<-a+v[i]+B1*X1+B2*X2
v[i]~dnorm(0,precv)}
# Prior distributions for "population" parameters
B1~dnorm(0,precB1)
B2~dnorm(0,precB2)
a~dnorm(0,preca)
preca<-~dgamma(0.001,0.001)
precv~dgamma(0.001,0.001)
precB1~dgamma(0.001,0.001)
precB2~dgamma(0.001,0.001)
}
```

Convergence of MCMC samplers is an important issue, as noted in Section 3.3.6. In all the examples above the samplers were monitored using a variety of diagnostics. The main diagnostic used on WinBUGS is the Brooks-Gelman-Rubin (BGR) diagnostic which examines the between- and within-sample variances for output streams. Figure A.14 displays the BGR diagnostic for two components of the logistic linear model for the South Carolina brain cancer example: the deviance and parameter a. The display suggests that the models converged early at least for these parameters.

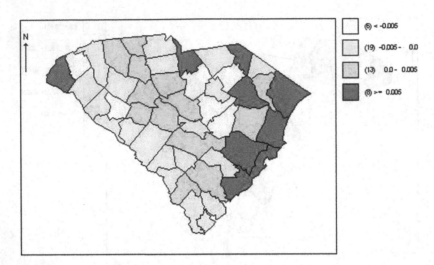

FIGURE A.13: Display of the posterior mean uncorrelated frailty (v[]) for
the logistic linear model by county in South Carolina for the five-year brain
cancer case example (1996 to 2000).

A.4   R2WinBUGS and R2OpenBUGS Functions

This function is very useful for executing R2WinBUGS programs from R. For
an overview see Sturtz et al. (2005). The main advantage of this lies in the
ability to further manipulate data within R before and after execution of
WinBUGS code. In addition, R2WinBUGS provides added facilities (such as a
different DIC calculation, and graphical summaries). R2OpenBUGS performs
in a similar manner except that OpenBUGS is called instead of WinBUGS.
The main advantage of this functionality is found in simulation. When multiple data sets must be repeatedly analysed then the ability to pass these to
WinBUGS from R and to continue processing of further simulation data sets
within a loop is a major advantage.

The main function call, once R2WinBUGS is loaded, is to a function called
bugs which activates WinBUGS and passes the relevant information for processing:

```
bugs(data, inits, parameters.to.save, model.file = "model.txt",
   n.chains = 3, n.iter = 2000, n.burnin = floor(n.iter/2),
   n.thin = max(1, floor(n.chains * (n.iter - n.burnin)/1000)),
   bin = (n.iter - n.burnin) / n.thin,
   digits = 5, codaPkg = FALSE,
```

FIGURE A.14: Display of Brooks-Gelman-Rubin (BGR) plots for the deviance and parameter a for the South Carolina county-based brain cancer logistic linear frailty model for 1996 to 2000. The plots are closely linear and parallel which suggests close convergence by 2000 iterations.

```
bugs.directory = "c:/WinBUGS/WinBUGS14/")
```

The main parameters for the function call are summarized below:

data: a named list (corresponding to the names used in the model) OR a vector or list of the names of objects OR a name of a file containing the data

inits: a list with n.chains elements, each element being a list of starting values OR a function that creates these values

parameters.to.save: a character vector of the names of parameters to be saved

model.file: file containing the WinBUGS code

n.chains: number of chains (defaults to 3)

n.iter: number of iterations (including burn-in)

n.burnin: number of burn-in iterations (default to half length of chain: n.iter/2)

Other parameters can be set to control, e.g. thinning rate (n.thin), DIC calculation (DIC).

A simple example of a call is as follows. The gamma-Poisson model where the relative risk is assumed to have mean $e_i\theta_i$ is considered below for the South

Carolina congenital anomaly mortality data. This data was first examined in Chapter 4. It consists of incident counts in 46 counties of South Carolina for one year. The expected rates associated with these counts are also available, computed from the statewide reference population. For purposes of data input to R2WinBUGS we need to set up a data structure which includes the outcome variable, the expected rate, and number of regions (y,e,m). The structure adat defined below is such a list structure:

```
adat<-list(m=46,
y=c(0,7,1,5,1,1,5,16,0,17,4,0,0,1,1,7,1,3,0,0,8,2,13,7,0,8,0,3,2,4,1,
11,0,1,2,3,3,8,6,14,3,11,6,0,1,5),
e=c(1.129778827,6.667008775,0.650279674,6.988864371,0.95571406,
1.123210345,5.908349156,8.539026017,0.601016062,18.92051111,
2.272694617,1.73736337,2.019808077,1.688099759,1.747216093,
3.221840201,1.835890594,5.221942834,0.978703751,1.254579976,
6.407553754,2.676656232,16.57884744,3.077333607,1.087083697,
7.606301637,1.018114641,2.15774619,2.844152512,2.955816698,
0.985272233,9.22871658,0.38097193,1.855596038,1.579719813,
1.579719813,2.647098065,4.791707292,4.144711859,15.70852363,
0.765228101,11.32077795,6.256478678,1.500898035,2.085492893,
7.297583004))
```

For initial values a list structure can be defined for this example that contains three initializations of the relevant parameters. The parameters to initialize are the predicted outcome values. To run R2WinBUGS the default is three chains. In this example, we can specify initial values for three chains as a list of lists with a,b,theta,ypred within each of three list elements (ab,ac,ad) which in turn, form the three component list ainits:

```
ab<-list(a=0.1,b=0.1,theta=c(1,1,1,1,1,1,1,1,1,1,1,1,1,1,1,1,1,1,1,1,1,
1,1,1,1,1,1,1,1,1,1,1,1,1,1,1,1,1,1,1,1,1,1,1,1,1),
ypred=c(0,7,1,5,1,1,5,16,0,15,4,0,0,1,1,7,1,3,0,0,8,2,13,7,0,
8,0,3,2,4,1,11,0,1,2,3,3,8,6,14,3,11,6,0,1,5))
ac<-list(a=0.12,b=0.12,theta=c(1.2,1,1,1,1,1,1,1,1,1,1.2,1,1,1,1,1,1,1,1,1,1,
1,1,1,1,1,1,1.2,1,1,1,1,1,1,1,1,1,1.5,1,1,1.6,1,1,1,1),
ypred=c(0,7,1,5,1,1,5,16,0,15,4,0,0,0,0,6,1,2,0,0,8,2,13,7,0,
8,0,3,2,4,1,11,0,1,2,3,3,8,6,14,3,11,6,0,1,5))
ad<-list(a=0.15,b=0.17,theta=c(1.4,1,1,1,1,1,1,1,1,1.8,1,1,1,1,1,1,1,1,1,1,1,
1,1,1,1,1,1,1.2,1,1,1,1,1,2.1,1,1,1,1,1,1,1,1,1.1,1,1),
ypred=c(0,7,1,5,0,0,0,0,0,0,4,0,0,1,1,7,1,3,0,0,8,2,13,7,0,8,0,
3,2,4,1,11,0,1,2,3,3,8,6,14,3,11,6,0,1,5))
ainits<-list(ab,ac,ad)
```

In the example, the parameters to monitor are the mean, residual, and predictive residual ("mu","r","rpred"), the data are in the object adat, the number of chains defaults to three, the inits for the three chains are in ainits, the number of iterations is 4000 and the model is found in text file "map_modelWB_pred.txt":

```
parameters<-c("mu","r","rpred")
map.sim<-bugs(adat,inits = ainits,
parameters,n.iter=4000,
model.file ="map_modelWB_pred.txt")
```

The model file map_modelWB_pred.txt contains the following model code:

```
model
{for (i in 1:m)
{# Poisson likelihood for observed counts
    y[i]~dpois(mu[i])
    ypred[i]~dpois(mu[i])
    mu[i]<-e[i]*theta[i]
    probexc[i]<-step(theta[i]-1)
    # Relative Risk
    theta[i]~dgamma(a,b)
    r[i]<-y[i]-mu[i]
    rpred[i]<-y[i]-ypred[i]}
# Prior distributions for "population" parameters
a~dexp(10)
b~dexp(10)
# Population mean and population variance
mean<-a/b
var<-a/pow(b,2)
}
```

Submitting the command map.sim<-bugs(....) in R calls WinBUGS with the parameters specified, and stores the results of the call in object map.sim. Upon return from WinBUGS, map.sim contains a range of information. For example, some useful components and summaries are as follows:

map.sim $mean yields the posterior mean estimate.

map.sim$sd yields the posterior standard deviations.

map.sim$mean$rpred yields the separate posterior mean for component rpred.

map.sim$median yields the posterior sample median.

map.sim$summary[,3] yields the 2.5 posterior sample percentile.

map.sim$summary[,7] yields the 97.5 posterior sample percentile.

map.sim $sims.matrix yields the complete output matrix.

Diagnostic plots of the output are available via use of the plot command applied to the bugs object. For example, the caterpillar and box plots

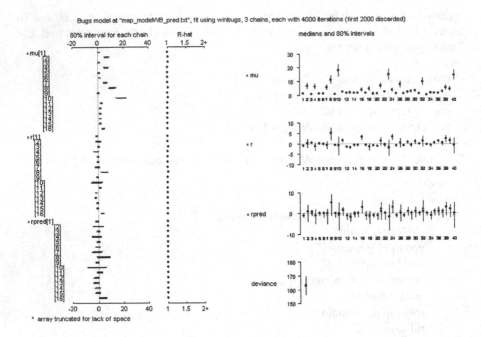

FIGURE A.15: Display of the plot command applied to the output object from the bugs(..) command. The output consists of caterpillar and box plots for the nomibnated output.

available on WinBUGS can be produced with the command plot(map.sim) (see Figure A.15). Alternatively simple plotting commands applied to elements of the output can be used. For example, plot(density(map.sim$summary [1:46,1]),main="") will produce a density estimate of the mean posterior distribution of parameter mu (see Figure A.16), whereas the commands

```
smat<-map.sim$sims.matrix
dfsmat<-data.frame(smat[,47:92])
boxplot(dfsmat)
```

will produce a multiple boxplot of the residuals in R for each unit (see Figure A.17).

FIGURE A.16: Display of the density estimate for the posterior mean for parameter mu from the bugs output object map.sim.

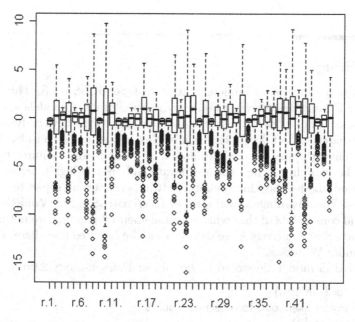

FIGURE A.17: Display of multiple boxplots of posterior mean residuals in r as output by R2WinBUGS.

A.5   OpenBUGS and JAGS

An open source version of BUGS has been under development for some time and a Windows implementation of the package is available which is designed to be equivalent in use to WinBUGS. OpenBUGS is available in its current version from http://www.openbugs.info/w/. OpenBUGS runs with MS Windows, Unix/Linux, or Macintosh (using e.g. Wine). The differences between OpenBUGS and WinBUGS are highlighted at the http://www.openbugs.info/w/OpenVsWin page. One major difference is the capability on OpenBUGS to choose the sampling algorithms used for MCMC. However, while most capabilities of WinBUGS are now available in Open-BUGS there is still a troubling issue of differences in results obtained from either package (for the same program) and hence whether the packages can be run to produce the same output.

JAGS (Just another gibbs sampler) is a program for analysis using MCMC simulation not wholly unlike BUGS. JAGS language constructs are close to BUGS but the package has the advantage that it can be run from R directly without calling an external package. WinBUGS code can usually be easily converted to JAGS. However not all distributions available on WinBUGS are available on JAGS (for example CAR models).

A.6   BRugs

The BRugs package is an implementation of OpenBUGS on R. This is a collection of R functions allowing users to analyse graphical models via MCMC techniques. Most of the R functions provide a link to the BRugs dynamic link library (shared object file). Essentially this allows users considerable freedom to build graphical models from scratch, but it also requires that each step in the modeling process is a separate call to a BRugs function. Hence compilation, model checking, monitor setting, etc., are all separate calls to R functions. The advantage of BRugs over R2WinBUGS is that WinBUGS is not called and constraints of that windows-implemented package are not imposed. This means that relatively large data sets can be handled and there is no need to run under Windows.

For a BUGS model file stored in "log-linear-Poisson-model2.txt":

```
model{for (i in 1:m)
 { # Poisson likelihood for observed counts
 y[i]~dpois(mu[i])
 mu[i]<-e[i]*theta[i]
```

```
# Relative Risk
log(theta[i])<-a+v[i]
v[i]~dnorm(0,precv)}
# Prior distributions for "population" parameters
a~dnorm(0,preca)
preca<-pow(sda,-2)
precv<-pow(sdv,-2)
sda~dunif(0,100)
sdv~dunif(0,100)}
```

Data stored in list format in "adat.txt":

```
list(m=46,
y=c(0,7,1,5,1,1,5,16,0,17,4,0,0,1,1,7,1,3,0,0,8,2,13,7,0,8,0,3,2,4,1,
11,0,1,2,3,3,8,6,14,3,11,6,0,1,5),
e=c(1.129778827,6.667008775,0.650279674,6.988864371,0.95571406,
1.123210345,5.908349156,8.539026017,0.601016062,18.92051111,
2.272694617,1.73736337,2.019808077,1.688099759,1.747216093,
3.221840201,1.835890594,5.221942834,0.978703751,1.254579976,
6.407553754,2.676656232,16.57884744,3.077333607,1.087083697,
7.606301637,1.018114641,2.15774619,2.844152512,2.955816698,
0.985272233,9.22871658,0.38097193,1.855596038,1.579719813,
1.579719813,2.647098065,4.791707292,4.144711859,15.70852363,
0.765228101,11.32077795,6.256478678,1.500898035,2.085492893,
7.297583004))
```

Initialization values in "inits.txt":

```
list(a=0.1,sda=0.1,sdv=0.1)
```

the following command sequence will fit the BUGS code model on R, with a burn-in of 10000 iterations with 2 chains monitored :

```
library(BRugs)
modelCheck("log-linear-Poisson-model2.txt")
modelData("adat.txt")
modelCompile(numChains=2)
modelInits(rep("inits.txt",2))
modelGenInits( )                          #......random effect initialisation
modelUpdate(10000)
samplesSet(c("theta","v","deviance"))
modelUpdate(2000)
samplesStats("*")
```

Note that output can be stored in a dataframe for later inspection:

```
asd<-samplesStats("*").
```

FIGURE A.18: BGR statistic plot for the Poisson log linear relative risk model applied to a 46-region data set from BRugs.

These functions can be installed into a purpose-built R function. Within the BRugs package there is a command already specified:

```
BRugsFit( "log-linear-Poisson-model2.txt", data="adat.txt", numChains=2,
parametersToSave=c("deviance","a"),nBurnin = 5000, nIter = 2000, nThin =
1)
```

This command will fit the model specified in "log-linear-Poisson-model2.txt" with a burn-in of 5000 and a sample of size 2000, and will display the deviance and estimate for parameter "a", as well as the DIC.

Note also that convergence checking can be done via the usual BGR statistic:

```
plotBgr("deviance",col=gray(0:3/3))
```

This command gives the BGR plot for the above model fit as in Figure A.18. The dark line is the ratio of between and within chain variance estimates.

Appendix B: Selected WinBUGS Code

A selection of WinBUGS models are presented here to provide a sample of how code can be specified for the model discussed. The ODC files for all the fitted models cited book are available for download from *http://academicdepartments.musc.edu/phs/research/lawson/*.

B.1   Code for Convolution Model (Chapter 5)

```
model
  {
  for( i in 1 : m ) {
  y[i] ~dpois(mu[i])
  mu[i] <- e[i] * rr[i]
  smr[i]<-(y[i]+eps2)/(e[i]+eps2)
  rr[i] <- exp(al0 +v[i]+u[i])
  v[i]~dnorm(0,1.0E-6)
  PP[i]<-step(rr[i]-1+eps)
  ypred[i] ~dpois(mu[i])
  PPL[i] <- pow(ypred[i]-y[i],2)
  PPL2[i] <- abs(ypred[i]-y[i])
  }
  mape <- mean(PPL2[])
  mspe <- mean(PPL[])
  # car normal
  u[1:m]~car.normal(adj[],weights[],num[],tau.u)
  for(k in 1:sumNumNeigh)
  {
      weights[k]<-1
  }
  al0 ~dflat()
  eps<-1.0E-6
  eps2~dnorm(0,1000)
  tau.u~dgamma(0.005,0.005)
  }
```

B.2   Code for Spatial Spline Model (Chapter 5)

This fits the model outlined in Section 5.7.2 for count data:

```
model{
for (i in 1 : 100){
ME[i]<-0}
a
for (i in 1:100){
for (j in 1:100){
SOM[i,j]<-tauS*OM[i,j]
}}
for (i in 1 : m){
y[i]~dpois(mu[i])
mu[i]<-e[i]*theta[i]
ssum[i]<-inprod(gam[],CDZ[i,])
log(theta[i])<-a0+a1*xc[i]+a2*yc[i]+ssum[i]}
gam[1:100]~dmnorm(ME[],SOM[,])
a0~dflat()
a1~dnorm(0,0.0001)
a2~dnorm(0,0.0001)
tauS<-1/(sigmaS*sigmaS)
sigmaS~dunif(0,10)}
```

B.3   Code for Spatial Autologistic Model
(Chapter 6)

This model was specified as a simple regression model on the neighbor sums:

```
model
{
for (i in 1:m)
{
 yb[i]~dbern(p[i])
 logit(p[i])<-alpha0+alpha*x[i]+beta*ysum[i]}
alpha0~dflat()
alpha~dnorm(0.0,0.1)
beta~dnorm(0.0,0.1)}
```

B.4 Code for Logistic Spatial Case Control Model (Chapter 6)

Logistic spatial case control model with a fully specified spatial covariance:

```
model {
for (i in 1:N){
ind[i]~dbern(p[i])
 logit(p[i])<-f[i]
f[i]<-gam1*dis[i]+gam0+v[i]+W[i]
v[i]~dnorm(0,0.001)
res[i]<-(ind[i]-p[i])/sqrt(p[i]*(1-p[i]))
prexR2[i]<-step(res[i]-2)}
gam0~dflat()
gam1~dnorm(0,0.01)
 for(i in 1:N){ mu[i]<-0}
 W[1:N] ~spatial.exp(mu[], x[], y[], tau, phi,1)
tau~dexp(0.1)
phi~dexp(0.1)
}
```

B.5 Code for PP Residual Model (Chapter 6)

This fits a heterogeneous Poisson process Bayesian model using the Dirichlet tesselation weights. The background risk is estimated externally (e.g. in R) and appears evaluated at the case event locations in the variable den[]

```
#w[] Dirichlet weights
# d[] distance from putative source
#I[] indicator variable denoting 'real' data point or 'dummy' data point
# x,y,w,den,I and d must be read in:
model{
C <- 10000
                for (i in 1:N) {
                f[i]<-1.+exp(bet1*d[i])
                    zeros[i] <- 0
                        log(lam[i])<-bet0+log(f[i])
                    log(L[i])<-I[i]*log(lam[i])-w[i]*den[i]*lam[i]
                    phi[i] <- -log(L[i]) + C
                    zeros[i] ~dpois(phi[i])
```

```
                         res[i]<-((1/w[i])-den[i]*lam[i])
                         x1[i]<-x[i]
                         y1[i]<-y[i]
                         probexR[i]<-step(res[i]-0)}
     bet0~dnorm(0,0.001)
     bet1~dnorm(0,0.001)}
```

B.5.1   Same Model with Uncorrelated Random Effect

```
model{
   C <- 10000
            for (i in 1:N) {
            f[i]<-1.+exp(bet1*d[i])
                  zeros[i] <- 0
                     log(lam[i])<-bet0+log(f[i])
                     log(L[i])<-I[i]*(log(lam[i])+v[i])-w[i]*den[i]*lam[i]*exp(v[i])
                     phi[i] <- -log(L[i]) + C
                     zeros[i] ~dpois(phi[i])
                     res[i]<-((1/w[i])-den[i]*lam[i])
                     x1[i]<-x[i]
                     y1[i]<-y[i]
                     probexR[i]<-step(res[i]-0)
                  v[i]~dnorm(0,tauv)
            }

   bet0~dnorm(0,0.001)
   bet1~dnorm(0,0.001)
   tauv<-0.1
      }
```

B.6   Code for Logistic Spatial Case-Control Model (Chapter 6)

This specification is for use with the R2winBUGS function:

```
# x,y and ind as inputs
# dis is distance
model
{
C <- 10000
for (i in 1:N){
```

```
f[i]<-(1+exp(-gam1*dis[i]))*exp(gam0+v[i])
zeros[i] <- 0
log(lam[i])<-log(f[i])
log(L[i])<-ind[i]*log(lam[i])-log(1.+lam[i])
phi[i] <- -log(L[i]) + C
zeros[i] ~dpois(phi[i])
mu[i]<-0
p[i]<-lam[i]/(1+lam[i])
x1[i]<-x[i]
y1[i]<-y[i]
v[i]~dnorm(0,0.1)
res[i]<-(ind[i]-p[i])/sqrt(p[i]*(1-p[i]))
prexR2[i]<-step(res[i]-2)
}
gam0~dnorm(0,0.1)
gam1~dnorm(0,0.1)
}
```

In file "logistic_case_control_P2.txt":

datalist elements: N, x, y, dis, ind
initial values: gam0, gam1, v
parameters: v, res, p, prexR2

B.6.0.1 R2WinBUGS Commands

The following commands set up the parameter vectors to report and the mapping tools. Use is made of the multivariate B spline package (MBA) for contouring:

```
>parameters<-c("res","v","prexR2","p")
>reslogist2<-bugs(data,inits,parameters,model.file="logistic_case_control_
   P2.txt", n.iter=10000)
```

The following R code will plot the mapped surface of $\Pr(r_i > 2)$:
```
>plot(data$x[data$ind==1],data$y[data$ind==1],xlab="easting",ylab=
   "northing")
>zedprexR2<-cbind(data$x,data$y,reslogist2$mean$prexR2)
>ressurfprex<-mba.surf(zedprexR2,20,20)$xyz.est
>contour(ressurfprex,levels=c(0.05,0.1,0.2),add=T)
```

B.7   Code for Poisson Residual Clustering Example (Chapter 6)

This is a standard Poisson risk model for count data with UH effect and residual and exceedances computed

```
model
{
 for( i in 1 : m ) {
 y[i] ~dpois(mu[i])
 mu[i] <- e[i] * rr[i]
 log(rr[i]) <- al0 +v[i]
 v[i]~dnorm(0,0.001)
 PP[i]<-step(rr[i]-1+eps)
 res[i]<-(y[i]-mu[i])/sqrt(mu[i])
 prexR1[i]<-step(res[i]-1)
 prexR2[i]<-step(res[i]-2)
 }
 al0~dflat()
 eps<-1.0E-6
}
```

B.8   Code for Proper CAR Model (Chapter 5)

This is a proper CAR model for Ohio respiratory cancer data from 1988:

```
model{
    for(i in 1 : 88) {
        m[i] <- 1/num[i]
    }
    cumsum[1] <- 0
    for(i in 2:(88+1)) {
    cumsum[i] <- sum(num[1:(i-1)])
    }
    for(k in 1 : sumNumNeigh) {
    for(i in 1:88) {
    pick[k,i] <- step(k - cumsum[i] - epsilon) * step(cumsum[i+1] - k) }
```

```
    C[k] <- 1 / inprod(num[], pick[k,]) # weight for each pair of neighbours
    }
    epsilon <- 0.0001
    for (i in 1 : 88) {
    y[i] ~dpois(mu[i])
    log(mu[i])<-log(e[i])+a0+a1*(pov[i]+ep1[i])+S[i]+v[i]
    RR[i] <- exp(a0+a1*(pov[i]+ep1[i])+S[i]+v[i]
    theta[i] <- a0+a1*(pov[i]+ep1[i])+S[i]+v[i]
    v[i]~dnorm(0,0.0001)
    ep1[i]~dnorm(0,0.0001)
    dis1[i]<-dist[i]}
    # Proper CAR prior distribution for spatial random effects:
    S[1:88] ~car.proper(theta[], C[], adj[], num[], m[], prec, gamma)
    # Other priors:
    prec <-1/pow(sdS,2)                    # prior on precision
    sdS~dunif(0,100)                    # uniform sd prior
    sigma <- sqrt(1 / prec)                     # standard deviation
    gamma.min <- min.bound(C[], adj[], num[], m[])
    gamma.max <- max.bound(C[], adj[], num[], m[])
    gamma ~dunif(0.5, gamma.max)
    a0~dflat()
    a1~dnorm(0,0.0001)
    #tau.v<-1/pow(sdv,2)
    #sdv~dunif(0,10)
    #tau.p1<-1/pow(sdp1,2)
    #sdp1~dunif(0,10)
    }
```

B.9 Code for Multiscale Model for PH and County-Level Data (Chapter 9)

The multiscale model for PH districts and counties in Georgia where u and v for districts are included in county effects:

```
model{
for( i in 1:18){
yph[i]~dpois(muph[i])
log(muph[i])<-log(eph[i])+log(thph[i])
thph[i]<-exp(aph0+uph[i]+vph[i])
vph[i]~dnorm(0,0.0001)
}
```

```
for( j in 1:159){
yc[j]~dpois(muc[j])
log(muc[j])<-log(ec[j])+log(thc[j])
thc[j]<-exp(ac0+uc[j]+vc[j]+uph[phc[j]]+vph[phc[j]])
vc[j]~dnorm(0,0.0001)
}
for (k in 1 :nsumph){weiph[k]<-1}
uph[1:18]~car.normal(adj[],weiph[],num[],tauph)
#tau.vph<-1/pow(sdvph,2)
tauph<-1/pow(sdph,2)
aph0~dflat()
#sdvph~dunif(0,100)
sdph~dunif(0,100)
for (o in 1 :nsumc){weic[o]<-1}
uc[1:159]~car.normal(adj2[],weic[],num2[],tauc)
#tau.vc<-1/pow(sdvc,2)
tauc<-1/pow(sdc,2)
ac0~dflat()
#sdvc~dunif(0,100)
sdc~dunif(0,100) }
```

B.10   Code for Shared Component Model for Georgia Asthma and COPD (Chapter 10)

The joint shared component model for asthma and COPD in Georgia counties is:

```
MODEL{
for(i in 1:N){
asthma[i]~dpois(Muas[i])
log(Muas[i])<-log(Easthma[i])+log(thetAS[i])
log(thetAS[i])<-b0+Was[i]+Wcom[i]*delta
COPD[i]~dpois(MuC[i])
log(MuC[i])<-log(ECOPD[i])+log(thetCOPD[i])
log(thetCOPD[i])<-b1+Wcom[i]/delta
angina1[i]<-angina[i]
Eangina1[i]<-Eangina[i]
}
delta<-exp(sdf)
sdf~dnorm(0,5.88)
for( i in 1:sumNumNeigh){weights[i]<-1}
```

```
Wcom[1:N]~car.normal(adj[],weights[],num[],tWcom)
Was[1:N]~car.normal(adj[],weights[],num[],tWas)
b0~dflat()
b1~dflat()
tWcom<-pow(sdWcom,-2)
sdWcom~dunif(0,100)
tWas<-pow(sdWas,-2)
sdWas~dunif(0,100) }
```

B.11 Code for Seizure Example with Spatial Effect (Chapter 11)

The seizure longitudinal example with a spatial contextual effect w is:

```
model{
for(i in 1:59){
 y[i,1]<-t1[i]
 y[i,2]<-t2[i]
 y[i,3]<-t3[i]
 y[i,4]<-t4[i]
}
ep[1]~dnorm(0,0.0001)
for (i in 1:59){
for (k in 1:4){
y[i,k]~dpois(mu[i,k])
log(mu[i,k])<-log(baseC[i])+alp0+alp1*group[i]+alp2*age[i]+ep[k]+w[ind[i]]
}}
for (k in 2:4){
ep[k]~dnorm(ep[k-1],alp3)}
alp0~dflat()
alp1~dnorm(0,0.0001)
alp2~dnorm(0,0.0001)
alp3~dgamma(0.001,0.001)
w[1:46]~car.normal(adj[],weights[],num[],tauW)
for (i in 1:sumNumNeigh){weights[i]<-1}
tauW<-pow(sdW,-2)
sdW~dunif(0,100)}}
```

B.12   Code for Knorr-Held Model for Space-Time Relative Risk Estimation (Chapter 12)

This is a space-time model for the VLBW example from Georgia with a time-dependent temporal effect and time-dependent interaction term:

```
model{
for (i in 1:159){
for (j in 1: 11){
vlbw[i,j]~dbin(p[i,j],birth[i,j])
logit(p[i,j])<-a0+v[i]+u[i]+g[j]+psi[i,j]
}
psi[i,1]~dnorm(0,taupsi)
for (j in 2:11){
psi[i,j]~dnorm(psi[i,j-1],taupsi)}
v[i]~dnorm(0,tauv)      }
u[1:159]~car.normal(adj[], weights[], num[], tauu)
    for(k in 1:sumNumNeigh) {
         weights[k] <- 1}
g[1]~dnorm(0,0.0001)
time1[1]<-time[1]-1993
for (j in 2: 11){
g[j]~dnorm(g[j-1],tau.g)
time1[j]<-time[j]-1993}
taupsi<-1/pow(sdpsi,2)
sdpsi~dunif(0,100)
tau.g<-1/pow(sdg,2)
sdg~dunif(0,100)
tauu<-1/pow(sdu,2)
sdu~dunif(0,100)
tauv<-1/pow(sdv,2)
sdv~dunif(0,100)
a0~dflat()}
```

B.13   Code for Space-Time Autologistic Model (Chapter 12)

This is the ST autologistic model with temporally dependent intercept and slope parameters for the neighbor sum terms:

```
model
{
for (i in 1:m)
{
 logit(p[i,1])<-a[1]+b[1]*ys1[i,1]
ypred[i,1]~dbern(p[i,1])
diff[i,1]<-pow(y[i,1]-ypred[i,1],2)
     for (k in 2:T)
     {
         y[i,k]~dbern(p[i,k])
 logit(p[i,k])<-a[k]+b[k]*ys1[i,k]+c[k]*ys1[i,(k-1)]
     ypred[i,k]~dbern(p[i,k])
     diff[i,k]<-pow(y[i,k]-ypred[i,k],2)
     }
}
for (k in 1:T){
 a[k] ~dnorm(0,0.01)
 b[k] ~dnorm(0,0.01)
 c[k] ~dnorm(0,0.01)
         }
         mspe<-mean(diff[,])
}
```

Appendix C: R Code for Thematic Mapping

The following R code was developed by Bo Cai and can be used to produce thematic maps from ArcView shapefiles on R. The function is called fillmap and must be called with a complete set of ArcView .shp, .sbn, .dbf, .shx files resident in the active work directory.

```
fillmap<-function(map, figtitle, y , n.col, bk="e", cuts,legendtxt="", legend-
pos=c(466000,4330000), titlepos=c(350000,4650000)){
  polylist<- Map2poly(map)
  if(bk=="q"){p <- seq(0,1, length=n.col+1)
  br <- round(quantile(y, probs=p),2)}
  if(bk=="e"){br <- round(seq(min(y), max(y), length=n.col+1),2)}
  if(bk=="c"){if (length(cuts)!= (n.col+1)) {cat("Cut off and color categories
do not match. ", "\n")
  break} else {br <- cuts} }
  # 0: dark 1: light light Current shading ranges from darkest to light gray white
(to distinguish with lakes).
  shading<-gray((n.col-1):0/(n.col-1))
  y.grp<-findInterval(y, vec=br, rightmost.closed = TRUE, all.inside = TRUE)
  y.shad<-shading[y.grp]
  plot.polylist(polylist,col=y.shad,axes=F)
  br<-round(br, 2)
  if (legendtxt=="")
  {
  cn<-length(y[y>=br[n.col]]) # number of regions in this intervals
    leg.txt<-paste("[",br[n.col],",",br[n.col+1],"]","(",cn,")",sep="")
    for(i in (n.col-1):1){
  cn<-length(y[(y>=br[i])&(y<=br[i+1])])
  leg.txt<-append(leg.txt,paste("[",br[i],",",br[i+1],")","(",cn,")",sep="")) }
  leg.txt<-rev(leg.txt)
  } else {leg.txt<-legendtxt}
  legend(legendpos[1],legendpos[2],legend=leg.txt,fill=shading,cex=0.7,ncol=1,
bty="n")
  text(titlepos[1],titlepos[2], figtitle, cex=1.2)}
```

FIGURE C.1: Georgia county-level plot: simulated data.

Calling the function:

  1) Load libraries:
  >library(sp)
  >library(maptools)

  2) To call the routine for the Georgia shapefile with 159 counties with values
generated from a Gaussian distribution with mean 1 and standard deviation
0.1:

```
>x<-readShapePoly("Georgia.shp")
>y<-rnorm(159,1,.1)
>maintitle<-paste("")
>fillmap(x,maintitle,y,n.col=5)
```

This produced Figure C.1.
    In addition, there is now a version of the fillmap function that can pro-
duce multiple thematic maps and is available from github. On R: in-
stall_github("carrollrm/fillmap") will allow the user to use fillmaps.

Appendix D: CAR Model Examples

We have a set of regions: $i = 1, ..., m$ and an effect that has an intrinsic CAR prior distribution.

The effect is, in vector form, vector $\boldsymbol{u}$, $i = 1, ..., m$. Hence u_i represents the effect in the i th region. In general,

$$\boldsymbol{u} \sim \boldsymbol{N}(\boldsymbol{x}'\beta, r[I - \phi C]^{-1}M)$$

where $\boldsymbol{x}'\beta$ is a linear predictor, and M is a known diagonal matrix chosen so that $r[I - \phi C]^{-1}M$ is symmetric and PD. Conditionally then

$$u_i|.... \sim N(\boldsymbol{x}_i'\beta + \phi \sum_{j \in \delta_i} c_{ij}(u_j - (\boldsymbol{x}'\beta)_j), rm_{ii})$$

where δ_i is a neighborhood of the i th region, and the conditional variance is $rm_{ii} > 0$. Note that for no trend this reduces to

$$u_i|.... \sim N(\phi \sum_{j \in \delta_i} c_{ij}(u_j), rm_{ii}).$$

For symmetry we have $c_{ij}m_{jj} = c_{ji}m_{ii}$, $i, j = 1, ..., m$ and for positive definiteness

$$\phi \in \{\phi_{\min}, \phi_{\max}\}$$

where $\phi_{\min} = \eta_1^{-1}$, and $\phi_{\max} = \eta_m^{-1}$ and η_1 and η_m are the smallest and largest eigenvalues of the matrix

$$M^{-1/2}CM^{1/2}.$$

Note also that the partial correlation ($Pcorr$) for a CAR model is given by $\phi\sqrt{c_{ij}c_{ji}}$.

Some common assumptions for the "no trend" case are given in Table D1.

CAR Model	c_{ij}	m_{ii}	ϕ	Pcorr $\phi\sqrt{c_{ij}c_{ji}}$	Mean	Variance
Intrinsic (WCAR)	$1/n_{\delta_i}$	$1/n_{\delta_i}$	1	$1/\sqrt{(n_{\delta_i}n_{\delta_j})}$	$\sum_{j\in\delta_i} u_j/n_{\delta_i}$	τ/n_{δ_i}
Proper (WCAR)	$1/n_{\delta_i}$	$1/n_{\delta_i}$	ϕ	$\phi/\sqrt{(n_{\delta_i}n_{\delta_j})}$	$\phi\sum_{j\in\delta_i} u_j/n_{\delta_i}$	τ/n_{δ_i}
Stern and Cressie	$(e_j/e_i)^{1/2}$	$1/e_i$	ϕ	ϕ	$\phi\sum_{j\in\delta_i} c_{ij}\cdot u_j$	τ/e_i
Cressie and Chan	$f\left(\frac{d_{ij}}{d_{\max}}\right)\left(\frac{n_{\delta_i}}{n_{\delta_j}}\right)^{1/2}$	$1/n_{\delta_i}$	ϕ	$\phi f\left(\frac{d_{ij}}{d_{\max}}\right)$	$\phi\sum_{j\in\delta_i} c_{ij}\cdot u_j$	τ/n_{δ_i}
ACAR Cressie (Cressie and Kapat, 2008)	$\left(\frac{n_{\delta_j}}{n_{\delta_i}}\right)^{1/2}$	$1/n_{\delta_i}$	ϕ	ϕ	$\phi\sum_{j\in\delta_i}\sqrt{n_{\delta_j}}\cdot u_j/\sqrt{n_{\delta_i}}$	τ/n_{δ_i}
HCAR Cressie (lattice)	1 if neighbor	1	ϕ	ϕ	$\phi\sum_{j\in\delta_i} u_j$	τ
Shared intrinsic (Lawson et al., 2015)	$n_{\delta_{ij}}/n_{\delta_i}$	$1/n_{\delta_i}$	1	$n_{\delta_{ij}}/\sqrt{n_{\delta_i}n_{\delta_j}}$	$\sum_{j\in\delta_i} n_{\delta_{ij}}\cdot u_j/n_{\delta_i}$	τ/n_{δ_i}
Shared proper	$n_{\delta_{ij}}/n_{\delta_i}$	$1/n_{\delta_i}$	ϕ	$\phi n_{\delta_{ij}}/\sqrt{n_{\delta_i}n_{\delta_j}}$	$\phi\sum_{j\in\delta_i} n_{\delta_{ij}}\cdot u_j/n_{\delta_i}$	τ/n_{δ_i}
GCM (Moraga and Lawson, 2012)	-	-	1	$n_{\delta_{ij}}/\sqrt{n_{\delta_i}n_{\delta_j}}$	$\sum_{j\in\delta_i} u_j/n_{\delta_i}$	τ/n_{δ_i}

Table D.1: Common assumptions for "no trend" case for CAR models

References

Abellan, J., S. Richardson, and N. Best (2008). Use of space-time models to investigate the stability of patterns of disease. *Environmental Health Perspectives 116*, 1111–1118. doi:10.1289/ehp.10814.

Agarwal, D., A. Gelfand, and S. Citron-Pousty (2002). Zero-inflated models with application to spatial count data. *Environmental and Ecological Statistics 9*, 341–355.

Al-Hadhrami, A. and A. B. Lawson (2011). Bayesian hierarchical modeling of latent period switching in small-area putative health hazard studies. *Statistical Methods in Medical Research 20*, 5–28.

Anderson, C., D. Lee, and N. Dean (2014). Identifying clusters in Bayesian disease mapping. *Biostatistics 15*, 457–469.

Anderson, C. and L. Ryan (2017). A comparison of spatio-temporal disease mapping approaches including an application to ischaemic heart disease in New South Wales, Australia. *International Journal of Environmental Research and Public Health 14*(2), 146.

Anderson, R. and R. May (1992). *Infectious Disease of Humans: Dynamics and Control*. New York: Oxford University Press.

Andersson, H. and T. Britton (2013). *Stochastic Epidemic Models and Their Statistical Analysis*. New York: Springer.

Anto, J. and J. Sunyer (1990). Epidemiological studies of asthma epidemics in Barcelona. *Chest 98*, 185–190.

Aregay, M., R. Carroll, A. B. Lawson, C. Faes, R. S. Kirby, and K. Watjou (2017). Zero inflated multiscale models for aggregated small area health data. *Environmetrics*. 29, 1, DOI: 10.1002/env.2477.

Aregay, M., A. B. Lawson, C. Faes, and R. Kirby (2014). Bayesian multiscale modeling for aggregated disease mapping data. In *Proceedings of Third ACM SIGSPATIAL International Workshop on the Use of GIS in Public Health*, New York, pp. 45–48. ACM.

Aregay, M., A. B. Lawson, C. Faes, R. Kirby, R. Carroll, and K. Watjou (2016a). Multiscale spatiotemporal models for aggregated small area health outcomes. *Environmetrics*. to appear.

Aregay, M., A. B. Lawson, C. Faes, R. S. Kirby, R. Carroll, and K. Watjou (2015). Impact of income on small area low birth weight incidence using multiscale models. *AIMS Public Health 2*, 667.

Aregay, M., A. B. Lawson, C. Faes, R. S. Kirby, R. Carroll, and K. Watjou (2016b). Multiscale measurement error models for aggregated small area health data. *Statistical Methods in Medical Research 25*(4), 1201–1223.

Aregay, M., A. B. Lawson, C. Faes, R. S. Kirby, R. Carroll, and K. Watjou (2016c). Spatial mixture multiscale modeling for aggregated health data. *Biometrical Journal 58*(5), 1091–1112.

Arya, S. P. (1998). *Air Pollution Metereology and Dispersion*. New York: Oxford University Press.

Azarian, T., S. Winn, S. Zaheer, J. Buehler, and R. Hopkins (2009). Utilization of syndromic surveillance with multiple data sources to enhance public health response. *Advances in Disease Surveillance 7*, 1–6.

Baddeley, A., J. Möller, and A. Pakes (2008). Properties of residuals for spatial point processes. *Annals of the Institute of Statistical Mathematics 60*, 627–649.

Baddeley, A. and R. Turner (2000). Practical maximum pseudolikelihood for spatial point patterns. *Australia and New Zealand Journal of Statistics 42*, 283–322.

Baddeley, A., R. Turner, J. Møller, and M. Hazelton (2005). Residual analysis for spatial point processes. *Journal of the Royal Statistical Society B 67*, 617–666.

Bailey, T. C., M. Carvalho, T. Lapa, W. Souza, and M. Brewer (2005). Modeling of under-detection of cases in disease surveillance. *Annals of Epidemiology 15*(5), 335–343.

Balakrishnan, S. and D. Madigan (2006). Algorithms for sparse linear classifiers in the massive data setting. *Journal of Machine Learning Research 1*.

Banerjee, S. (2016). Spatial survival models. In A. B. Lawson, S. Banerjee, R. Haining, and L. Ugarte (Eds.), *Handbook of Spatial Epidemiology*. New York: CRC Press.

Banerjee, S., B. Carlin, and A. Gelfand (2014). *Hierarchical Modelling and Analysis for Spatial Data* (2nd ed.). New York: CRC Press.

Banerjee, S. and B. P. Carlin (2003). Semiparametric spatio-temporal frailty modeling. *Environmetrics 14*, 523–535.

Banerjee, S., B. P. Carlin, and A. E. Gelfand (2004). *Hierarchical Modeling and Analysis for Spatial Data*. London: Chapman & Hall/CRC Press.

Banerjee, S., A. Gelfand, A. Finley, and H. Sang (2008). Gaussian predictive process models for large spatial data sets. *Journal of the Royal Statistical Society 70*, 825–848.

Banerjee, S., M. M. Wall, and B. P. Carlin (2003). Frailty modeling for spatially correlated survival data, with application to infant mortality in Minnesota. *Biostatistics 4*, 123–142.

Bankman, I. (Ed.) (2007). *Handbook of Medical Image Processing and Analysis* (2nd ed.), Volume 1. New York: Academic Press.

Barbieri, M. and J. Berger (2004). Optimal predictive model selection. *Annals of Statistics 32*, 870–897.

Barndorff-Nielsen, O. E., W. S. Kendall, and M. N. M. van Lieshout (1999). *Stochastic Geometry: Likelihood and Computation*. Boca Raton: Chapman & Hall.

Bastos, L. and D. Gamerman (2006). Dynamical survival models with spatial frailty. *Lifetime Data Analysis 12*, 441–460.

Becker, N. G. (1989). *Analysis of Infectious Disease Data*. Chapman & Hall.

Becker, N. G. (2015). *Modeling to Inform Infectious Disease Control*. New York: CRC Press.

Berman, M. and T. R. Turner (1992). Approximating point process likelihoods with GLIM. *Applied Statistics 41*, 31–38.

Bernardinelli, L., D. G. Clayton, C. Pascutto, C. Montomoli, M. Ghislandi, and M. Songini (1995). Bayesian analysis of space-time variation in disease risk. *Statistics in Medicine 14*, 2433–2443.

Bernardo, J. M. and A. F. M. Smith (1994). *Bayesian Theory*. New York: Wiley.

Berrocal, V. J., A. E. Gelfand, and D. M. Holland (2010). A bivariate space-time downscaler under space and time misalignment. *Annals of Applied Statistics 4*(4), 1942–1975.

Berzuini, C., N. G. Best, W. R. Gilks, and C. Larissa (1997). Dynamic conditional independence models and Markov chain Monte Carlo methods. *Journal of the American Statistical Association 92*, 1403–1412.

Besag, J. (1975). Statistical analysis of non-lattice data. *The Statistician 24*, 179–195.

Besag, J. (1986). On the statistical analysis of dirty pictures (with discussion). *Journal of the Royal Statistical Society, Series B 48*, 259–302.

Besag, J. and P. J. Green (1993). Spatial statistics and Bayesian computation. *Journal of the Royal Statistical Society, Series B 55*, 25–37.

Besag, J. and J. Newell (1991). The detection of clusters in rare diseases. *Journal of the Royal Statistical Society, Series A 154*, 143–155.

Besag, J. and J. Tantrum (2003). Likelihood analysis of binary data in space and time. In P. Green, N. Hjort, and S. Richardson (Eds.), *Highly Structured Stochastic Systems*, Chapter 9A, pp. 289–295. New York: Oxford University Press.

Besag, J., J. York, and A. Mollié (1991). Bayesian image restoration with two applications in spatial statistics. *Annals of the Institute of Statistical Mathematics 43*, 1–59.

Best, N., S. Richardson, and A. Thomson (2005). A comparison of Bayesian spatial models for disease mapping. *Statistical Methods in Medical Research 14*, 35–59.

Best, N. G. and J. C. Wakefield (1999). Accounting for inaccuracies in population counts and case registration in cancer mapping studies. *Journal of the Royal Statistical Society A 162*, 363–382.

Bishop, C. (2006). *Pattern Recognition and Machine Learning*. New York: Springer.

Blangiardo, M. and M. Cameletti (2015). *Spatial and Spatio-Temporal Bayesian Models with R-INLA*. New York: Wiley.

Blangiardo, M. and M. Cameletti (2016). Computational issues and R packages for spatial data analysis. In A. B. Lawson, S. Banerjee, R. Haining, and L. Ugart (Eds.), *Handbook of Spatial Epidemiology*, pp. 417–448. New York: CRC Press.

Blangiardo, M., M. Cameletti, G. Baio, and H. Rue (2013). Spatial and spatio-temporal models with R-INLA. *Spatial and Spatio-temporal Epidemiology 4*, 33–49.

Blangiardo, M., S. Richardson, J. Gulliver, and A. Hansell (2011). A Bayesian analysis of the impact of air pollution episodes on cardio-respiratory hospital admissions in the Greater London area. *Statistical Methods in Medical Research 20*, 69–80.

Bobb, J. F., L. Valeri, B. Claus Henn, D. C. Christiani, R. O. Wright, M. Mazumdar, J. J. Godleski, and B. A. Coull (2015). Bayesian kernel machine regression for estimating the health effects of multi-pollutant mixtures. *Biostatistics 16*(3), 493–508.

Boehning, D., E. Dietz, and P. Schlattmann (2000). Space-time mixture modelling of public health data. *Statistics in Medicine 19*, 2333–2344.

Boehning, D., E. Dietz, P. Schlattmann, L. Mendonca, and U. Kirchner (1999). The zero-inflated Poisson model and the decayed, missing and filled teeth index in dental epidemiology. *Journal of the Royal Statistical Society A 162*, 195–209.

Bottai, M., M. Geraci, and A. Lawson (2007). Testing for unusual aggregation of health risk in semiparametric models. *Statistics in Medicine.* DoI: 10/1002/sim.3126.

Bowman, A. and A. Azzalini (1997). *Applied Smoothing Techniques for Data Analysis: The Kernel Approach with S-Plus Illustrations*. New York: Oxford University Press.

Breslow, N. and D. G. Clayton (1993). Approximate inference in generalised linear mixed models. *Journal of the American Statistical Association 88*, 9–25.

Brix, A. and P. Diggle (2001). Spatio-temporal prediction for log-Gaussian Cox processes. *Journal of the Royal Statistical Society B 63*, 823–841.

Brookmeyer, R. (2006). Modeling an outbreak of anthrax. In R. Peck, G. Casella, G. Cobb, R. Hoerl, D. Nolan, R. Starbuck, and H. Stern (Eds.), *Statistics: A Guide to the Unknown* (4th ed.)., pp. 197–210. New York: Thomson.

Brooks, S., A. Gelman, G. Jones, and X. Meng (Eds.) (2011). *Handbook of Markov Chain Monte Carlo*. New York: CRC Press.

Brooks, S. and A. E. Gelman (1998). General methods for monitoring convergence of iterative simulations. *Journal of Computational and Graphical Statistics 7*, 434–455.

Brooks, S., P. Guidici, and A. Philippe (2003). Nonparametric convergence assessment for MCMC model selection. *Journal of Computational and Graphical Statistics 12*, 1–22.

Brooks, S. P. (1998). Quantitative convergence assessment for MCMC via Cusums. *Statistics and Computing 8*, 267–274.

Browne, W. J. (2012). MCMC estimation in MLwiN, v2.25. Technical report, Centre for Multilevel Modelling, University of Bristol.

Buckeridge, D., D. Owens, P. Switzer, J. Frank, and M. Musen (2006). Evaluating detection of an inhalational anthrax outbreak. *Emerging Infectious Diseases 12*, 1942–1949.

Buckland, S. T., D. R. Anderson, K. P. Burnham, J. L. Laake, D. L. Borchers, and L. Thomas (2001). *Introduction to Distance Sampling*. New York: Oxford University Press.

Burnham, K. and D. Anderson (2010). *Model Selection and Multimodel Inference* (2nd ed.). New York: Springer.

Byers, S. and A. Raftery (2002). Bayesian estimation and segmentation of spatial point processes using Voronoi tilings. In A. B. Lawson and D. Denison (Eds.), *Spatial Cluster Modelling*, Chapter 6. New York: CRC Press.

Cai, B. and D. Dunson (2008). Bayesian variable selection in generalised linear mixed models. In D. Dunson (Ed.), *Random Effect and Latent Variable Model Selection*, pp. 63–91. New York: Springer.

Cai, B., A. B. Lawson, M. M. Hossain, and J. Choi (2012). Bayesian latent structure models with space-time dependent covariates. *Statistical Modelling 12*, 145–164.

Cai, B., A. B. Lawson, M. M. Hossain, J. Choi, R. Kirby, and J. Liu (2013). Bayesian semi-parametric model with spatially-temporally varying coefficient selection. *Statistics in Medicine 32*, 3670–3685.

Calder, C. A. (2007). Dynamic factor process convolution models for multivariate space–time data with application to air quality assessment. *Environmental and Ecological Statistics 14*(3), 229–247.

Calder, C. A. (2008). A dynamic process convolution approach to modeling ambient particulate matter concentrations. *Environmetrics 19*(1), 39–48.

Calder, C. A., C. Holloman, and D. Higdon (2002). Exploring space-time structure in ozone concentration using a dynamic process convolution model. In *Bayesian Case Studies VI*. New York: Springer.

Caragea, P. and M. Kaiser (2006). Covariates and time in the autologistic model. Technical Report 2006-19, Iowa State University, Department of Statistics.

Carlin, B. P. and S. Banerjee (2003). Hierarchical multivariate CAR models for spatio-temporally correlated survival data. In J. B. Bernardo, D. A. P., J. O. Berger, and M. West (Eds.), *Bayesian Statistics 7*. Oxford: Oxford University Press.

Carlin, B. P. and J. Hodges (1999). Hierarchical proportional hazards regression models for highly stratified data. *Biometrics 55*, 1162–1170.

Carlin, B. P. and T. Louis (2000). *Bayes and Empirical Bayes Methods for Data Analysis* (2nd ed.). Boca Raton: Chapman & Hall/CRC Press.

Carlin, B. P. and T. Louis (2008). *Bayesian Methods for Data Analysis* (3rd ed.). New York: CRC Press.

Carroll, R., A. B. Lawson, C. Faes, R. Kirby, and M. Aregay (2015). Comparing INLA and OpenBUGS for hierarchical Poisson modeling in disease mapping. *Spatial and Spatio-temporal Epidemiology 14-15*, 45–54.

Carroll, R., A. B. Lawson, C. Faes, R. S. Kirby, M. Aregay, and K. Watjou (2015). Spatially-dependent Bayesian model selection for disease mapping. *Statistical Methods in Medical Research.* DOI 10.1177/0962280215627298 PMCID: 5374035.

Carroll, R., A. B. Lawson, C. Faes, R. S. Kirby, M. Aregay, and K. Watjou (2016). Spatio-temporal Bayesian model selection for disease mapping. *Environmetrics 27*(8), 466–478. doi: 10.1002/env.2410.

Carroll, R., D. Ruppert, L. Stefanski, and C. Crainiceanu (2006). *Measurement Error in Nonlinear Models* (2nd ed.). New York: Chapman & Hall.

Cassella, G. and E. I. George (1992). Explaining the Gibbs sampler. *The American Statistician 46*, 167–174.

Celeux, G., F. Forbes, C. Robert, and M. Titterington (2006). Deviance information criteria for missing data models. *Bayesian Analysis 1*, 651–674.

Chaix, B., M. Rosvall, J. Lynch, and J. Merlo (2006). Disentangling contextual effects on cause specific mortality in a longitudinal 23-year follow-up study: impact of population density or socio-economic environment? *International Journal of Epidemiology 35*, 633–643.

Chan, T. C., C. C. King, M. Y. Yen, P. Chiang, C. S. Huang, and C. K. Hsiao (2010). Probabilistic daily ILI syndromic surveillance with a spatio-temporal Bayesian hierarchical model. *Plos One 5(7).* e11626.

Chang, H. (2016). Data assimilation for environemental pollution fields. In A. B. Lawson, S. Banerjee, R. Haining, and L. Ugart (Eds.), *Handbook of Spatial Epidemiology,* pp. 289–302. New York: CRC Press.

Charras-Garrido, M., D. Abrial, and J. de Goer (2012). Classification method for disease risk mapping based on discrete hidden Markov random fields. *Biostatistics 13*, 241–255.

Chen, C., J. Wakefield, and T. Lumley (2014). The use of sample weights in Bayesian hierarchical models for small area estimation. *Spatial and Spatio-temporal Epidemiology 11*, 33–43.

Chen, M., Q. Shao, and J. Ibrahim (2000). *Monte Carlo Methods in Bayesian Computation.* New York: Springer.

Chen, R., R. Connelly, and N. Mantel (1993). Analysing post-alarm data in a monitoring system, in order to accept or reject the alarm. *Statistics in Medicine 12*, 1807–1812.

Chen, R., J. Iscovich, and U. Goldbourt (1997). Clustering of leukaemia cases in a city in Israel. *Statistics in Medicine 16*, 1873–1887.

Choi, J. and A. B. Lawson (2011). Evaluation of Bayesian spatial-temporal latent models in small area health data. *Environmetrics 22*(8), 1008–1022.

Choi, J. and A. B. Lawson (2016). Bayesian spatially-dependent variable selection for small area health modeling. *Statistical Methods in Medical Research*. DOI: 10.1177/0962280215627184.

Chopin, N. (2002). A sequential particle filter method for static models. *Biometrika 89*, 539–552.

Chung, D. and S. Keles (2010). Sparse partial least squares classification for high dimensional data. *Statistical Applications in Genetics and Molecular Biology 9*(1). doi:10.2202/1544-6115.1492.

Clark, A. B. and A. B. Lawson (2002). Spatio-temporal cluster modelling of small area health data. In A. B. Lawson and D. Denison (Eds.), *Spatial Cluster Modelling*, Chapter 14, pp. 235–258. New York: CRC Press.

Clayton, D. G. and L. Bernardinelli (1992). Bayesian methods for mapping disease risk. In P. Elliott, J. Cuzick, D. English, and R. Stern (Eds.), *Geographical and Environmental Epidemiology: Methods for Small-Area Studies*. Oxford: Oxford University Press.

Clayton, D. G., L. Bernardinelli, and C. Montomoli (1993). Spatial correlation in ecological analysis. *International Journal of Epidemiology 22*, 1193–1202.

Clayton, D. G. and J. Kaldor (1987). Empirical Bayes estimates of age-standardised relative risks for use in disease mapping. *Biometrics 43*, 671–691.

Cliff, A. D. and J. K. Ord (1981). *Spatial Processes: Models and Applications*. London: Pion.

Congdon, P. (2003). *Applied Bayesian Modelling*. New York: Wiley.

Congdon, P. (2005). *Bayesian Models for Categorical Data*. New York: Wiley.

Congdon, P. (2007). *Bayesian Statistical Modelling* (2nd ed.). New York: Wiley.

Congdon, P. (2010). *Applied Bayesian Hierarchical Methods*. New York: CRC Press.

Congdon, P. (2014). *Applied Bayesian Modelling* (2nd ed.). New York: Wiley.

Congdon, P., M. Almog, S. Curtis, and R. Ellerman (2007). A spatial structural equation modelling framework for health count responses. *Statistics in Medicine 26*, 5267–5284.

Cook, R. and J. Lawless (2007). *The Statistical Analysis of Recurrent Events*. New York: Springer.

Cooner, F., S. Banerjee, and M. McBean (2006). Modelling geographically referenced survival data with a cure fraction. *Statistical Methods in Medical research 15*(4), 307–324.

Corberan, A., A. B. Lawson, and R. Boaz (2017). Integration of animal health and public health surveillance sources to exhaustively inform the risk of zoonosis: An application to leishmaniasis data in Brazil. Consultation Report to Ministry of Health in Brazil, authorized by PAHO.

Corberán-Vallet, A. (2012). Prospective surveillance of multivariate spatial disease data. *Statistical Methods in Medical Research 21*(5), 457–477. DOI:10.1177/0962280212446319.

Corberán-Vallet, A. and A. B. Lawson (2011). Conditional predictive inference for online surveillance of spatial disease incidence. *Statistics in Medicine 30*, 3095–3116.

Corberán-Vallet, A. and A. B. Lawson (2016). Spatial health surveillance. In A. B. Lawson, S. Banerjee, R. Haining, and L. Ugart (Eds.), *Handbook of Spatial Epidemiology*, pp. 501–520. New York: CRC Press.

Costain, D. (2009). Bayesian partitioning for modeling and mapping spatial case-control data. *Biometrics 65*, 1123–1132.

Cowles, M. K. (2004). Review of WinBUGS 1.4. *The American Statistician 58*, 330–336.

Cressie, N. A. C. (1993). *Statistics for Spatial Data* (revised ed.). New York: Wiley.

Cressie, N. A. C. (1996). Change of support and the modifiable areal unit problem. *Geographical Systems 2*, 83–101.

Cressie, N. A. C. and P. Kapat (2008). Some diagnostics for Markov random fields. *Journal of Computational and Graphical Statistics 17*, 726–749.

Cressie, N. A. C. and A. Mugglin (2000). Spatio-temporal hierarchical modelling of an infectious disease from (simulated) count data. In J. Bethlehem and P. van der Heijden (Eds.), *Compstat 2000*. Heidelberg: Physica Verlag.

Cressie, N. A. C., S. Richardson, and I. Jaussent (2004). Ecological bias: use of maximum-entropy approximations. *Australia–New Zealand Journal of Statistics 46*, 233–255.

Cressie, N. A. C. and C. Wikle (2011). *Statistics for Spatio-Temporal Data*. New York: Wiley.

Cressie, N. C. and A. Lawson (2000). Hierarchical probability models and Bayesian analysis of mine locations. *Advances in Applied Probability 32*(2), 315–330.

Cromley, E. and S. McLafferty (2011). *GIS and Public Health* (2nd ed.). New York: Guildford Press.

Dabney, A. and J. Wakefield (2005). Issues in the mapping of two diseases. *Statistical Methods in Medical Research 14*, 83–112.

Dagne, G. A. (2004). Hierarchical Bayesian analysis of correlated zero-inflated count data. *Biometrical Journal 46*(6), 653–663.

Daley, D. and J. Gani (1999). *Epidemic Modelling: An Introduction.* New York: Cambridge University Press.

Davison, A. C. and D. V. Hinkley (1997). *Bootstrap Methods and their Application.* London: Cambridge University Press.

Deardon, R., S. Brooks, B. Grenfell, M. Keeling, M. Tildesley, N. Savill, D. Shaw, and M. Woolhouse (2010). Inference for individual-level models of infectious diseases in large populations. *Statistica Sinica 20*, 239–261.

Dellaportas, P., J. Forster, and I. Ntzoufras (2002). On Bayesian model and variable selection using MCMC. *Statistics and Computing 12*, 27–36.

Denison, D., N. Adams, C. Holmes, and D. Hand (2002). Bayesian partition modelling. *Computational Statistics and Data Analysis 38*, 475–485.

Denison, D. and C. Holmes (2001). Bayesian partitioning for estimating disease risk. *Biometrics 57*, 143–149.

Denison, D., C. Holmes, B. Mallick, and A. Smith (2002). *Bayesian Methods for Nonlinear Classification and Regression.* New York: Wiley.

Dey, D., M-H. Chen, and H. Chang (1997). Bayesian approach for nonlinear random effects models. *Biometrics 53*, 1239–1252.

Diggle, P. (2007). Spatio-temporal point processes: Methods and applications. In B. Finkenstadt, L. Held, and V. Isham (Eds.), *Statistical Methods for Spatio-Temporal Systems*, Chapter 1, pp. 1–45. London: CRC Press.

Diggle, P. and P. Elliott (1995). Statistical issues in the analysis of disease risk near point sources using individual or spatially aggregated data. *Journal of Epidemiology and Community Health 49*, s20–s27.

Diggle, P., T. Fiksel, Y. Ogata, D. Stoyan, and M. Tanemura (1994). On parameter estimation for pairwise interaction processes. *International Statistical Review 62*, 99–117.

Diggle, P. and E. Gabriel (2010). Spatio-temporal point processes. In A.Gelfand, P. Diggle, M. Fuentes, and P. Guttorp (Eds.), *Handbook of Spatial Statistics*, Chapter 25. New York: CRC Press.

Diggle, P., P. Heagerty, K.-Y. Liang, and S. Zeger (2002). *Analysis of Longitudinal Data* (2nd ed.). New York: Oxford University Press.

Diggle, P., L. Knorr-Held, B. Rowlingson, T. Su, P. Hawtin, and T. Bryant (2004). On-line monitoring of public health surveillance data. In R. Brookmeyer and D. Stroup (Eds.), *Monitoring the Health of Populations: Statistical Principles and Methods for Public Health Surveillance*, Chapter 9. London: Oxford University Press.

Diggle, P., S. Morris, P. Elliott, and G. Shaddick (1997). Regression modelling of disease risk in relation to point sources. *Journal of the Royal Statistical Society, A 160*, 491–505.

Diggle, P. and P. Ribeiro Jr. (2007). *Model-based Geostatistics*. New York: Springer.

Diggle, P. and B. Rowlingson (1994). A conditional approach to point process modelling of elevated risk. *Journal of the Royal Statistical Society A 157*, 433–440.

Diggle, P., P. Zheng, and P. Durr (2005). Nonparametric estimation of spatial segregation in a multivariate point process: bovine tuberculosis in Cornwall, UK. *Journal of the Royal Statistical Society C 54*, 645–658.

Diggle, P. J. (1985). A kernel method for smoothing point process data. *Journal of the Royal Statistical Society C 34*, 138–147.

Diggle, P. J. (1990). A point process modelling approach to raised incidence of a rare phenomenon in the vicinity of a prespecified point. *Journal of the Royal Statistical Society A 153*, 349–362.

Diggle, P. J. (2003). *Statistical Analysis of Spatial Point Patterns* (2nd ed.). London: Arnold.

Diggle, P. J. (2005). Spatio-temporal point processes, partial likelihood, foot and mouth disease. *Statistical Methods in Medical Research 15*, 325–336.

Diggle, P. J., S. Morris, and J. C. Wakefield (2000). Point-source modelling using matched case-control data. *Biostatistics 1*, 1–17.

Diggle, P. J., J. Tawn, and R. Moyeed (1998). Model-based geostatistics. *Journal of the Royal Statistical Society C 47*, 299–350.

Diva, U., D. K. Dey, and S. Banerjee (2008). Parametric models for spatially correlated survival data for individuals with multiple cancers. *Statistics in Medicine 27*(12), 2127–2144.

Doucet, A., N. deFreitas, and N. Gordon (Eds.) (2001). *Sequential Monte Carlo Methods in Practice*. New York: Springer.

Duan, J., M. Guidani, and A. Gelfand (2007). Generalized spatial Dirichlet process models. *Biometrika 94*, 809–825.

Durr, P. and A. Gatrell (Eds.) (2004). *GIS and Spatial Analysis in Veterinary Science*. London: Oxford University Press.

Durr, P., N. Tait, and A. B. Lawson (2005). Bayesian hierarchical modelling to enhance the epidemiological value of abattoir surveys for bovine fasciolosis. *Preventive Veterinary Medicine 71*, 157–172.

Eberley, L. and B. P. Carlin (2000). Identifiability and convergence issues for Markov chain Monte Carlo fitting of spatial models. *Statistics in Medicine 19*, 2279–2294.

Elliott, P., J. Cuzick, D. English, and R. Stern (Eds.) (1992). *Geographical and Environmental Epidemiology:Methods for Small-Area Studies*. New York: Oxford University Press.

Elliott, P., J. C. Wakefield, N. G. Best, and D. J. Briggs (Eds.) (2000). *Spatial Epidemiology: Methods and Applications*. London: Oxford University Press.

Esman, N. A. and G. M. Marsh (1996). Applications and limitations of air dispersion modeling in environmental epidemiology. *Journal of Exposure Analysis and Environmental Epidemiology 6*, 339–353.

Fahrmeir, L. and S. Lang (2001). Bayesian inference for generalized additive mixed models based on Markov random field priors. *Journal of the Royal Statistical Society 50*, 201–220.

Faraway, J. (2006). *Extending the Linear Model with R*. New York: Chapman & Hall/CRC Press.

Farnsworth, M., J. Hoeting, N. Hobbs, and M. Miller (2006). Linking chronic wasting disease to mule deer movement: a hierarchical Bayesian approach. *Ecological Applications 16*, 1026–1036.

Fernandez, C. and P. Green (2002). Modelling spatially correlated data via mixtures: a Bayesian approach. *Journal of the Royal Statistical Society B 64*, 805–826.

Ferreira, J., D. Denison, and C. Holmes (2002). Partition modelling. In A. B. Lawson and D. Denison (Eds.), *Spatial Cluster Modelling*, Chapter 7, pp. 125–145. New York: CRC Press.

Fonseca, T. C. O. and M. A. R. Ferreira (2017). Dynamic multiscale spatiotemporal models for Poisson data. *Journal of the American Statistical Association 112*(517), 215–234.

Forbes, F., M. Charras-Garrido, L. Azizi, S. Doyle, and D. Abrial (2013). Spatial risk mapping for rare disease with hidden Markov fields and variational EM. *Annals of Applied Statistics 7*, 1192–1216.

Fotheringham, A. S., C. Brunsdon, and M. Charlton (2002). *Geographically Weighted Regression: The Analysis of Spatially Varying Relationships*. New York: Wiley.

French, J. and M. Wand (2004). Generalized additive models for cancer mapping with incomplete covariates. *Biostatistics 5*, 177–191.

Freni-Sterrantino, A., M. Ventrucci, and H. Rue (2017). A note on intrinsic conditional autoregressive models for disconnected graphs. *ArXiv e-prints*. arXiv:1705.04854 [stat.ME].

Fricker Jr, R. (2011). Some methodological issues in biosurveillance. *Statistics in Medicine 30*, 403–415.

Fricker Jr, R. (2013). *Introduction to Statistical Methods for Biosurveillance: With an Emphasis on Syndromic Surveillance*. New York: Cambridge University Press.

Frisen, M. (1992). Evaluations of methods for statistical surveillance. *Statistics in Medicine 11*, 1489–1502.

Frisen, M. and J. D. Mare (1991). Optimal surveillance. *Biometrika 78*, 271–280.

Frisén, M. and C. Sonesson (2005). Optimal surveillance. In A. B. Lawson and K. Kleinman (Eds.), *Spatial and Syndromic Surveillance for Public Health*, Chapter 3. New York: Wiley.

Gamerman, D. and H. Lopes (2006). *Markov Chain Monte Carlo: Stochastic Simulation for Bayesian Inference*. New York: CRC Press.

Gangnon, R. (2006). Impact of prior choice on local Bayes factors for cluster detection. *Statistics in Medicine 25*, 883–895.

Gangnon, R. and M. Clayton (2000). Bayesian detection and modeling of spatial disease clustering. *Biometrics 56*, 922–935.

Gardner, M. J. (1989). Review of reported increases of childhood cancer rates in the vicinity of nuclear installations in the UK. *Journal of the Royal Statistical Society 152*, 307–325.

Gelfand, A. (2010). Misaligned spatial data: The change of support problem. In A.Gelfand, P. Diggle, M. Fuentes, and P. Guttorp (Eds.), *Handbook of Spatial Statistics*, Chapter 29. New York: CRC Press.

Gelfand, A. and D. Dey (1994). Bayesian model choice: Asymptotics and exact calculations. *Journal of the Royal Statistical Society B 56*, 501–514.

Gelfand, A., P. Diggle, M. Fuentes, and P. Guttorp (Eds.) (2010). *Handbook of Spatial Statistics*. New York: CRC Press.

Gelfand, A. and S. Ghosh (1998). Model choice: A minimum posterior predictive loss approach. *Biometrika 85*, 1–11.

Gelfand, A., A. Kottas, and S. MacEachern (2005). Bayesian nonparametric spatial modeling with Dirichlet process mixing. *Journal of the American Statistical Association 100*, 1021–1035.

Gelfand, A. and P. Vounatsou (2003). Proper multivariate conditional autoregressive models for spatial data. *Biostatistics 4*, 11–25.

Gelfand, A. E. and S. Banerjee (2015). Bayesian wombling: finding rapid change in spatial maps. *Wiley Interdisciplinary Reviews: Computational Statistics* 7(5), 307–315.

Gelman, A. (2006). Prior distributions for variance parameters in hierarchical models. *Bayesian Analysis 1*, 515–533.

Gelman, A., J. B. Carlin, H. S. Stern, and D. Rubin (2004). *Bayesian Data Analysis*. London: Chapman & HAll/CRC Press.

Gelman, A. and J. Hill (2007). *Data Analysis Using Regression and Multilevel-Hierarchical Models*. New York: Cambridge University Press.

Gelman, A., J. Huang, and A. Vehtari (2014). Understanding predictive information criteria in Bayesian models. *Statistics and Computing 24*, 997–1016.

Gelman, A. and K. Shirley (2011). Inference and monitoring convergence. In S. Brooks, A. Gelman, G. Jones, and X. Meng (Eds.), *Handbook of Markov Chain Monte Carlo*, Chapter 6. New York: CRC Press.

Gelman, A. E. and D. Rubin (1992). Inference from iterative simulation using multiple sequences (with discussion). *Statistical Science 7*, 457–511.

Geweke, J. (1992). Evaluating the accuracy of sampling -based approaches to the calculation of posterior moments. In J. Bernardo, J. Berger, A. Dawid, and A. Smith (Eds.), *Bayesian Statistics 4*. Oxford University Press.

Geyer, C. (2011). Importance sampling, simulated tempering, and umbrella sampling. In S. Brooks, A. Gelman, G. Jones, and X. Meng (Eds.), *Handbook of Markov Chain Monte Carlo*, Chapter 11. New York: CRC Press.

Ghosh, M. and J. N. K. Rao (1994). Small area estimation: an appraisal. *Statistical Science 9*, 55–93.

Ghosh, S., P. Mukhopadhyay, and J.-C. Lu (2006). Bayesian analysis of zero-inflated regression models. *Journal of Statistical Planning and Inference 136*, 1360–1375.

Gilks, W. R., D. G. Clayton, D. J. Spiegelhalter, N. G. Best, A. J. McNeil, L. D. Sharples, and A. J. Kirby (1993). Modelling complexity: Applications of Gibbs sampling in medicine. *Journal of the Royal Statistical Society B 55*, 39–52.

Gilks, W. R., S. Richardson, and D. J. Spiegelhalter (Eds.) (1996). *Markov Chain Monte Carlo in Practice*. London: Chapman & Hall.

Gneiting, T., M. Genton, and P. Guttorp (2007). Geostatistical space-time models, stationarity, separability, and full symmetry. In B. Finkenstadt, L. Held, and V. Isham (Eds.), *Statistical Methods for Spatio-Temporal Systems*, Chapter 4, pp. 151–175. London: CRC Press.

Godsill, S. (2001). On the relationship between Markov chain Monte Carlo methods for model uncertainty. *Journal of Computational and Graphical Statistics 10*, 230–248.

Goldstein, H. and A. Leyland (2001). Further topics in multilevel modelling: Context and composition. In A. Leyland and H. Goldstein (Eds.), *Multilevel Modelling of Health Statistics*, Chapter 12.4, pp. 181–186. New York: Wiley.

Greco, F., A. B. Lawson, D. Cocchi, and T. Temples (2005). Voronoi tesselation and other methods for misaligned small area health data. *Environmental and Ecological Statistics 12*, 379–395.

Green, P. J. (1995). Reversible jump MCMC computation and Bayesian model determination. *Biometrika 82*, 711–732.

Green, P. J. and S. Richardson (2002). Hidden Markov models and disease mapping. *Journal of the American Statistical Association 97*, 1055–1070.

Greenland, S. (1992). Divergent biases in ecologic and individual-level studies. *Statistics in Medicine 11*, 1209–1223.

Greenland, S. and J. Robins (1994). Ecologic studies: biases, misconceptions and counterexamples. *American Journal of Sociology 139*, 747–759.

Griffin, J. and M. Steel (2006). Order-based dependent Dirichlet processes. *Journal of the American Statistical Association 101*, 179–194.

Grigg, O. and V. Farewell (2004). A risk-adjusted sets method for monitoring adverse medical outcomes. *Statistics in Medicine 23*, 1593–1602.

Grigg, O. and D. Spiegelhalter (2007). A simple risk-adjusted exponentially weighted moving average. *Journal of the American Statistical Association 102*, 140–152.

Guo, X. and B. P. Carlin (2004). Separate and joint modeling of longitudinal and event time data using standard computer packages. *The American Statistician 58*(1), 16–24.

Gustafson, P. (1998). Flexible Bayesian modelling for survival data. *Lifetime Data Analysis 4*, 281–299.

Gustafson, P. (2004). *Measurement Error and Misclassification in Statistics and Epidemiology*. London: Chapman & Hall.

Gutmann, M. U. and J. Corander (2016). Bayesian optimization for likelihood-free inference of simulator-based statistical models. *Journal of Machine Learning Research 17*(125), 1–47.

Heagerty, P. and S. Lele (1998). A composite likelihood approach to binary spatial data. *Journal of the American Statistical Association 93*, 1099–1111.

Heaton, M., D. L. Banks, J. Zou, A. Karr, G. Datta, J. Lynch, and F. Verae (2012). A spatio-temporal absorbing state model for disease and syndromic surveillance. *Statistics in Medicine*. DOI:10.1002/sim.5350.

Hegarty, A. and D. Barry (2008). Bayesian disease mapping using product partition models. *Statistics in Medicine 27*(19), 3868–3893.

Heisterkamp, S., G. Doornbos, and N. Nagelkerke (2000). Assessing health impact of environmental pollution sources using space-time models. *Statistics in Medicine 19*, 2569–2578.

Held, L., L. Hofmann, M. Hohle, and V. Schmid (2006). A two-component model for counts of infectious diseases. *Biostatistics 7*, 422–437.

Held, L., I. Natario, S. Fenton, H. Rue, and N. Becker (2005). Towards joint disease mapping. *Statistical Methods in Medical Research 14*, 61–82.

Henderson, R., S. Shimakura, and D. Gorst (2002). Modeling spatial variation in leukaemia survival data. *Journal of the American Statistical Association 97*, 965–972.

Higdon, D. (2002). Space and space-time modeling using process convolutions. In C. Anderson (Ed.), *Quantitative Methods for Current Environmental Issues*. London: Springer.

Hodges, J. S. and B. Reich (2010). Adding spatially correlated errors can mess up the fixed effect you love. *The American Statistician 64*, 325–334.

Hoffman, M. D. and A. Gelman (2014). The no-u-turn sampler: Adaptively setting path lengths in Hamiltonian Monte Carlo. *Journal of Machine Learning Research 15*, 1351–1381.

Höhle, M., E. Jorgensen, and P. O'Neill (2005). Inference in disease transmission experiments by using stochastic epidemic models. *Journal of the Royal Statistical Society C 54*, 349–366.

Hossain, M. and A. Lawson (2008). Approximate methods in Bayesian point process spatial models. *Computational Statistics and Data Analysis 53*, 2831–2842.

Hossain, M. and A. B. Lawson (2005). Local likelihood disease clustering: Development and evaluation. *Environmental and Ecological Statistics 12*, 259–273.

Hossain, M. and A. B. Lawson (2006). Cluster detection diagnostics for small area health data with reference to evaluation of local likelihood models. *Statistics in Medicine 25*, 771–786.

Hossain, M. and A. B. Lawson (2016). Mixtures and latent structure in spatial epidemiology. In A. B. Lawson, S. Banerjee, R. Haining, and L. Ugart (Eds.), *Handbook of Spatial Epidemiology*, pp. 349–362. New York: CRC Press.

Hossain, M., A. B. Lawson, B. Cai, J. Choi, J. Liu, and R. Kirby (2012). Space-time stick-breaking processes for small area disease cluster estimation. *Environmental and Ecological Statistics*, 1–17.

Hossain, M. M. and A. B. Lawson (2010). Space-time Bayesian small area disease risk models: Development and evaluation with a focus on cluster detection. *Environmental and Ecological Statistics 17*, 73–95.

Ibrahaim, J., M. Chen, and D. Sinha (2000). *Bayesian Survival Analysis*. New York: Springer.

Illian, J., H. Sorbye, and H. Rue (2012). A toolbox for fitting complex spatial point process models using integrated nested Laplace approximation (INLA). *Annals of Applied Statistics 6*(4), 1499–1530.

Inskip, H., V. Beral, P. Fraser, and P. Haskey (1983). Methods for age-adjustment of rates. *Statistics in Medicine 2*, 483–493.

Isham, V. and G. Medley (Eds.) (1996). *Models for Infectious Human Diseases*. Cambridge: Cambridge University Press.

Ishwaran, H. and L. James (2001). Gibbs sampling methods for stick-breaking priors. *Journal of the American Statistical Association 96*, 161–173.

Ishwaran, H. and L. James (2002). Approximate Dirichlet process computing in finite normal mixtures: Smoothing and prior information. *Journal of Computational and Graphical Statistics 11*, 1–26.

Jin, X., S. Banerjee, and B. Carlin (2008). Order-free coregionalized areal data models with application to multiple disease mapping. *Journal of the Royal Statistical Society B*, to appear.

Jin, X., B. Carlin, and S. Banerjee (2005). Generalized hierarchical multivariate CAR models for areal data. *Biometrics 61*, 950–961.

Jung, I. (2016). Scan tests. In A. B. Lawson, S. Banerjee, R. Haining, and L. Ugart (Eds.), *Handbook of Spatial Epidemiology*, pp. 179–194. New York: CRC Press.

Katz, R., L. May, J. Baker, and E. Test (2011). Redefining syndromic surveillance. *Journal of Epidemiology and Global Health 1*, 21–31.

Kauermann, G. and J. D. Opsomer (2003). Local likelihood estimation in generalized additive models. *Scandinavian Journal of Statistics 30*, 317–337.

Keeling, M. and P. Rohani (2007). *Modeling Infectious Diseases in Humans and Animals*. New York: Princeton University Press.

Kelsall, J. and P. Diggle (1998). Spatial variation in risk of disease: a nonparametric binary regression approach. *Applied Statistics 47*, 559–573.

Kelsall, J. and J. Wakefield (2002). Modelling spatial variation in disease risk: A geostatistical approach. *Journal of the American Statitsical Association 97*, 692–701.

Khan, D., L. M. Rossen, B. Hamilton, E. Dienes, Y. He, and R. Wei (2017). Spatiotemporal trends in teen birth rates in the USA, 2003–2012. *Journal of the Royal Statistical Society: A (Statistics in Society)*, n/a–n/a.

Kim, H., D. Sun, and R. Tsutakawa (2001). A bivariate Bayes method for improving the estimates of mortality rates with a twofold autoregressive model. *Journal of the American Statistical Association 96*, 1506–1521.

Kim, S., M. Tadesse, and M. Vannucci (2006). Variable selection in clustering via Dirichlet process mixture models. *Biometrika 93*, 877–893.

Kipnis, V., R. J. Carroll, L. S. Fredman, and L. Li (1999). A new dietary measurement error model and its application to the estimation of relative risk: Application to four validation studies. *American Journal of Epidemiology 150*, 642–651.

Kipnis, V., D. Midthune, L. S. Freedman, S. Bingham, A. Schatzkin, A. Subar, and R. J. Carroll (2001). Empirical evidence of correlated biases in dietary assessment instruments and its implications. *American Journal of Epidemiology 153*, 394–403.

Kipnis, V., A. Subar, D. Midthune, L. S. Freedman, L. Ballard-Barbash, R. Troiano, S. Bingham, D. Schoeller, A. Schatzkin, and R. J. Carroll (2003). The structure of dietary measurement error: Results of the open biomarker study. *American Journal of Epidemiology 158*, 14–21.

Kirkpatrick, S., C. D. Gelatt, and M. P. Vecchi (1983). Optimization by simulated annealing. *Science 220*(4598), 671–680.

Knorr-Held, L. (2000). Bayesian modelling of inseparable space-time variation in disease risk. *Statistics in Medicine 19*, 2555–2567.

Knorr-Held, L. and J. Besag (1998). Modelling risk from a disease in time and space. *Statistics in Medicine 17*, 2045–2060.

Knorr-Held, L. and N. G. Best (2001). A shared component model for detecting joint and selective clustering of two diseases. *Journal of the Royal Statistical Society 164*, 73–85.

Knorr-Held, L. and G. Rasser (2000). Bayesian detection of clusters and discontinuities in disease maps. *Biometrics 56*, 13–21.

Knorr-Held, L. and S. Richardson (2003). A hierarchical model for space-time surveillance data on meningococcal disease incidence. *Journal of the Royal Statistical Society 52*, 169–183.

Koenker, R. (2005). *Quantile Regression*. New York: Cambridge University Press.

Koenker, R. and G. S. Bassett (1978). Regression quantiles. *Econometrica 46*, 33–50.

Kottas, A. (2016). Bayesian nonparametric modeling for disease incidence data. In A. B. Lawson, S. Banerjee, R. Haining, and L. Ugart (Eds.), *Handbook of Spatial Epidemiology*, pp. 363–374. New York: CRC Press.

Kottas, A., J. Duan, and A. Gelfand (2008). Modelling disease incidence data with spatial and spatio-temporal Dirichlet process mixtures. *Biometrical Journal 50*(2), 29–42.

Kulldorff, M. and N. Nagarwalla (1995). Spatial disease clusters: detection and inference. *Statistics in Medicine 14*, 799–810.

Kunsch, H. (1987). Intrinsic autoregressions and related models on the two-dimensional lattice. *Biometrika 74*, 517–524.

Kuo, L. and B. Mallick (1998). Variable selection for regression models. *Sankhya 60*, 65–81.

Lachin, J. L. (2000). Statistical considerations in the intent-to-treat principle. *Controlled Clinical Trials 21*, 167–189.

Lambert, D. (1992). Zero-inflated Poisson regression, with application to defects in manufacturing. *Technometrics 34*, 1–14.

Lambert, P., A. Sutton, P. Burton, K. Abrams, and P. Jones (2005). How vague is vague? A simulation study of the impact of the use of vague prior distributions in MCMC using WinBUGS. *Statistics in Medicine 24*, 2401–2428.

Lawson, A. and J. Choi (2016). Spatiotemporal disease mapping. In A. B. Lawson, S. Banerjee, R. Haining, and L. Ugart (Eds.), *Handbook of Spatial Epidemiology*, pp. 335–348. New York: CRC Press.

Lawson, A. B. (1992a). GLIM and normalising constant models in spatial and directional data analysis. *Computational Statistics and Data Analysis 13*, 331–348.

Lawson, A. B. (1992b). On fitting non-stationary Markov point processes on GLIM. In Y. Dodge and J. Whittacker (Eds.), *Computational Statistics I*. Heidelberg: Physica Verlag.

Lawson, A. B. (1993a). A deviance residual for heterogeneous spatial Poisson processes. *Biometrics 49*, 889–897.

Lawson, A. B. (1993b). On the analysis of mortality events around a prespecified fixed point. *Journal of the Royal Statistical Society A 156*, 363–377.

Lawson, A. B. (1995). Markov chain Monte Carlo methods for putative pollution source problems in environmental epidemiology. *Statistics in Medicine 14*, 2473–2486.

Lawson, A. B. (1996). Use of deprivation indices in small area studies: Letter. *Journal of Epidemiology and Community Health 50*, 689–690.

Lawson, A. B. (2000). Cluster modelling of disease incidence via RJMCMC methods: a comparative evaluation. *Statistics in Medicine 19*, 2361–2376.

Lawson, A. B. (2001). *Statistical Methods in Spatial Epidemiology*. New York: Wiley.

Lawson, A. B. (2002). The analysis of putative sources of health hazard. In D. Zimmerman (Ed.), *Encyclopedia of Life Support Systems*. Paris: UNESCO.

Lawson, A. B. (2004). Some issues in the spatio-temporal analysis of public health surveillance data. In R. Brookmeyer and D. Stroup (Eds.), *Monitoring the Health of Populations: Statistical Principles and Methods for Public Health Surveillance*, Chapter 11, pp. 289–314. New York: Oxford University Press.

Lawson, A. B. (2005). Spatial and spatio-temporal disease analysis. In A. B. Lawson and K. Kleinman (Eds.), *Spatial and Syndromic Surveillance for Public Health*, Chapter 4. New York: Wiley.

Lawson, A. B. (2006a). Disease cluster detection: a critique and a Bayesian proposal. *Statistics in Medicine 25*, 897–916.

Lawson, A. B. (2006b). *Statistical Methods in Spatial Epidemiology* (2nd ed.). New York: Wiley.

Lawson, A. B. (2012). Bayesian point event modeling in spatial and environmental epidemiology: a review. *Statistical Methods in Medical Research 21*, 509–529.

Lawson, A. B. (2013). *Bayesian Disease Mapping: Hierarchical Modeling in Spatial Epidemiology* (2nd ed.). New York: CRC Press.

Lawson, A. B., S. Banerjee, R. Haining, and L. Ugarte (Eds.) (2016). *Handbook of Spatial Epidemiology*. New York: CRC Press.

Lawson, A. B., A. Biggeri, D. Boehning, E. Lesaffre, J.-F. Viel, A. Clark, P. Schlattmann, and F. Divino (2000). Disease mapping models: An empirical evaluation. *Statistics in Medicine 19*, 2217–2242. Special issue.

Lawson, A. B., A. Biggeri, and E. Dreassi (1999). Edge effects in disease mapping. In A. B. Lawson, D. B'ohning, E. Lesaffre, A. Biggeri, J.-F. Viel, and R. Bertollini (Eds.), *Disease Mapping and Risk Assessment for Public Health*, Chapter 6, pp. 85–97. Chichester: Wiley.

Lawson, A. B., D. Böhning, E. Lessafre, A. Biggeri, J. F. Viel, and R. Bertollini (Eds.) (1999). *Disease Mapping and Risk Assessment for Public Health.* Chichester: Wiley.

Lawson, A. B., W. J. Browne, and C. L. Vidal-Rodiero (2003). *Disease Mapping with WinBUGS and MLwiN.* New York: Wiley.

Lawson, A. B., R. Carroll, C. Faes, R. S. Kirby, M. Aregay, and K. Watjou (2017). Spatio-temporal multivariate mixture models for Bayesian model selection in disease mapping. *Environmetrics.* DOI: 10.1002/env.2465.

Lawson, A. B., J. Choi, B. Cai, M. M. Hossain, R. Kirby, and J. Liu (2012). Bayesian 2-stage space-time mixture modeling with spatial misalignment of the exposure in small area health data. *Journal of Agricultural, Biological and Environmental Statistics 17*, 417–441.

Lawson, A. B., J. Choi, and J. Zhang (2014). Prior choice in discrete latent modeling of spatially referenced cancer survival. *Statistical Methods in Medical Research 23*, 183–200.

Lawson, A. B. and A. Clark (1999a). Markov chain Monte Carlo methods for clustering in case event and count data in spatial epidemiology. In M. E. Halloran and D. Berry (Eds.), *Statistics and Epidemiology: Environment and Clinical Trials*, pp. 193–218. New York: Springer.

Lawson, A. B. and A. Clark (1999b). Markov chain Monte Carlo methods for putative sources of hazard and general clustering. In A. B. Lawson, D. Böhning, E. Lesaffre, A. Biggeri, J.-F. Viel, and R. Bertollini (Eds.), *Disease Mapping and Risk Assessment for Public Health*, Chapter 9, pp. 119–141. New York: Wiley.

Lawson, A. B. and A. Clark (2002). Spatial mixture relative risk models applied to disease mapping. *Statistics in Medicine 21*, 359–370.

Lawson, A. B., A. B. Clark, and C. L. Vidal-Rodeiro (2004). Developments in general and syndromic surveillance for small area health data. *Journal of Applied Statistics 31*, 951–966.

Lawson, A. B. and N. Cressie (2000). Spatial statistical methods for environmental epidemiology. In C. R. Rao and P. K. Sen (Eds.), *Handbook of Statistics: Bio-Environmental and Public Health Statistics*, Volume 18, pp. 357–396. Amsterdam: Elsevier.

Lawson, A. B. and D. Denison (2002). Spatial cluster modelling: an overview. In A. B. Lawson and D. Denison (Eds.), *Spatial Cluster Modelling*, Chapter 1, pp. 1–19. New York: CRC Press.

Lawson, A. B. and K. Kleinman (Eds.) (2005). *Spatial and Syndromic Surveillance for Public Health.* New York: Wiley.

Lawson, A. B. and P. Leimich (2000). Approaches to space-time modelling of infectious disease behaviour. *IMA journal of Mathematics Applied to Medicine and Biology 17*, 1–13.

Lawson, A. B., G. Onicescu, and C. Ellerbe (2011). Foot and mouth disease revisited: Re-analysis using Bayesian spatial susceptible-infectious-removed models. *Spatial and Spatio-Temporal Epidemiology 2*, 185–194.

Lawson, A. B. and C. Rotejanaprasert (2014). Childhood brain cancer in Florida: A Bayesian clustering approach. *Statistics and Public Policy 1*, 99–107.

Lawson, A. B., C. Rotejanaprasert, P. Moraga, and J. Choi (2015). A shared neighbour conditional autoregressive model for small area spatial data. *Environmetrics 26*, 383–392.

Lawson, A. B. and H. R. Song (2009). Semiparametric space-time survival modeling of chronic wasting disease in deer. *Environmental and Ecological Statistics*. DOI 10.1007/s10651-009-0118-z.

Lawson, A. B. and H.-R. Song (2010). Bayesian hierarchical modeling of the dynamics of spatio-temporal influenza season outbreaks. *Spatial and Spatio-temporal Epidemiology 1*, 187–195.

Lawson, A. B., H.-R. Song, B. Cai, M. M. Hossain, and K. Huang (2010). Space-time latent component modeling of geo-referenced health data. *Statistics in Medicine 29*, 2012–2017.

Lawson, A. B. and J.-F. Viel (1995). Tests for directional space-time interaction in epidemiological data. *Statistics in Medicine 14*, 2383–2392.

Lawson, A. B. and F. Williams (2000). Spatial competing risk models in disease mapping. *Statistics in Medicine 19*, 2451–2468.

Lawson, A. B. and F. L. R. Williams (2001). *An Introductory Guide to Disease Mapping*. New York: Wiley.

Lawson, A. B. and H. Zhou (2005). Spatial statistical modeling of disease outbreaks with particular reference to the UK FMD epidemic of 2001. *Preventive Veterinary Medicine 71*, 141–156.

Le Strat, Y. (2005). Overview of temporal surveillance. In *Spatial and Syndromic Surveillance for Public Health*, Chapter 2, pp. 13–29. New York: Wiley.

Lee, D. and A. B. Lawson (2016). Quantifying the spatial inequality and temporal trends in maternal smoking rates in Glasgow. *Annals of Applied Statistics 10*(3), 1427–1446.

Lee, D., S. Mukhopadhyay, A. Rushworth, and S. K. Sahu (2017). A rigorous statistical framework for spatio-temporal pollution prediction and estimation of its long-term impact on health. *Biostatistics 18*(2), 370–385.

Lee, D. and T. Neocleous (2010). Bayesian quantile regression for count data with application to environmental epidemiology. *Applied Statistics 59*(5), 905–920.

Lee, E. C., J. M. Asher, S. Goldlust, J. D. Kraemer, A. B. Lawson, and S. Bansal (2016). Mind the scales: Harnessing spatial big data for infectious disease surveillance and inference. *Journal of Infectious Diseases 214*(suppl_4), S409.

Leonard, T. and H. Hsu (1999). *Bayesian Methods*. London: Cambridge University Press.

Leroux, B., X. Lei, and N. Breslow (2000). Estimation of disease rates in small areas: A new mixed model for spatial dependence. In M. Halloran and D. Berry (Eds.), *Statistical Models in Epidemiology, the Environment and Clinical Trials*, pp. 135–178. New York: Springer.

Lesaffre, E. and A. B. Lawson (2012). *Bayesian Biostatistics*. New York: Wiley.

Leyland, A. and H. Goldstein (Eds.) (2001). *Multilevel Modelling in Health Statistics*. London: Wiley.

Li, G., N. Best, A. L. Hansell, I. Ahmed, and S. Richardson (2012). Bay STDetect: detecting unusual temporal patterns in small area data via Bayesian model choice. *Biostatistics 13*(4), 695–710.

Liang, S., S. Banerjee, and B. Carlin (2009). Bayesian wombling for spatial point processes. *Biometrics 65*, 1243–1253.

Liang, S., B. Carlin, and A. Gelfand (2009). Analysis of Minnesota colon and rectum cancer point patterns with spatial and nonspatial covariate information. *Annals of Applied Statistics 3*, 943–962.

Lindgren, F. and H. Rue (2015). Bayesian spatial modelling with R-INLA. *Journal of Statistical Software 63*, 1–25.

Lindgren, F., H. Rue, and J. Lindstrom (2011). An explicit link between Gaussian fields and Gaussian Markov random fields: the stochastic partial differential equation approach. *Journal of the Royal Statistical Society B 74*, 423–498.

Lindsay, B. G. (1988). Composite likelihood methods. *Contemporary Mathematics 80*, 221–238.

Little, R. and D. Rubin (2002). *Statistical Analysis with Missing Data*. New York: Wiley.

Liu, J. S. (2001). *Monte Carlo Strategies in Scientific Computing*. New York: Springer.

Liu, X., M. Wall, and J. Hodges (2005). Generalised spatial structural equation models. *Biostatistics 6*, 539–557.

Louie, M. M. and E. D. Kolaczyk (2006). Multiscale detction of localised anomalous structure in aggregate disease incidence data. *Statistics in Medicine 25*, 787–810.

Lu, H. and B. Carlin (2005). Bayesian areal wombling for geographical boundary analysis. *Geographical Analysis 37*, 265–285.

Lunn, D., C. Jackson, N. Best, A. Thomas, and D. Spiegelhalter (2012). *The BUGS Book*. New York: CRC Press.

Ma., B., A. B. Lawson, and Y. Liu (2007). Evaluation of Bayesian models for focused clustering in health data. *Environmetrics 18*, 1–16.

Ma, H., B. Carlin, and S. Banerjee (2010). Hierarchical and joint site edge methods for Medicare hospice service region boundary analysis. *Biometrics 66*.

Machado, J. and J. Silva (2005). Quantiles for counts. *Journal of the American Statistical Association 100*, 1226–1237.

Macnab, Y. (2007). Spline smoothing in Bayesian disease mapping. *Environmetrics 18*, 727–744.

MacNab, Y. C. (2016). Linear models of coregionalization for multivariate lattice data: Order-dependent and order-free cMCARs. *Statistical Methods in Medical Research 25*(4), 1118–1144.

Maheswaran, R. and M. Craglia (Eds.) (2004). *GIS in Public Health Practice*. New York: CRC press.

Maindonald, J. and J. Braun (2003). *Data Analysis and Graphics Using R*. New York: Cambridge University Press.

Mantel, N. (1967). The detection of disease clustering and a generalised regression approach. *Cancer Research 27*, 209–220.

Marin, J.-M., P. Pudlo, C. Robert, and R. Ryder (2012). Approximate Bayesian computational methods. *Statistics and Computing 22*, 1167–1180.

Marin, J.-M. and C. Robert (2007). *Bayesian Core: A Practical Approach to Computational Bayesian Statistics*. New York: Springer.

Marshall, E. and D. Spiegelhalter (2003). Approximate cross-vaidatory predictive checks in disease mapping models. *Statistics in Medicine 22*, 1649–1660.

Martinez-Beneito, M., D. Conessa, A. Lopez-Quilez, and A. Lopez-Maside (2008). Bayesian Markov switching models for the early detection of influenza epidemics. *Statistics in Medicine 27*, 4455–4468.

Martinez-Beneito, M. A., P. Botella-Rocamora, and S. Banerjee (2017). Towards a multidimensional approach to Bayesian disease mapping. *Bayesian Anal. 12*(1), 239–259.

McCullagh, P. and J. Nelder (1989). *Generalised Linear Models* (2nd ed.). London: Chapman & Hall.

McKeague, I. and M. Loiseaux (2002). Perfect sampling for point process cluster modelling. In A. B. Lawson and D. Denison (Eds.), *Spatial Cluster Modelling*, Chapter 5. New York: CRC Press.

Mercer, L., J. Wakefield, C. Chen, and T. Lumley (2014). A comparison of spatial smoothing methods for small area estimation with sampling weights. *Spatial Statistics 8*, 69–85.

Mira, A., J. Moller, and G. Roberts (2001). Perfect slice samplers. *Journal of the Royal Statistical Society B 63*, 583–606.

Molenberghs, G. and G. Verbeke (2005). *Models for Discrete Longitudinal Data*. New York: Springer.

Molitor, J., M. Papathomas, M. Jerrett, and S. Richardson (2010). Bayesian profile regression with an application to the national survey of children's health. *Biostatistics 11*(3), 484–498.

Móller, J. and O. Skare (2001). Coloured Voronoi tesselations for Baycsian image analysis and reservoir modelling. *Statistical Modelling 1*, 213–232.

Møller, J., A. Syversveen, and R. P. Waagepetersen (1998). Log Gaussian Cox processes. *Scandinavian Journal of Statistics 25*, 451–482.

Móller, J. and R. Waagepetersen (1998). Markov-connected component fields. *Advances in Applied Probability 30*, 1–35.

Móller, J. and R. Waagpetersen (2004). *Statistical Inference and Simulation for Spatial Point Processes*. New York: CRC Press.

Mollison, D. (Ed.) (1995). *Epidemic Models: Their Structure and Relation to Data*. London: Cambridge University Press.

Moraga, P. and A. Lawson (2012). Gaussian component mixtures and car models in Bayesian disease mapping. *Computational Statistics and Data Analysis 56*, 1417–1433.

Morel, P. D., A. Doucet, and A. Jasra (2012). An adaptive sequential Monte Carlo method for approximate Bayesian computation. *Statistics and Computing 22*, 1009–1020.

Morgan, O., M. Vreiheid, and H. Dolk (2004). Risk of low birth weight near EUROHAZCON hazardous waste landfill sites in England. *Archives of Environmental Health 59*, 149–151.

Morton, A. and B. Finkenstadt (2005). Discrete time modelling of disease incidence time series by using Markov chain Monte Carlo methods. *Journal of the Royal Statistical Society C 54*, 575–594.

Mugglin, A., N. Cressie, and I. Gemmel (2002). Hierarchical statistical modelling of influenza epidemic dynamics in space and time. *Statistics in Medicine 21*, 2703–2721.

Napier, G., D. Lee, C. Robertson, and A. Lawson (2017). A Bayesian space-time model for clustering areal units based on their disease trends. Submitted.

Nathoo, F. S., M. L. Lesperance, A. B. Lawson, and C. B. Dean (2013). Comparing variational Bayes with Markov chain Monte Carlo for Bayesian computation in neuroimaging. *Statistical Methods in Medical Research 22*(4), 398–423.

Neal, P. and G. Roberts (2004). Statistical inference and model selection for the 1861 Hagelloch measles epidemic. *Biostatistics 5*, 249–261.

Neal, R. (2011). MCMC using Hamiltonian dynamics. In S. Brooks, A. Gelman, G. Jones, and X. Meng (Eds.), *Handbook of Markov Chain Monte Carlo*, Chapter 5. New York: CRC Press.

Neal, R. M. (2003). Slice sampling. *Annals of Statistics 31*, 1–34.

Nieuwenhuijsen, M. (Ed.) (2015). *Exposure Assessment in Environmental Epidemiology* (2nd ed.). New York: Oxford University Press.

Nott, D. J. and T. Rydén (1999). Pairwise likelihood methods for inference in image models. *Biometrika 86*, 661–676.

Nsubuga, P., M. White, S. Thacker, M. Anderson, S. Blount, C. Broome, M. C. Chiller, T. Espitia, V. Imtiaz, R. Sosin, D. Stroup, D.Tauxe, R. Vijayaraghavan, and M. Trostle (2006). Public health surveillance: A tool for targeting and monitoring interventions. In D. T. Jamison, J. G. Breman, A. R. Measham, G. Alleyne, M. Claeson, D. B. Evans, P. Jha, A. Mills, and P. Musgrove (Eds.), *Disease Control Priorities in Developing Countries* (2nd ed.)., pp. 997–1015. New York: Oxford University Press.

Ntzoufras, I. (2009). *Bayesian Modelling Using WinBUGS*. New York: Wiley.

Oden, N., R. Sokal, M. Fortin, and H. Goebl (1993). Categorical wombling: Detecting regions of significant change in spatially located categorical variables. *Geographical Analysis 25*(4), 315–336.

O'Hara, R. B. and M. J. Sillanpää (2009). A review of Bayesian variable selection methods: What, how, and which. *Bayesian Analysis 4*, 85–118.

O'Loughlin, J. (2002). The electoral geography of weimar germany: Exploratory spatial data analyses (ESDA) of Protestant support for the Nazi party. *Political Analysis 10*(3), 217–243.

Onicescu, G. and A. B. Lawson (2016). Bayesian modeling and inference. In A. B. Lawson, S. Banerjee, R. Haining, and L. Ugart (Eds.), *Handbook of Spatial Epidemiology*, pp. 133–160. New York: CRC Press.

Onicescu, G., A. B. Lawson, J. Zhang, M. Gebregziabher, K. Wallace, and J. M. Eberth (2017a). Bayesian accelerated failure time model for space-time dependency in a geographically augmented survival model. *Statistical Methods in Medical Research*. PMID:26220537.

Onicescu, G., A. B. Lawson, J. Zhang, M. Gebregziabher, K. Wallace, and J. M. Eberth (2017b). Spatially explicit survival modeling for small area cancer data. *Journal of Applied Statistics 0*, 1–18.

Park, E. S., P. Hopke, M.-S. Oh, E. Symanski, D. Han, and C. Spiegelman (2014). Assessment of source-specific health effects associated with an unknown number of major sources of multiple air pollutants: A unified Bayesian approach. *Biostatistics 15*.

Park, T. and G. Casella (2008). The Bayesian lasso. *Journal of the American Statistical Association 103*(482), 681–686.

Paul, M. and L. Held (2011). Predictive assessment of a non-linear random effects model for multivariate time series of infectious disease counts. *Statistics in Medicine 30*, 1118–1136.

Paul, M., L. Held, and A. Toschke (2008). Multivariate modelling of infectious disease surveillance data. *Statistics in Medicine 27*, 6250–6267.

Peng, R. and F. Dominici (2008). *Statistical Methods for Environmental Epidemiology with R: A Case Study in Air Pollution and Health*. New York: Springer.

Petralias, A. and P. Dellaportas (2013). An MCMC model search algorithm for regression problems. *Journal of Statistical Computation and Simulation 83*(9), 1722–1740.

Pfeiffer, D. U., T. P. Robinson, M. Stevenson, K. B. Stevens, D. J. Rogers, and A. C. A. Clements (Eds.) (2008). *Spatial Analysis in Epidemiology*. New York: Springer.

Pike, F. and L. Weissfeld (2013). Joint modeling of censored longitudinal and event time data. *Journal of Applied Statistics 40*(1), 17–27.

Plummer, M. and D. G. Clayton (1996). Estimation of population exposure in ecological studies. *Journal of the Royal Statistical Society B 58*, 113–126.

Propp, J. and D. Wilson (1996). Exact sampling with coupled Markov chains. *Random Structures and Algortithms 9*, 223–252.

Raghunathan, T. E., D. Xie, N. Schenker, V. L. Parsons, W. W. Davis, K. W. Dodd, and E. J. Feuer (2007). Combining information from surveys to estimate county-level prevalence rates of cancer risk factors and screening. *Journal of the American Statistical Association 102*, 474–486.

Rao, J. K. N. (2003). *Small Area Estimation.* New York: Wiley.

Rao, J. N. K. and I. Molina (2015). *Small Area Estimation* (2nd ed.). New York: Wiley.

Reich, B. (2016). Quantile regression for epidemiological applications. In A. B. Lawson, S. Banerjee, R. Haining, and L. Ugart (Eds.), *Handbook of Spatial Epidemiology*, pp. 239–250. New York: CRC Press.

Reich, B., J. Hodges, and V. Zadnik (2006). Effects of residual smoothing on the posterior of the fixed effects in disease-mapping models. *Biometrics 62*, 1197–1206.

Richardson, S., J. Abellan, and N. Best (2006). Bayesian spatio-temporal analysis of joint patterns of male and female lung cancer risks in Yorkshire (UK). *Statistical Methods in Medical Research 15*, 97–118.

Richardson, S., A. Thomson, N. Best, and P. Elliott (2004). Interpreting posterior relative risk estimates in disease mapping studies. *Environmental Health Perspectives 112*, 1016–1025.

Ripley, B. D. (1981). *Spatial Statistics.* New York: Wiley.

Ripley, B. D. (1987). *Stochastic Simulation.* New York: Wiley.

Ripley, B. D. (1988). *Statistical Inference for Spatial Processes.* New York: Cambridge University Press.

Robert, C. and G. Casella (2005). *Monte Carlo Statistical Methods* (2nd ed.). New York: Springer.

Rockova, V. and E. George (2014). Negotiating multicollinearity with spike-and-slab priors. *Metron 72*, 217–229.

Rodriguez, A., D. Dunson, and A. Gelfand (2009). Bayesian nonparametric functional data analysis through density estimation. *Biometrika 96*, 149–162.

Rossow, H., J. Ollgren, J. Hytnen, H. Rissanen, O. Huitu, H. Henttonen, M. Kuusi, and O. Vapalahti (2015). Incidence and seroprevalence of tularaemia in Finland, 1995 to 2013: regional epidemics with cyclic pattern. *Eurosurveillance 20*(33).

Rotejanaprasert, C. (2014). Evaluation of cluster recovery for small area relative risk models. *Statistical Methods in Medical Research 23*(6), 531–551.

Rotejanaprasert, C. and A. Lawson (2016). Bayesian prospective detection of small area health anomalies using Kullback-Lleibler divergence. *Statistical Methods in Medical Research. 27*(4), 1076–1087.

Rotejanaprasert, C., A. Lawson, S. Bolick-Aldrich, and D. Hurley (2016). Spatial Bayesian surveillance for small area case event data. *Statistical Methods in Medical Research 25*(4), 1101–1117. PMID: 27566768.

Rotejanaprasert, C., A. B. Lawson, H. Rossow, J. Sane, O. Huitu, H. Henttonen, and V. D. R. Vilas (2017). Towards integrated surveillance of zoonoses: Spatiotemporal joint modeling of rodent population data and human tularemia cases in Finland. *BMC Research Methodology*.

Rowlingson, B., P. Diggle, and T.-L. Su (2005). Point process methodology for on-line spatio-temporal disease surveillance. *Environmetrics 16*, 423–434.

Royle, A. and M. Berliner (1999). A hierarchical approach to multivariate spatial modeling and prediction. *Journal of Agricultural, Biological and Environmental Statistics 4*, 29–56.

Rue, H. and L. Held (2005). *Gaussian Markov Random Fields: Theory and Applications*. New York: Chapman & Hall/CRC Press.

Rue, H., S. Martino, and N. Chopin (2009). Approximate Bayesian inference for latent Gaussian models by using integrated nested Laplace approximations. *Journal of the Royal statistical Society B 71*, 319–392.

Rue, H., A. Reibler, S. Sorbye, J. Illian, D. Simpson, and F. Lindgren (2016). Bayesian computing with INLA: a review. *arXiv*. https://arxiv.org/abs/1604.00860.

Rushworth, A., D. Lee, and R. Mitchell (2014). A spatio-temporal model for estimating the long-term effects of air pollution on respiratory hospital admissions in Greater London. *Spatial and Spatio-temporal Epidemiology 10*(Supplement C), 29–38.

Rushworth, A., D. Lee, and C. Sarran (2017). An adaptive spatiotemporal smoothing model for estimating trends and step changes in disease risk. *Journal of the Royal Statistical Society: C 66*(1), 141–157.

Salway, R. and J. Wakefield (2005). Sources of bias in ecological studies of non-rare events. *Environmental and Ecological Statistics 12*, 321–347.

Sattenspiel, L. and A. Lloyd (2009). *The Geographic Spread of Infectious Diseases: Models and Applications*. New York: Princeton University Press.

Savoca, M., D. Ludwig, S. Jones, K. Clodfelter, J. Sloop, L. Bollhalter, and A. Bertoni (2017). Geographical information systems to assess external validity in randomized trials. *American Journal of Preventive Medicine 53*, 252–259.

Scheel, I., P. Green, and J. Rougier (2011). A graphical diagnostic for identifying influential model choices in Bayesian hierarchical models. *Scandinavian Journal of Statistics 38*, 529–550.

Schlattman, P. and D. Böhning (1993). Mixture models and disease mapping. *Statistics in Medicine 12*, 1943–1950.

Schrodle, B. and L. Held (2011). Spatio-temporal disease mapping using INLA. *Environmetrics 22*, 725–734.

Schrodle, B., L. Held, A. Rieber, and J. Danuser (2011). Using integrated nested Laplace approximations for the evaluation of veterinary surveillance data from Switzerland: a case study. *Applied Statistics 60*, 261–279.

Shmeuli, G. and H. Burkom (2010). Statistical challenges facing early outbreak detection in biosurveillance. *Technometrics 52*, 39–51.

Simpson, D., F. Lindgren, and H. Rue (2012). In order to make spatial statistics computationally feasible, we need to forget about the covariance function. *Environmetrics 23*, 65–74.

Simpson, D. P., H. Rue, T. G. Martins, A. Riebler, and S. H. Sørbye (2015). Penalising model component complexity: A principled, practical approach to constructing priors. *ArXiv e-prints*.

Smith, A. F. M. and A. E. Gelfand (1992). Bayesian statistics without tears: A sampling-resampling perspective. *American Statistician 46*, 84–88.

Smith, A. F. M. and G. Roberts (1993). Bayesian computation via the Gibbs sampler and related Markov chain Monte Carlo methods. *Journal of the Royal Statistical Society B 55*, 3–23.

Snow, J. (1854). *On the Mode of Communication of Cholera* (2nd ed.). London: Churchill Livingstone.

Sonesson, C. and D. Bock (2003). A review and discussion of prospective statistical surveillance in public health. *Journal of the Royal Statistical Society A 166*, 5–21.

Sosin, D. (2003). Draft framework for evaluating syndromic surveillance systems. *Journal of Urban Health 80*, i8–i13. Supplement.

Spiegelhalter, D., N. Best, B. Carlin, and A. van der Linde (2014). The deviance information criterion: 12 years on. *Journal of the Royal Statistical Society B 76*, 485–493.

Spiegelhalter, D., A. Thomas, N. Best, and D. Lunn (2007). *WinBUGS manual*. Cambridge,UK: MRC Biostatistics Unit. Version 1.4.3.

Spiegelhalter, D. J., N. G. Best, B. P. Carlin, and A. van der Linde (2002). Bayesian deviance, the effective number of parameters and the comparison of arbitrarily complex models. *Journal of the Royal Statistical Society B 64*, 583–640.

Spiegelhalter, D. J., N. G. Best, W. R. Gilks, and H. Inskip (1996). Hepatitis B: a case study in MCMC methods. In W. R. Gilks, S. Richardson, and D. J. Spiegelhalter (Eds.), *Markov Chain Monte Carlo in Practice*. London: Chapman & Hall.

Steele, F., H. Goldstein, and W. Browne (2004). A general multilevel multistate competing risks model for event history data, with an application to a study of contraceptive use dynamics. *Statistical Modelling 4*, 145–159.

Stern, H. and Y. Jeon (2004). Applying structural equation models with incomplete data. In A. Gelman and X.-L. Meng (Eds.), *Applied Bayesain Modeling and Causal Inference from Incomplete-Data Perspectives*, Chapter 30, pp. 331–342. London: Wiley.

Stern, H. S. and N. A. C. Cressie (1999). Inference for extremes in disease mapping. In A. B. Lawson, A. Biggeri, D. Boehning, E. Lesaffre, J. F. Viel, and R. Bertollini (Eds.), *Disease Mapping and Risk Assessment for Public Health*, Chapter 5. New York: Wiley.

Stern, H. S. and N. A. C. Cressie (2000). Posterior predictive model checks for disease mapping models. *Statistics in Medicine 19*, 2377–2397.

Stevenson, M., R. Morris, A. B. Lawson, J. Wilesmith, J. M. Ryan, and R. Jackson (2005). Area level risks for BSE in British cattle before and after the July 1988 meat and bone meal feed ban. *Preventive Veterinary Medicine 69*, 129–144.

Stevenson, M. A., R. S. Morris, A. B. Lawson, J. W. Wilesmith, J. B. M. Ryan, and R. A. Jackson (2001). Spatial regression analysis of the bovine spongiform encephalopathy epidemic in Great Britain. In A. C. Gatrell and P. Durr (Eds.), *The GISVET Conference: Application of Geographical Information Systems and Spatial Analysis to Veterinary Science*, Lancaster.

Stevenson, M. A., J. W. Wilesmith, J. B. M. Ryan, R. S. Morris, A. B. Lawson, D. U. Pfeiffer, and D. Lin (2000). Descriptive spatial analysis of the epidemic of bovine spongiform encephalopathy in Great Britain to June 1997. *Veterinary Record 147*, 379–384.

Strat, Y. L. and F. Carrat (1999). Monitoring epidemiologic surveillance data using hidden Markov models. *Statistics in Medicine 18*, 3463–3478.

Sturtz, S., U. Ligges, and A. Gelman (2005). R2WinBUGS: A package for running WinBUGS from R. *Journal of Statistical Software 12*, 1–16.

Sun, D., R. Tsutakawa, H. Kim, and Z. He (2000). Spatio-temporal interaction with disease mapping. *Statistics in Medicine 19*, 2015–2035.

Sutton, S. (2005). *The modeling of spatially-referenced recurrent event data in a South Carolina population.* Ph.D. thesis, University of South Carolina.

Tanner, M. A. (1996). *Tools for Statistical Inference* (3rd ed.). New York: Springer.

Taylor, B., T. Davies, B. Rowlingson, and P. Diggle (2015). Bayesian inference and data augmentation schemes for spatial, spatiotemporal and multivariate log-Gaussian Cox processes in r. *Journal of Statistical Software 63*(7), 1–47.

Taylor, B. and P. Diggle (2014). INLA or MCMC? A tutorial and comparative evaluation for spatial prediction in log-Gaussian Cox processes. *Journal of Statistical Computation and Simulation 84*(10), 2266–2284.

Teng, M., F. Nathoo, and T. Johnson (2017). Bayesian computation for log-Gaussian Cox processes: a comparative analysis of methods. *Journal of Statistical Computation and Simulation 87*, 2227–2252.

Thacker, S. (1994). Historical development. In S. Teusch and R. Churchill (Eds.), *Principles and Practice of Public Health Surveillance*, pp. 3–17. New York: Oxford University Press.

Thacker, S. and R. Berkelman (1992). History of public health surveillance. In W. Halperin, L. Baker, and R. Monson (Eds.), *Public Health Surveillance*, pp. 1–15. New York: Van Nostrand Reinhold.

Thomas, A., N. Best, D. Lunn, R. Arnold, and D. Spiegelhalter (2004). Geobugs User Manual v 1.2. www.mrc-bsu.cam.ac.uk/bugs.

Thomas, D. (2009). *Statistical Methods on Environmental Epidemiology.* London: Oxford University Press.

Tibshirani, R. and T. Hastie (1987). Local likelihood estimation. *Journal of the American Statistical Association 82*, 559–568.

Tierney, L. and J. Kadane (1986). Accurate approximations for posterior moments and marginal densities. *Journal of the American Statistical Association 81*, 82–86.

Tzala, E. and N. Best (2008). Bayesian latent variable modelling of multivariate spatio-temporal variation in cancer mortality. *Statistical Methods in Medical Research 17*, 97–118.

Unkel, S., C. P. Farrington, P. Garthwaite, C. Robertson, and N. Andrews (2012). Statistical methods for the prospective detection of infectious disease outbreaks: a review. *Journal of the Royal Statistical Society 175*, 49–82.

van Lieshout, M. and A. Baddeley (2002). Extrapolating and interpolating spatial patterns. In A. B. Lawson and D. Denison (Eds.), *Spatial Cluster Modelling*, Chapter 4, pp. 61–86. New York: CRC Press.

Vandendijck, Y., C. Faes, R. Kirby, A. Lawson, and N. Hens (2016). Model-based inference for small area estimation with sampling weights. *Spatial Statistics 18, Part B*, 455–473.

Varin, C., G. Høst, and Ø. Skare (2005). Pairwise likelihood infernce in spatial generalized linear mixed models. *Computational Statistics and Data Analysis 49*, 1173–1191.

Verbeke, G. and G. Molenberghs (2000). *Linear Mixed Models for Longitudinal Data*. New York: Springer.

Vidal-Rodeiro, C. and A. B. Lawson (2006a). Monitoring changes in spatio-temporal maps of disease. *Biometrical Journal 48*, 463–480.

Vidal-Rodeiro, C. and A. B. Lawson (2006b). Online updating of space-time disease surveillance models via particle filters. *Statistical Methods in Medical Research 15*, 423–444.

Vidal-Rodiero, C. L. and A. B. Lawson (2005). An evaluation of edge effects in disease mapping. *Computational Statistics and Data Analysis 49*, 45–62.

Voss, D. (2004). Using ecological inference for contextual research. In G. King, O. Rosen, and M. Tanner (Eds.), *Ecological Inference: New Methodological Strategies*, Chapter 3. New York: Cambridge University Press.

Wackernagel, H. (2003). *Multivariate Geostatistics* (3rd ed.). New York: Springer.

Wakefield, J., S. Haneuse, A. Dobra, and E. Teeple (2011). Bayes computation for ecological inference. *Statistics in Medicine 30*, 1381–1396.

Wakefield, J. and A. Kim (2013). A Bayesian model for cluster detection. *Biostatistics 14*, 752–765.

Wakefield, J. and G. Shaddick (2006). Health-exposure modeling and the ecological fallacy. *Biostatistics 7*, 438–455.

Wakefield, J. and T. Smith (2016). Ecological modeling: General issues. In A. B. Lawson, S. Banerjee, R. Haining, and L. Ugart (Eds.), *Handbook of Spatial Epidemiology*, pp. 99–118. New York: CRC Press.

Wakefield, J. C. and S. Morris (2001). The Bayesian modeling of disease risk in relation to a point source. *Journal of the American Statistical Association 96*, 77–91.

Waller, L. and C. Gotway (2004). *Applied Spatial Statistics for Public Health Data.* New York: Wiley.

Waller, L. A., B. P. Carlin, H. Xia, and A. E. Gelfand (1997). Hierarchical spatio-temporal mapping of disease rates. *Journal of the American Statistical Association 92*, 607–617.

Wang, F. and M. Wall (2003). Generalized common spatial factor model. *Biostatistics 4*, 569–582.

Wang, J. and R. Swendsen (1990). Cluster Monte Carlo algorithms. *Physica A: Statistical Mechanics and Its Applications 167*(3), 565–579.

Watjou, K., C. Faes, A. Lawson, R. S. Kirby, M. Aregay, R. Carroll, and Y. Vandendijck (2017). Spatial small area smoothing models for handling survey data with nonresponse. *Statistics in Medicine 36*(23), 3708–3745.

Wikle, C. K. (2002). Spatial modelling of count data: A case study in modelling breeding bird survey data on large spatial domains. In A. B. Lawson and D. G. T. Denison (Eds.), *Spatial Cluster Modelling*, Chapter 11. London: CRC Press.

Williams, F., A. Lawson, and O. Lloyd (1992). Low sex ratios of births in areas at risk from air pollution from incinerators, as shown by geographical analysis and 3-dimensional mapping. *International Journal of Epidemiology 21*, 311–319.

Wolpert, R. L. and K. Ickstadt (1998). Poisson/gamma random field models for spatial statistics. *Biometrika 85*, 251–267.

Woodall, W. (2006). The use of control charts in health-care and public-health surveillance. *Journal of Quality Technology 38*, 89–104.

Woodall, W., O. Grigg, and H. Burkom (2010). Research issues and ideas on health-related monitoring. In H. J. Lenz (Ed.), *Frontiers in Statistical Quality Control 9*, pp. 145–155. Berlin: Springer.

Xia, H. and B. P. Carlin (1998). Spatio-temporal models with errors in covariates: Mapping Ohio lung cancer mortality. *Statistics in Medicine 17*, 2025–2043.

Xia, H., B. P. Carlin, and L. A. Waller (1997). Hierarchical models for mapping Ohio lung cancer rates. *Environmetrics 8*, 107–120.

Yu, B. and P. Mykland (1998). Looking at Markov samplers through cusum path plots: A simple diagnostic idea. *Statistics and Computing 8*, 275–286.

Yu, K. and R. Moyeed (2001). Bayesian quantile regression. *Statistics and Probability Letters 54*, 437–447.

Zhang, J. and A. Lawson (2011). Bayesian parametric accelerated failure time spatial model and its application to prostate cancer. *Journal of Applied Statistics 38*, 591–603.

Zhang, S., D. Sun, C. He, and M. Schootman (2006). A Bayesian semi-parametric model for colorectal cancer incidences. *Statistics in Medicine 25*, 285–309.

Zhou, H. and A. B. Lawson (2008). EWMA smoothing and Bayesian spatial modeling for health surveillance. *Statistics in Medicine 27*, 5907–5928.

Zhou, H., A. B. Lawson, J. Hebert, E. Slate, and E. Hill (2008a). A Bayesian hierarchical modeling approach for studying the factors affecting the stage at diagnosis of prostate cancer. *Statistics in Medicine 27*, 1468–89.

Zhou, H., A. B. Lawson, J. Hebert, E. Slate, and E. Hill (2008b). Joint spatial survival modelling for the date of diagnosis and the vital outcome for prostate cancer. *Statistics in Medicine 27*, 3612–28.

Zhu, L., B. Carlin, and A. Gelfand (2003). Hierarchical regression with mis-aligned spatial data: relating ambient ozone and pediatric asthma ER visits in Atlanta. *Environmetrics 14*, 537–557.

Zhu, L. and B. P. Carlin (2000). Comparing hierarchical models for spatio-temporally misaligned data using the DIC criterion. *Statistics in Medicine 19*, 2265–2278.

Index